James Wels
ΔΤΔ

$$\log = \ln$$

$$\log_{10} = \log \text{ base } 10$$

$$e^{\ln x} = X$$

if $\ln_e x = c$

then $e^c = X$

$$a = \frac{d^2x}{dt^2} = \frac{dv}{dt} = v\frac{dv}{dx}$$

$$e^{\ln u} = u$$

$$\ln_e a = b$$

$$e^b = a \quad + \text{ from } \quad e^{\ln a} = a$$

$$\ln u = X$$

$$u = e^X$$

TABLES OF INTEGRALS
AND OTHER MATHEMATICAL DATA

TABLES OF INTEGRALS
AND OTHER MATHEMATICAL DATA

BY

HERBERT BRISTOL DWIGHT, D.Sc

PROFESSOR OF ELECTRICAL MACHINERY
MASSACHUSETTS INSTITUTE OF TECHNOLOGY

Revised Edition

NEW YORK

THE MACMILLAN COMPANY

PREFACE TO THE FIRST EDITION

The first study of any portion of mathematics should not be done from a synopsis of compact results, such as this collection. The references, although they are far from complete, will be helpful, it is hoped, in showing where the derivation of the results is given or where further similar results may be found. A list of numbered references is given at the end of the book. These are referred to in the text as "Ref. 7, p. 32," etc., the page number being that of the publication to which reference is made.

Letters are considered to represent real quantities unless otherwise stated. Where the square root of a quantity is indicated, the positive value is to be taken, unless otherwise indicated. Two vertical lines enclosing a quantity represent the absolute or numerical value of that quantity, that is, the modulus of the quantity. The absolute value is a positive quantity. Thus, $\log |-3| = \log 3$.

The constant of integration is to be understood after each integral. The integrals may usually be checked by differentiating.

In algebraic expressions, the symbol log represents natural or Napierian logarithms, that is, logarithms to the base e. When any other base is intended, it will be indicated in the usual manner. When an integral contains the logarithm of a certain quantity, integration should not be carried from a negative to a positive value of that quantity. If the quantity is negative, the logarithm of the absolute value of the quantity may be used, since $\log (-1) = (2k+1)\pi i$ will be part of the constant of integration (see **409.03**). Accordingly, in many cases, the logarithm of an absolute value is shown, in giving an integral, so as to indicate that it applies to real values, both positive and negative.

Inverse trigonometric functions are to be understood as referring to the principal values.

Suggestions and criticisms as to the material of this book and as to errors that may be in it, will be welcomed.

v

The author desires to acknowledge valuable suggestions from Professors P. Franklin, W. H. Timbie, L. F. Woodruff, and F. S. Woods, of Massachusetts Institute of Technology.

H. B. Dwight.

Cambridge, Mass.
December, 1933.

PREFACE TO THE SECOND EDITION

A considerable number of items have been added, including groups of integrals involving

$$(ax^2 + bx + c)^{1/2}, \quad \frac{1}{a + b \sin x} \quad \text{and} \quad \frac{1}{a + b \cos x},$$

also additional material on inverse functions of complex quantities and on Bessel functions. A probability integral table (No. 1045) has been included.

It is desired to express appreciation for valuable suggestions from Professor Wm. R. Smythe of California Institute of Technology and for the continued help and interest of Professor Philip Franklin of the Department of Mathematics, Massachusetts Institute of Technology.

Herbert B. Dwight.

Cambridge, Mass.

CONTENTS

CONTENTS

APPENDIX

TABLES OF INTEGRALS
AND OTHER MATHEMATICAL DATA

TABLES OF INTEGRALS AND OTHER MATHEMATICAL DATA

ALGEBRAIC FUNCTIONS

1. $(1 + x)^n = 1 + nx + \dfrac{n(n-1)}{2!} x^2 + \dfrac{n(n-1)(n-2)}{3!} x^3$
$$+ \cdots + \frac{n!}{(n-r)!\,r!} x^r + \cdots.$$

Note that, here and elsewhere, we take $0! = 1$. If n is a positive integer, the expression consists of a finite number of terms. If n is not a positive integer, the series is convergent for $x^2 < 1$; and if $n > 0$, the series is convergent also for $x^2 = 1$. [Ref. 21, p. 88.]

2. The coefficient of x^r in No. 1 is denoted by $\dbinom{n}{r}$ or $_nC_r$.

Values are given in the following table.

TABLE OF BINOMIAL COEFFICIENTS

$_nC_r$: Values of n in left column; values of r in top row

	0	1	2	3	4	5	6	7	8	9	10
1	1	1									
2	1	2	1				N.B. Sum of any two adjacent numbers in same row is equal to number just below the right-hand one of them.				
3	1	3	3	1							
4	1	4	6	4	1						
5	1	5	10	10	5	1					
6	1	6	15	20	15	6	1				
7	1	7	21	35	35	21	7	1			
8	1	8	28	56	70	56	28	8	1		
9	1	9	36	84	126	126	84	36	9	1	
10	1	10	45	120	210	252	210	120	45	10	1

For a large table see Ref. 59, v. 1, second section, p. 69.

3. $(1 - x)^n = 1 - nx + \dfrac{n(n-1)}{2!} x^2 - \dfrac{n(n-1)(n-2)}{3!} x^3$
$$+ \cdots + (-1)^r \frac{n!}{(n-r)!\,r!} x^r + \cdots.$$

[See Table 2 and note under No. 1.]

4. $(a \pm x)^n = a^n \left(1 \pm \dfrac{x}{a} \right)^n.$

4.2. $(1 \pm x)^2 = 1 \pm 2x + x^2.$

4.3. $(1 \pm x)^3 = 1 \pm 3x + 3x^2 \pm x^3.$

4.4. $(1 \pm x)^4 = 1 \pm 4x + 6x^2 \pm 4x^3 + x^4,$

and so forth, using coefficients from Table 2.

5.1. $(1 \pm x)^{1/4} = 1 \pm \dfrac{1}{4} x - \dfrac{1 \cdot 3}{4 \cdot 8} x^2 \pm \dfrac{1 \cdot 3 \cdot 7}{4 \cdot 8 \cdot 12} x^3$

$$- \frac{1 \cdot 3 \cdot 7 \cdot 11}{4 \cdot 8 \cdot 12 \cdot 16} x^4 \pm \cdots, \qquad [x^2 \le 1].$$

5.2. $(1 \pm x)^{1/3} = 1 \pm \dfrac{1}{3} x - \dfrac{1 \cdot 2}{3 \cdot 6} x^2 \pm \dfrac{1 \cdot 2 \cdot 5}{3 \cdot 6 \cdot 9} x^3$

$$- \frac{1 \cdot 2 \cdot 5 \cdot 8}{3 \cdot 6 \cdot 9 \cdot 12} x^4 \pm \cdots, \qquad [x^2 \le 1].$$

5.3. $(1 \pm x)^{1/2} = 1 \pm \dfrac{1}{2} x - \dfrac{1 \cdot 1}{2 \cdot 4} x^2 \pm \dfrac{1 \cdot 1 \cdot 3}{2 \cdot 4 \cdot 6} x^3$

$$- \frac{1 \cdot 1 \cdot 3 \cdot 5}{2 \cdot 4 \cdot 6 \cdot 8} x^4 \pm \cdots, \qquad [x^2 \le 1].$$

5.4. $(1 \pm x)^{3/2} = 1 \pm \dfrac{3}{2} x + \dfrac{3 \cdot 1}{2 \cdot 4} x^2 \mp \dfrac{3 \cdot 1 \cdot 1}{2 \cdot 4 \cdot 6} x^3$

$$+ \frac{3 \cdot 1 \cdot 1 \cdot 3}{2 \cdot 4 \cdot 6 \cdot 8} x^4 \mp \frac{3 \cdot 1 \cdot 1 \cdot 3 \cdot 5}{2 \cdot 4 \cdot 6 \cdot 8 \cdot 10} x^5 + \cdots, \qquad [x^2 \le 1].$$

5.5. $(1 \pm x)^{5/2} = 1 \pm \dfrac{5}{2} x + \dfrac{5 \cdot 3}{2 \cdot 4} x^2 \pm \dfrac{5 \cdot 3 \cdot 1}{2 \cdot 4 \cdot 6} x^3$

$$- \frac{5 \cdot 3 \cdot 1 \cdot 1}{2 \cdot 4 \cdot 6 \cdot 8} x^4 \pm \frac{5 \cdot 3 \cdot 1 \cdot 1 \cdot 3}{2 \cdot 4 \cdot 6 \cdot 8 \cdot 10} x^5 - \cdots, \qquad [x^2 \le 1].$$

6. $(1 + x)^{-n} = 1 - nx + \dfrac{n(n + 1)}{2!} x^2 - \dfrac{n(n + 1)(n + 2)}{3!} x^3$

$$+ \cdots + (-1)^r \frac{(n + r - 1)!}{(n - 1)! \, r!} x^r + \cdots, \qquad [x^2 < 1].$$

7. $(1 - x)^{-n} = 1 + nx + \dfrac{n(n + 1)}{2!} x^2 + \dfrac{n(n + 1)(n + 2)}{3!} x^3$

$$+ \cdots + \frac{(n + r - 1)!}{(n - 1)! \, r!} x^r + \cdots, \qquad [x^2 < 1].$$

8. $(a \pm x)^{-n} = a^{-n} \left(1 \pm \dfrac{x}{a}\right)^{-n}, \qquad\qquad [x^2 < a^2].$

9.01. $\quad (1 \pm x)^{-1/4} = 1 \mp \dfrac{1}{4}x + \dfrac{1 \cdot 5}{4 \cdot 8}x^2 \mp \dfrac{1 \cdot 5 \cdot 9}{4 \cdot 8 \cdot 12}x^3$

$$+ \dfrac{1 \cdot 5 \cdot 9 \cdot 13}{4 \cdot 8 \cdot 12 \cdot 16}x^4 \mp \cdots, \qquad [x^2 < 1].$$

9.02. $\quad (1 \pm x)^{-1/3} = 1 \mp \dfrac{1}{3}x + \dfrac{1 \cdot 4}{3 \cdot 6}x^2 \mp \dfrac{1 \cdot 4 \cdot 7}{3 \cdot 6 \cdot 9}x^3$

$$+ \dfrac{1 \cdot 4 \cdot 7 \cdot 10}{3 \cdot 6 \cdot 9 \cdot 12}x^4 \mp \cdots, \qquad [x^2 < 1].$$

9.03. $\quad (1 \pm x)^{-1/2} = 1 \mp \dfrac{1}{2}x + \dfrac{1 \cdot 3}{2 \cdot 4}x^2 \mp \dfrac{1 \cdot 3 \cdot 5}{2 \cdot 4 \cdot 6}x^3$

$$+ \dfrac{1 \cdot 3 \cdot 5 \cdot 7}{2 \cdot 4 \cdot 6 \cdot 8}x^4 \mp \cdots, \qquad [x^2 < 1].$$

9.04. $\quad (1 \pm x)^{-1} \; = 1 \mp x + x^2 \mp x^3 + x^4 \mp \cdots, \qquad [x^2 < 1].$

9.05. $\quad (1 \pm x)^{-3/2} = 1 \mp \dfrac{3}{2}x + \dfrac{3 \cdot 5}{2 \cdot 4}x^2 \mp \dfrac{3 \cdot 5 \cdot 7}{2 \cdot 4 \cdot 6}x^3$

$$+ \dfrac{3 \cdot 5 \cdot 7 \cdot 9}{2 \cdot 4 \cdot 6 \cdot 8}x^4 \mp \cdots, \qquad [x^2 < 1].$$

9.06. $\quad (1 \pm x)^{-2} \; = 1 \mp 2x + 3x^2 \mp 4x^3 + 5x^4 \mp \cdots,$

$$[x^2 < 1].$$

9.07. $\quad (1 \pm x)^{-5/2} = 1 \mp \dfrac{5}{2}x + \dfrac{5 \cdot 7}{2 \cdot 4}x^2 \mp \dfrac{5 \cdot 7 \cdot 9}{2 \cdot 4 \cdot 6}x^3$

$$+ \dfrac{5 \cdot 7 \cdot 9 \cdot 11}{2 \cdot 4 \cdot 6 \cdot 8}x^4 \mp \cdots, \qquad [x^2 < 1].$$

9.08. $\quad (1 \pm x)^{-3} \; = 1 \mp \dfrac{1}{1 \cdot 2}\{2 \cdot 3x \mp 3 \cdot 4x^2 + 4 \cdot 5x^3$

$$\mp 5 \cdot 6x^4 + \cdots\}, \qquad [x^2 < 1].$$

9.09. $\quad (1 \pm x)^{-4} \; = 1 \mp \dfrac{1}{1 \cdot 2 \cdot 3}\{2 \cdot 3 \cdot 4x \mp 3 \cdot 4 \cdot 5x^2 + 4 \cdot 5 \cdot 6x^3$

$$\mp 5 \cdot 6 \cdot 7x^4 + \cdots\}, \qquad [x^2 < 1].$$

9.10. $\quad (1 \pm x)^{-5} \; = 1 \mp \dfrac{1}{1 \cdot 2 \cdot 3 \cdot 4}\{2 \cdot 3 \cdot 4 \cdot 5x \mp 3 \cdot 4 \cdot 5 \cdot 6x^2$

$$+ 4 \cdot 5 \cdot 6 \cdot 7x^3 \mp 5 \cdot 6 \cdot 7 \cdot 8x^4 + \cdots\}, \qquad [x^2 < 1].$$

10.

$2! =$	2
$3! =$	6
$4! =$	24
$5! =$	120
$6! =$	720
$7! =$	$5\ 040$
$8! =$	$40\ 320$
$9! =$	$362\ 880$
$10! =$	$3\ 628\ 800$
$11! =$	$39\ 916\ 800$

10.1.

$1/2!$	$= .5$
$1/3!$	$= .166\ 666\ 7$
$1/4!$	$= .041\ 666\ 7$
$1/5!$	$= .008\ 333\ 3$
$1/6!$	$= .001\ 388\ 9$
$1/7!$	$= .000\ 198\ 4$
$1/8!$	$= .000\ 024\ 80$
$1/9!$	$= .000\ 002\ 756$
$1/10!$	$= .000\ 000\ 275\ 6$
$1/11!$	$= .000\ 000\ 025\ 05$

For a large table see Ref. 59, v. 1, second section, pp. 58–68.

11.
$$\lim_{n \to \infty} \frac{n!}{n^n e^{-n} \sqrt{n}} = \sqrt{(2\pi)}.$$

This gives approximate values of $n!$ for large values of n. When $n = 12$ the value given by the formula is $0.007(n!)$ too large and when $n = 20$ it is $0.004(n!)$ too large. [Ref. 21, p. 74. See also 851.4 and 850.4.]

12.

$2^2 = 4.$	$2^6 = 64.$	$2^{10} = 1024.$
$2^3 = 8.$	$2^7 = 128.$	$2^{11} = 2048.$
$2^4 = 16.$	$2^8 = 256.$	$2^{12} = 4096.$
$2^5 = 32.$	$2^9 = 512.$	$2^{13} = 8192.$

15.1. $(a + b + c)^2 \equiv a^2 + b^2 + c^2 + 2ab + 2bc + 2ca.$

[The sign $\equiv$ expresses an identity.]

15.2. $(a + b - c)^2 \equiv a^2 + b^2 + c^2 + 2ab - 2bc - 2ca.$

15.3. $(a - b - c)^2 \equiv a^2 + b^2 + c^2 - 2ab + 2bc - 2ca.$

16. $(a + b + c + d)^2 \equiv a^2 + b^2 + c^2 + d^2 + 2ab + 2ac$
$$+ 2ad + 2bc + 2bd + 2cd.$$

17. $(a + b + c)^3 \equiv a^3 + b^3 + c^3 + 6abc$
$$+ 3(a^2b + ab^2 + b^2c + bc^2 + c^2a + ca^2).$$

20.1. $a + x \equiv (a^2 - x^2)/(a - x).$

20.11. $1 + x \equiv (1 - x^2)/(1 - x).$

20.2. $a^2 + ax + x^2 \equiv (a^3 - x^3)/(a - x).$

20.3. $a^3 + a^2x + ax^2 + x^3 \equiv (a^4 - \dot{x}^4)/(a - x)$
$$\equiv (a^2 + x^2)(a + x).$$

20.4. $a^4 + a^3x + a^2x^2 + ax^3 + x^4 \equiv (a^5 - x^5)/(a - x).$

20.5. $a^5 + a^4x + a^3x^2 + a^2x^3 + ax^4 + x^5$
$$\equiv (a^6 - x^6)/(a - x) \equiv (a^3 + x^3)(a^2 + ax + x^2).$$

21.1. $a - x \equiv (a^2 - x^2)/(a + x).$

21.2. $a^2 - ax + x^2 \equiv (a^3 + x^3)/(a + x).$

21.3. $a^3 - a^2x + ax^2 - x^3 \equiv (a^4 - x^4)/(a + x)$
$$\equiv (a^2 + x^2)(a - x).$$

21.4. $a^4 - a^3x + a^2x^2 - ax^3 + x^4 \equiv (a^5 + x^5)/(a + x).$

21.5. $a^5 - a^4x + a^3x^2 - a^2x^3 + ax^4 - x^5$
$$\equiv (a^6 - x^6)/(a + x) \equiv (a^3 - x^3)(a^2 - ax + x^2).$$

22. $a^4 + a^2x^2 + x^4 \equiv (a^6 - x^6)/(a^2 - x^2)$
$$\equiv (a^2 + ax + x^2)(a^2 - ax + x^2).$$

22.1. $a^4 - a^2x^2 + x^4 \equiv (a^6 + x^6)/(a^2 + x^2).$

23. $a^4 + x^4 \equiv (a^2 + x^2)^2 - 2a^2x^2$
$$\equiv (a^2 + ax\sqrt{2} + x^2)(a^2 - ax\sqrt{2} + x^2).$$

25. **Arithmetic Progression** of the first order (first differences constant), to n terms,

$$a + (a + d) + (a + 2d) + (a + 3d) + \cdots + \{a + (n - 1)d\}$$

$$\equiv na + \frac{1}{2}n(n - 1)d$$

$$\equiv \frac{n}{2} \text{ (1st term + nth term).}$$

26. **Geometric Progression,** to n terms,

$$a + ar + ar^2 + ar^3 + \cdots + ar^{n-1} \equiv a(1 - r^n)/(1 - r)$$
$$\equiv a(r^n - 1)/(r - 1).$$

26.1. If $r^2 < 1$, the limit of the sum of an infinite number of terms is $a/(1 - r)$.

27. The reciprocals of the terms of a series in arithmetic progression of the first order are in **Harmonic Progression.** Thus

$$\frac{1}{a}, \qquad \frac{1}{a + d}, \qquad \frac{1}{a + 2d}, \qquad \cdots \qquad \frac{1}{a + (n - 1)d}$$

are in Harmonic Progression.

28.1. The **Arithmetic Mean** of n quantities is

$$\frac{1}{n}(a_1 + a_2 + a_3 + \cdots + a_n).$$

28.2. The **Geometric Mean** of n quantities is

$$(a_1\, a_2\, a_3\, \cdots\, a_n)^{1/n}.$$

28.3. Let the **Harmonic Mean** of n quantities be H. Then

$$\frac{1}{H} = \frac{1}{n}\left(\frac{1}{a_1} + \frac{1}{a_2} + \frac{1}{a_3} + \cdots + \frac{1}{a_n}\right).$$

28.4. The arithmetic mean of a number of positive quantities is $\geqq$ their geometric mean, which in turn is $\geqq$ their harmonic mean.

29. Arithmetic Progression of the kth order (kth differences constant).

Series: $u_1, u_2, u_3, \cdots u_n$.

First differences: $d_1', d_2', d_3', \cdots$
where $d_1' = u_2 - u_1$, $d_2' = u_3 - u_2$, etc.

Second differences: $d_1'', d_2'', d_3'', \cdots$
where $d_1'' = d_2' - d_1'$, etc.

Sum of n terms of the series

$$= \frac{n!}{(n-1)!\,1!}u_1 + \frac{n!}{(n-2)!\,2!}d_1' + \frac{n!}{(n-3)!\,3!}d_1'' + \cdots$$

29.01. If a numerical table consists of values u_n of a function at equal intervals h of the argument, as follows,

$$f(a) = u_1, \quad f(a+h) = u_2, \quad f(a+2h) = u_3, \quad \text{etc.,}$$

then

$$f(a + ph) = u_1 + pd_1' + \frac{p(p-1)}{2!}d_1''$$
$$+ \frac{p(p-1)(p-2)}{3!}d_1''' + \cdots$$

where $p < 1$ and where d_1', d_1'', etc., are given by 29. The coefficients of d_1', d_1'', d_1''', etc., are called Gregory-Newton

Interpolation Coefficients. For numerical values of these coefficients see Ref. 44, v. 1, pp. 102–109 and Ref. 45, pp. 184–185.

29.1. $1 + 2 + 3 + \cdots + n = \dfrac{n}{2}(n + 1).$

29.2. $1^2 + 2^2 + 3^2 + \cdots + n^2 = \dfrac{n}{6}(n + 1)(2n + 1)$

$$= \dfrac{n}{6}(2n^2 + 3n + 1).$$

29.3. $1^3 + 2^3 + 3^3 + \cdots + n^3 = \dfrac{n^2}{4}(n + 1)^2$

$$= \dfrac{n^2}{4}(n^2 + 2n + 1).$$

29.4. $1^4 + 2^4 + 3^4 + \cdots + n^4$

$$= \dfrac{n}{30}(n + 1)(2n + 1)(3n^2 + 3n - 1)$$

$$= \dfrac{n}{30}(6n^4 + 15n^3 + 10n^2 - 1).$$

29.9. $\displaystyle\sum_{u=1}^{n} u^p = \dfrac{n^{p+1}}{p + 1} + \dfrac{n^p}{2} + \dfrac{B_1}{2!}pn^{p-1}$

$$- \dfrac{B_2}{4!}p(p - 1)(p - 2)n^{p-3} + \cdots,$$

omitting terms in n^0 and those that follow.

For values of B_1, B_2, $\cdots$, see **45**.

The above results may be used to find the sum of a series whose nth term is made up of n, n^2, n^3, etc.

30.1. $1 + 3 + 5 + 7 + 9 + \cdots + (2n - 1) = n^2.$

30.2. $1 + 8 + 16 + 24 + 32 + \cdots + 8(n - 1) = (2n - 1)^2.$

33.1. $1 + 3x + 5x^2 + 7x^3 + \cdots = \dfrac{1 + x}{(1 - x)^2}.$

33.2. $1 + ax + (a + b)x^2 + (a + 2b)x^3 + \cdots$

$$= 1 + \dfrac{ax + (b - a)x^2}{(1 - x)^2}.$$

33.3. $1 + 2^2x + 3^2x^2 + 4^2x^3 + \cdots = \dfrac{1 + x}{(1 - x)^3}.$

33.4. $1 + 3^2x + 5^2x^2 + 7^2x^3 + \cdots = \dfrac{1 + 6x + x^2}{(1 - x)^3}.$

[Contributed by W. V. Lyon. Ref. 43, p. 448.]

35. $\dfrac{1}{a} - \dfrac{1}{a + b} + \dfrac{1}{a + 2b} - \dfrac{1}{a + 3b} + \cdots = \displaystyle\int_0^1 \dfrac{x^{a-1}}{1 + x^b}\, dx,$

$[a, b > 0].$

35.1. $1 - \dfrac{1}{3} + \dfrac{1}{5} - \dfrac{1}{7} + \dfrac{1}{9} - \cdots = \dfrac{\pi}{4}.$ [See **120** and **48.1.**]

35.2. $1 - \dfrac{1}{4} + \dfrac{1}{7} - \dfrac{1}{10} + \dfrac{1}{13} - \cdots = \dfrac{1}{3}\left(\dfrac{\pi}{\sqrt{3}} + \log_\epsilon 2\right).$

[See **165.01.**]

35.3. $\dfrac{1}{2} - \dfrac{1}{5} + \dfrac{1}{8} - \dfrac{1}{11} + \dfrac{1}{14} - \cdots = \dfrac{1}{3}\left(\dfrac{\pi}{\sqrt{3}} - \log_\epsilon 2\right).$

[See **165.11.**]

35.4. $1 - \dfrac{1}{5} + \dfrac{1}{9} - \dfrac{1}{13} + \dfrac{1}{17} - \cdots$

$= \dfrac{1}{4\sqrt{2}}\{\pi + 2\log_\epsilon(\sqrt{2} + 1)\}.$ [See **170.**]

[Ref. 34, p. 161, Ex. 1.]

38.   If there is a **power series** for $f(h)$, it is

$$f(h) = f(0) + hf'(0) + \dfrac{h^2}{2!} f''(0) + \dfrac{h^3}{3!} f'''(0) + \cdots.$$

[MACLAURIN'S SERIES.]

38.1. $f(h) = f(0) + hf'(0) + \dfrac{h^2}{2!}f''(0) + \dfrac{h^3}{3!}f'''(0) + \cdots$

$+ \dfrac{h^{n-1}}{(n-1)!} f^{(n-1)}(0) + R_n,$

where, for a suitable value of θ between 0 and 1,

$$R_n = \dfrac{h^n}{n!}f^{(n)}(\theta h), \quad \text{or} \quad \dfrac{h^n}{(n-1)!}(1-\theta)^{n-1}f^{(n)}(\theta h).$$

39. $f(x + h) = f(x) + hf'(x) + \dfrac{h^2}{2!}f''(x) + \dfrac{h^3}{3!}f'''(x) + \cdots.$

[TAYLOR'S SERIES.]

39.1. $f(x + h) = f(x) + hf'(x) + \dfrac{h^2}{2!} f''(x) + \cdots$

$$+ \frac{h^{n-1}}{(n-1)!} f^{(n-1)}(x) + R_n,$$

where, for a suitable value of θ between 0 and 1,

$$R_n = \frac{h^n}{n!} f^{(n)}(x + \theta h), \quad \text{or} \quad \frac{h^n}{(n-1)!} (1 - \theta)^{n-1} f^{(n)}(x + \theta h).$$

40. $f(x + h, y + k) = f(x, y) + \left\{ h \dfrac{\partial f(x, y)}{\partial x} + k \dfrac{\partial f(x, y)}{\partial y} \right\}$

$$+ \frac{1}{2!} \left\{ h^2 \frac{\partial^2 f(x, y)}{\partial x^2} + 2hk \frac{\partial^2 f(x, y)}{\partial x\, \partial y} + k^2 \frac{\partial^2 f(x, y)}{\partial y^2} \right\}$$

$$+ \frac{1}{3!} \left\{ h^3 \frac{\partial^3 f(x, y)}{\partial x^3} + 3h^2 k \frac{\partial^3 f(x, y)}{\partial x^2 \partial y} + 3hk^2 \frac{\partial^3 f(x, y)}{\partial x\, \partial y^2} \right.$$

$$\left. + k^3 \frac{\partial^3 f(x, y)}{\partial y^3} \right\} + \cdots + R_n$$

where, for suitable values of θ_1 and θ_2 between 0 and 1,

$$R_n = \frac{1}{n!} \left\{ h^n \frac{\partial^n}{d x^n} + n h^{n-1} k \frac{\partial^n}{\partial x^{n-1} \partial y} \right.$$

$$+ \frac{n(n-1)}{2!} h^{n-2} k^2 \frac{\partial^n}{\partial x^{n-2} \partial y^2} + \cdots$$

$$\left. + k^n \frac{\partial^n}{\partial y^n} \right\} f(x + \theta_1 h, y + \theta_2 k). \quad [\text{Ref. 5, No. 807.}]$$

42.1. A number is divisible by 3 if the sum of the figures is divisible by 3.

42.2. A number is divisible by 9 if the sum of the figures is divisible by 9.

42.3. A number is divisible by 2^n if the number consisting of the last n figures is divisible by 2^n.

$P(x,y)\,dx + Q(x,y)\,dy = 0 \qquad \dfrac{\partial P}{\partial y} = \dfrac{\partial Q}{\partial x} \to \text{exact}$

$f(x,y) = \int P(x,y)\,dx + f(y)$

$f_y = \int \frac{\partial}{\partial y}(P(x,y)\,dx) + \frac{\partial}{\partial y} f(y) = Q(x,y)$

$f(y) = \int \frac{\partial}{\partial y}(f(y))$

$f(x,y) = \int P(x,y)\,dx + \int \frac{\partial}{\partial y}(f(y))$

Bernoulli's Numbers and Euler's Numbers

45.

BERNOULLI'S NUMBERS	$\text{Log}_{10} B_n$	EULER'S NUMBERS	$\text{Log}_{10} E_n$
$B_1 = \dfrac{1}{6}$	$\bar{1}.221\ 8487$	$E_1 = 1$	0
$B_2 = \dfrac{1}{30}$	$\bar{2}.522\ 8787$	$E_2 = 5$	$0.698\ 9700$
$B_3 = \dfrac{1}{42}$	$\bar{2}.376\ 7507$	$E_3 = 61$	$1.785\ 3298$
$B_4 = \dfrac{1}{30}$	$\bar{2}.522\ 8787$	$E_4 = 1,385$	$3.141\ 4498$
$B_5 = \dfrac{5}{66}$	$\bar{2}.879\ 4261$	$E_5 = 50,521$	$4.703\ 4719$
$B_6 = \dfrac{691}{2730}$	$\bar{1}.403\ 3154$	$E_6 = 2,702,765$	$6.431\ 8083$
$B_7 = \dfrac{7}{6}$	$0.066\ 9468$	$E_7 = 199,360,981$	$8.299\ 6402$
$B_8 = \dfrac{3617}{510}$	$0.850\ 7783$		
$B_9 = \dfrac{43,867}{798}$	$1.740\ 1350$		
$B_{10} = \dfrac{174,611}{330}$	$2.723\ 5577$		
$B_{11} = \dfrac{854,513}{138}$	$3.791\ 8396$		

For large tables see Ref. 27, pp. 176, 178; Ref. 34, pp. 234, 260; Ref. 44, v. 2, p. 230–242 and 294–302; and Ref. 59, (v. 1), second section, pp. 83–89.

The above notation is used in Ref. 27 and 34 and in "American Standard Mathematical Symbols," *Report of 1928*, Ref. 28. There are several different notations in use and, as stated in the above report, it is desirable when using the letters B and E for the above series of numbers, to give 47.1 and 47.4 as definitions, or to state explicitly the values of the first few numbers, as $B_1 = 1/6$, $B_2 = 1/30$, $B_3 = 1/42$, etc., $E_1 = 1$, $E_2 = 5$, $E_3 = 61$, etc.

46.1.
$$E_n = \frac{(2n)!}{(2n-2)!\,2!}\,E_{n-1} - \frac{(2n)!}{(2n-4)!\,4!}\,E_{n-2}$$
$$+ \cdots + (-1)^{n-1}$$

taking $0! = 1$ and $E_0 = 1$.

46.2.
$$B_n = \frac{2n}{2^{2n}(2^{2n}-1)}\left[\frac{(2n-1)!}{(2n-2)!\,1!}\,E_{n-1} - \frac{(2n-1)!}{(2n-4)!\,3!}\,E_{n-2}\right.$$
$$\left. + \cdots + (-1)^{n-1}\right].$$

47.1. $B_n = \dfrac{(2n)!}{\pi^{2n}2^{2n-1}}\left[1 + \dfrac{1}{2^{2n}} + \dfrac{1}{3^{2n}} + \dfrac{1}{4^{2n}} + \cdots\right].$

47.2. $B_n = \dfrac{(2n)!}{\pi^{2n}(2^{2n-1}-1)}\left[1 - \dfrac{1}{2^{2n}} + \dfrac{1}{3^{2n}} - \dfrac{1}{4^{2n}} + \cdots\right].$

47.3. $B_n = \dfrac{2(2n)!}{\pi^{2n}(2^{2n}-1)}\left[1 + \dfrac{1}{3^{2n}} + \dfrac{1}{5^{2n}} + \dfrac{1}{7^{2n}} + \cdots\right].$

47.4. $E_n = \dfrac{2^{2n+2}(2n)!}{\pi^{2n+1}}\left[1 - \dfrac{1}{3^{2n+1}} + \dfrac{1}{5^{2n+1}} - \dfrac{1}{7^{2n+1}} + \cdots\right].$

48.1. $1 - \dfrac{1}{3} + \dfrac{1}{5} - \dfrac{1}{7} + \cdots = \dfrac{E_0\pi}{4} = \dfrac{\pi}{4}.$

48.2. $1 + \dfrac{1}{2^2} + \dfrac{1}{3^2} + \dfrac{1}{4^2} + \cdots = B_1\pi^2 = \dfrac{\pi^2}{6}.$

48.3. $1 - \dfrac{1}{2^2} + \dfrac{1}{3^2} - \dfrac{1}{4^2} + \cdots = \dfrac{B_1\pi^2}{2} = \dfrac{\pi^2}{12}.$

48.4. $1 + \dfrac{1}{3^2} + \dfrac{1}{5^2} + \dfrac{1}{7^2} + \cdots = \dfrac{3B_1\pi^2}{4} = \dfrac{\pi^2}{8}.$

48.5. $1 + \dfrac{1}{2^4} + \dfrac{1}{3^4} + \dfrac{1}{4^4} + \cdots = \dfrac{B_2\pi^4}{3} = \dfrac{\pi^4}{90}.$

Reversion of Series

50. Let a known series be

$$y = ax + bx^2 + cx^3 + dx^4 + ex^5 + fx^6 + gx^7 + \cdots, \quad [a \neq 0],$$

to find the coefficients of the series

$$x = Ay + By^2 + Cy^3 + Dy^4 + Ey^5 + Fy^6 + Gy^7 + \cdots.$$

$A = \dfrac{1}{a}.$ $\qquad B = -\dfrac{b}{a^3}.$ $\qquad C = \dfrac{1}{a^5}(2b^2 - ac).$

$D = \dfrac{1}{a^7}(5abc - a^2d - 5b^3).$

$E = \dfrac{1}{a^9}(6a^2bd + 3a^2c^2 + 14b^4 - a^3e - 21ab^2c).$

$F = \dfrac{1}{a^{11}}(7a^3be + 7a^3cd + 84ab^3c - a^4f - 28a^2b^2d$
$$- 28a^2bc^2 - 42b^5).$$

$G = \dfrac{1}{a^{13}}(8a^4bf + 8a^4ce + 4a^4d^2 + 120a^2b^3d + 180a^2b^2c^2$
$$+ 132b^6 - a^5g - 36a^3b^2e - 72a^3bcd - 12a^3c^3 - 330ab^4c).$$

[See Ref. 23, p. 11, Ref. 31, p. 116 and *Philosophical Magazine*, vol. 19 (1910), p. 366, for additional coefficients.]

Powers of $S = a + bx + cx^2 + dx^3 + ex^4 + fx^5 \cdots$

51.1. $S^2 = a^2 + 2abx + (b^2 + 2ac)x^2 + 2(ad + bc)x^3$
$$+ (c^2 + 2ae + 2bd)x^4 + 2(af + be + cd)x^5 \cdots.$$

51.2. $S^{1/2} = a^{1/2}\left[1 + \dfrac{1}{2}\dfrac{b}{a}x + \left(\dfrac{1}{2}\dfrac{c}{a} - \dfrac{1}{8}\dfrac{b^2}{a^2}\right)x^2\right.$
$$+ \left(\dfrac{1}{2}\dfrac{d}{a} - \dfrac{1}{4}\dfrac{bc}{a^2} + \dfrac{1}{16}\dfrac{b^3}{a^3}\right)x^3$$
$$\left.+ \left(\dfrac{1}{2}\dfrac{e}{a} - \dfrac{1}{4}\dfrac{bd}{a^2} - \dfrac{1}{8}\dfrac{c^2}{a^2} + \dfrac{3}{16}\dfrac{b^2c}{a^3} - \dfrac{5}{128}\dfrac{b^4}{a^4}\right)x^4 \cdots\right].$$

51.3. $S^{-1/2} = a^{-1/2}\left[1 - \dfrac{1}{2}\dfrac{b}{a}x + \left(\dfrac{3}{8}\dfrac{b^2}{a^2} - \dfrac{1}{2}\dfrac{c}{a}\right)x^2\right.$
$$+ \left(\dfrac{3}{4}\dfrac{bc}{a^2} - \dfrac{1}{2}\dfrac{d}{a} - \dfrac{5}{16}\dfrac{b^3}{a^3}\right)x^3$$
$$\left.+ \left(\dfrac{3}{4}\dfrac{bd}{a^2} + \dfrac{3}{8}\dfrac{c^2}{a^2} - \dfrac{1}{2}\dfrac{e}{a} - \dfrac{15}{16}\dfrac{b^2c}{a^3} + \dfrac{35}{128}\dfrac{b^4}{a^4}\right)x^4 \cdots\right].$$

51.4. $S^{-1} = a^{-1}\left[1 - \dfrac{b}{a}x + \left(\dfrac{b^2}{a^2} - \dfrac{c}{a}\right)x^2 + \left(\dfrac{2bc}{a^2} - \dfrac{d}{a} - \dfrac{b^3}{a^3}\right)x^3\right.$
$$\left.+ \left(\dfrac{2bd}{a^2} + \dfrac{c^2}{a^2} - \dfrac{e}{a} - 3\dfrac{b^2c}{a^3} + \dfrac{b^4}{a^4}\right)x^4 \cdots\right].$$

51.5. $S^{-2} = a^{-2}\left[1 - 2\dfrac{b}{a}x + \left(3\dfrac{b^2}{a^2} - 2\dfrac{c}{a}\right)x^2\right.$
$$+ \left(6\dfrac{bc}{a^2} - 2\dfrac{d}{a} - 4\dfrac{b^3}{a^3}\right)x^3$$
$$\left.+ \left(6\dfrac{bd}{a^2} + 3\dfrac{c^2}{a^2} - 2\dfrac{e}{a} - 12\dfrac{b^2c}{a^3} + 5\dfrac{b^4}{a^4}\right)x^4 \cdots\right].$$

Roots of Quadratic Equation

55.1. The roots of $ax^2 + bx + c = 0$ are
$$\alpha = \frac{-b + \sqrt{(b^2 - 4ac)}}{2a} = \frac{-2c}{b + \sqrt{(b^2 - 4ac)}},$$
$$\beta = \frac{-b - \sqrt{(b^2 - 4ac)}}{2a} = \frac{-2c}{b - \sqrt{(b^2 - 4ac)}}.$$

The difference of two quantities is inconvenient to compute with precision and in such a case the alternative formula using the numerical sum of two quantities should be used.   [Ref. 41, p. 306.]

55.2. If one root α has been computed precisely, use

$$\beta = \frac{c}{a\alpha}.$$

Square Roots of Complex Quantity

58.1. $\sqrt{(x + iy)} = \pm \left[\sqrt{\left(\frac{r + x}{2}\right)} + i \sqrt{\left(\frac{r - x}{2}\right)} \right].$

58.2. $\sqrt{(x - iy)} = \pm \left[\sqrt{\left(\frac{r + x}{2}\right)} - i \sqrt{\left(\frac{r - x}{2}\right)} \right],$

where x may be positive or negative,

  y is positive
  $r = + \sqrt{(x^2 + y^2)}$
  $i = \sqrt{(-1)}.$

The positive square roots of $(r + x)/2$ and $(r - x)/2$ are to be used.
[Ref. 61, p. 260.]

58.3. An alternative method is to put $x + iy$ in the form

$$re^{i(\theta + 2\pi k)} \qquad \text{(see 604.05)}$$

where $r = \sqrt{(x^2 + y^2)}$, $\cos \theta = x/r$, $\sin \theta = y/r$, and k is an integer or 0. Then

$$\sqrt{(x + iy)} = \sqrt{(re^{i\theta})} = \pm \sqrt{r}\, e^{i\theta/2}$$
$$= \pm \sqrt{r}\left(\cos \frac{\theta}{2} + i \sin \frac{\theta}{2}\right).$$

I. F.

$\Theta(x,y) = e^{\int a(x)\,dx + b(y)\,dy}$

$\frac{\partial P}{\partial y} - \frac{\partial Q}{\partial x} = a(x)\, Q(x,y) - b(x)\, P(x,y)$

ALGEBRAIC FUNCTIONS—DERIVATIVES

60. $\dfrac{d(au)}{dx} = a\dfrac{du}{dx}$ where a is a constant.

61. $\dfrac{d(u+v)}{dx} = \dfrac{du}{dx} + \dfrac{dv}{dx}$.

62. $\dfrac{d(uv)}{dx} = u\dfrac{dv}{dx} + v\dfrac{du}{dx}$.

63. $\dfrac{d(uvw)}{dx} = uv\dfrac{dw}{dx} + vw\dfrac{du}{dx} + wu\dfrac{dv}{dx}$.

64. $\dfrac{d(x^n)}{dx} = nx^{n-1}$.

64.1. $\dfrac{d\sqrt{x}}{dx} = \dfrac{1}{2\sqrt{x}}$.

64.2. $\dfrac{d(1/x)}{dx} = -\dfrac{1}{x^2}$.

65. $\dfrac{d(u/v)}{dx} = \dfrac{1}{v}\dfrac{du}{dx} - \dfrac{u}{v^2}\dfrac{dv}{dx} = \dfrac{v\dfrac{du}{dx} - u\dfrac{dv}{dx}}{v^2}$.

66. $\dfrac{df(u)}{dx} = \dfrac{df(u)}{du} \cdot \dfrac{du}{dx}$.

67. $\dfrac{d^2f(u)}{dx^2} = \dfrac{df(u)}{du} \cdot \dfrac{d^2u}{dx^2} + \dfrac{d^2f(u)}{du^2} \cdot \left(\dfrac{du}{dx}\right)^2$.

68. $\dfrac{d^n(uv)}{dx^n} = v\dfrac{d^nu}{dx^n} + n\dfrac{dv}{dx}\dfrac{d^{n-1}u}{dx^{n-1}} + \dfrac{n(n-1)}{2!}\dfrac{d^2v}{dx^2}\dfrac{d^{n-2}u}{dx^{n-2}}$

$$+ \cdots + \dfrac{n!}{(n-k)!\,k!}\dfrac{d^kv}{dx^k}\dfrac{d^{n-k}u}{dx^{n-k}} + \cdots + \dfrac{ud^nv}{dx^n}.$$

69.1. $\dfrac{d}{dq}\displaystyle\int_p^q f(x)dx = f(q),$ $[p$ constant$]$.

69.2. $\dfrac{d}{dp}\displaystyle\int_p^q f(x)dx = -f(p),$ $[q$ constant$]$.

69.3. $\dfrac{d}{dc}\displaystyle\int_p^q f(x,c)dx = \int_\tau^q \dfrac{\partial}{\partial c}f(x,c)dx + f(q,c)\dfrac{dq}{dc} - f(p,c)\dfrac{dp}{dc}$.

72. If $\varphi(a) = 0$ and $\psi(a) = 0$, or if $\varphi(a) = \infty$ and $\psi(a) = \infty$, then

$$\lim_{x \to a} \frac{\varphi(x)}{\psi(x)} = \frac{\varphi'(a)}{\psi'(a)} .$$

If, also, $\varphi'(a) = 0$ and $\psi'(a) = 0$, or if $\varphi'(a) = \infty$ and $\psi'(a) = \infty$, then

$$\lim_{x \to a} \frac{\varphi(x)}{\psi(x)} = \frac{\varphi''(a)}{\psi''(a)} , \quad \text{and so on.}$$

72.1. If a function takes the form $0 \times \infty$ or $\infty - \infty$, it may, by an algebraic or other change, be made to take the form $0/0$ or ∞/∞.

72.2. If a function takes the form 0^0, ∞^0 or 1^∞, it may be made to take the form $0 \times \infty$ and therefore $0/0$ or ∞/∞ by first taking logarithms.   [Ref. 8, Chap. 42.]

79. General Formula for Integration by Parts.

$$\int u \, dv = uv - \int v \, du,$$

or

$$\int u \, dv = uv - \int v \, \frac{du}{dv} \, dv.$$

RATIONAL ALGEBRAIC FUNCTIONS—INTEGRALS

The constant of integration is to be understood with all integrals.

Integrals Involving x^n

80. $\displaystyle\int dx = x.$ **81.2.** $\displaystyle\int x^2 dx = \frac{x^3}{3}\cdot$

81.1. $\displaystyle\int x\,dx = \frac{x^2}{2}\cdot$ **81.9.** $\displaystyle\int x^n dx = \frac{x^{n+1}}{n+1},\quad [n \neq -1].$

82.1. $\displaystyle\int \frac{dx}{x} = \log_\epsilon |x|.$ [See note preceding 600.]

Integration in this case should not be carried from a negative to a positive value of x. If x is negative, use $\log |x|$, since $\log(-1) \equiv (2k+1)\pi i$ will be part of the constant of integration.   [See **409.03.**]

FIG. 82.1.   Graphs of $y = 1/x$ and $y = \log_\epsilon |x|$, where x is real.

82.2. $\displaystyle\int \frac{dx}{x^2} = -\frac{1}{x}\cdot$ **82.4.** $\displaystyle\int \frac{dx}{x^4} = -\frac{1}{3x^3}\cdot$

82.3. $\displaystyle\int \frac{dx}{x^3} = -\frac{1}{2x^2}\cdot$ **82.5.** $\displaystyle\int \frac{dx}{x^5} = -\frac{1}{4x^4}\cdot$

82.9. $\displaystyle\int \frac{dx}{x^n} = -\frac{1}{(n-1)x^{n-1}},$ $[n \neq 1].$

Integrals Involving $X = a + bx$

83. $\displaystyle\int (a + bx)^n dx = \frac{1}{b}\int X^n dX = \frac{X^{n+1}}{b(n + 1)}, \qquad [n \neq -1].$

84.1. $\displaystyle\int x^m(a + bx)^n dx$ may be integrated term-by-term after expanding $(a + bx)^n$ by the binomial theorem, when n is a positive integer.

84.2. If $m < n$, or if n is fractional, it may be shorter to use

$$\int x^m X^n dx = \frac{1}{b^{m+1}}\int (X - a)^m X^n dX$$

and expand $(X - a)^m$ by the binomial theorem, when m is a positive integer.

85. On integrals of rational algebraic fractions, see the topic partial fractions in text books, *e.g.*, Chapter II, Reference 7.

89. General formula for **90** to **95**:

$$\int \frac{x^m dx}{X^n} = \frac{1}{b^{m+1}}\int \frac{(X - a)^m dX}{X^n}$$

$$= \frac{1}{b^{m+1}}\left[\sum_{s=0}^{m} \frac{m!(- a)^s X^{m-n-s+1}}{(m - s)!\, s!\, (m - n - s + 1)} \right],$$

except where $m - n - s + 1 = 0$, in which case the corresponding term in the square brackets is

$$\frac{m!\, (- a)^{m-n+1}}{(m - n + 1)!\, (n - 1)!} \log |X|,$$

the letters representing real quantities. [Ref. 2, p. 7.] Integration should not be carried from a negative to a positive value of X in the case of $\log |X|$. If X is negative, use $\log |X|$ since $\log (- 1) \equiv (2k + 1)\pi i$ will be part of the constant of integration.

90. $\displaystyle\int \frac{dx}{X^n} = \frac{- 1}{(n - 1)bX^{n-1}}, \qquad\qquad [n \neq 1].$

90.1. $\displaystyle\int \frac{dx}{X} = \frac{1}{b}\log |X|.$ [See note on $\log |X|$ under **89**.]

90.2. $\displaystyle\int \frac{dx}{X^2} = -\frac{1}{bX}.$ **90.3.** $\displaystyle\int \frac{dx}{X^3} = -\frac{1}{2bX^2}.$

90.4. $\displaystyle\int \frac{dx}{X^4} = -\frac{1}{3bX^3}.$ **90.5.** $\displaystyle\int \frac{dx}{X^5} = -\frac{1}{4bX^4}.$

91. $\displaystyle\int \frac{x\,dx}{X^n} = \frac{1}{b^2}\left[\frac{-1}{(n-2)X^{n-2}} + \frac{a}{(n-1)X^{n-1}}\right],$

[except where any one of the exponents of X is 0, see 89].

91.1. $\displaystyle\int \frac{x\,dx}{X} = \frac{1}{b^2}[X - a\log|X|].$ [If $X < 0$, use $\log|X|$, see **89**.]

91.2. $\displaystyle\int \frac{x\,dx}{X^2} = \frac{1}{b^2}\left[\log|X| + \frac{a}{X}\right].$

91.3. $\displaystyle\int \frac{x\,dx}{X^3} = \frac{1}{b^2}\left[-\frac{1}{X} + \frac{a}{2X^2}\right].$

91.4. $\displaystyle\int \frac{x\,dx}{X^4} = \frac{1}{b^2}\left[-\frac{1}{2X^2} + \frac{a}{3X^3}\right].$

91.5. $\displaystyle\int \frac{x\,dx}{X^5} = \frac{1}{b^2}\left[-\frac{1}{3X^3} + \frac{a}{4X^4}\right].$

92. $\displaystyle\int \frac{x^2 dx}{X^n} = \frac{1}{b^3}\left[\frac{-1}{(n-3)X^{n-3}} + \frac{2a}{(n-2)X^{n-2}} - \frac{a^2}{(n-1)X^{n-1}}\right],$

[except where any one of the exponents of X is 0, see 89].

92.1. $\displaystyle\int \frac{x^2 dx}{X} = \frac{1}{b^3}\left[\frac{X^2}{2} - 2aX + a^2\log|X|\right].$

An alternative expression, which differs by a constant, is

$$\frac{x^2}{2b} - \frac{ax}{b^2} + \frac{a^2}{b^3}\log|a + bx|.$$

92.2. $\displaystyle\int \frac{x^2 dx}{X^2} = \frac{1}{b^3}\left[X - 2a\log|X| - \frac{a^2}{X}\right].$

92.3. $\displaystyle\int \frac{x^2 dx}{X^3} = \frac{1}{b^3}\left[\log|X| + \frac{2a}{X} - \frac{a^2}{2X^2}\right].$

92.4. $\displaystyle\int \frac{x^2 dx}{X^4} = \frac{1}{b^3}\left[-\frac{1}{X} + \frac{2a}{2X^2} - \frac{a^2}{3X^3}\right].$

92.5. $\displaystyle\int \frac{x^2 dx}{X^5} = \frac{1}{b^3}\left[-\frac{1}{2X^2} + \frac{2a}{3X^3} - \frac{a^2}{4X^4}\right].$

92.6. $\displaystyle\int \frac{x^2 dx}{X^6} = \frac{1}{b^3}\left[-\frac{1}{3X^3} + \frac{2a}{4X^4} - \frac{a^2}{5X^5}\right].$

92.7. $\displaystyle\int \frac{x^2 dx}{X^7} = \frac{1}{b^3}\left[-\frac{1}{4X^4} + \frac{2a}{5X^5} - \frac{a^2}{6X^6}\right].$

93. $\displaystyle\int \frac{x^3 dx}{X^n} = \frac{1}{b^4}\left[\frac{-1}{(n-4)X^{n-4}} + \frac{3a}{(n-3)X^{n-3}}\right.$
$$\left. -\frac{3a^2}{(n-2)X^{n-2}} + \frac{a^3}{(n-1)X^{n-1}}\right],$$

[except where any one of the exponents of X is 0, see 89].

93.1. $\displaystyle\int \frac{x^3 dx}{X} = \frac{1}{b^4}\left[\frac{X^3}{3} - \frac{3aX^2}{2} + 3a^2X - a^3 \log |X|\right]$
$$= \frac{x^3}{3b} - \frac{ax^2}{2b^2} + \frac{a^2 x}{b^3} - \frac{a^3}{b^4}\log |a + bx| + \text{constant.}$$

93.2. $\displaystyle\int \frac{x^3 dx}{X^2} = \frac{1}{b^4}\left[\frac{X^2}{2} - 3aX + 3a^2 \log |X| + \frac{a^3}{X}\right].$

93.3. $\displaystyle\int \frac{x^3 dx}{X^3} = \frac{1}{b^4}\left[X - 3a \log |X| - \frac{3a^2}{X} + \frac{a^3}{2X^2}\right].$

93.4. $\displaystyle\int \frac{x^3 dx}{X^4} = \frac{1}{b^4}\left[\log |X| + \frac{3a}{X} - \frac{3a^2}{2X^2} + \frac{a^3}{3X^3}\right].$

93.5. $\displaystyle\int \frac{x^3 dx}{X^5} = \frac{1}{b^4}\left[-\frac{1}{X} + \frac{3a}{2X^2} - \frac{3a^2}{3X^3} + \frac{a^3}{4X^4}\right].$

93.6. $\displaystyle\int \frac{x^3 dx}{X^6} = \frac{1}{b^4}\left[-\frac{1}{2X^2} + \frac{3a}{3X^3} - \frac{3a^2}{4X^4} + \frac{a^3}{5X^5}\right].$

93.7. $\displaystyle\int \frac{x^3 dx}{X^7} = \frac{1}{b^4}\left[-\frac{1}{3X^3} + \frac{3a}{4X^4} - \frac{3a^2}{5X^5} + \frac{a^3}{6X^6}\right].$

94. $\displaystyle\int \frac{x^4 dx}{X^n} = \frac{1}{b^5}\left[\frac{-1}{(n-5)X^{n-5}} + \frac{4a}{(n-4)X^{n-4}}\right.$
$$\left. -\frac{6a^2}{(n-3)X^{n-3}} + \frac{4a^3}{(n-2)X^{n-2}} - \frac{a^4}{(n-1)X^{n-1}}\right],$$

[except where any one of the exponents of X is 0, see 89].

94.1. $\displaystyle\int \frac{x^4 dx}{X} = \frac{1}{b^5}\left[\frac{X^4}{4} - \frac{4aX^3}{3} + \frac{6a^2X^2}{2} - 4a^3X + a^4 \log |X|\right]$
$$= \frac{x^4}{4b} - \frac{ax^3}{3b^2} + \frac{a^2 x^2}{2b^3} - \frac{a^3 x}{b^4} + \frac{a^4}{b^5}\log |a + bx| + \text{const.}$$

94.2. $\displaystyle\int \frac{x^4 dx}{X^2} = \frac{1}{b^5}\left[\frac{X^3}{3} - \frac{4aX^2}{2} + 6a^2X - 4a^3 \log |X| - \frac{a^4}{X}\right].$

94.3. $\displaystyle\int \frac{x^4 dx}{X^3} = \frac{1}{b^5}\left[\frac{X^2}{2} - 4aX + 6a^2\log|X| + \frac{4a^3}{X} - \frac{a^4}{2X^2}\right].$

94.4. $\displaystyle\int \frac{x^4 dx}{X^4} = \frac{1}{b^5}\left[X - 4a\log|X| - \frac{6a^2}{X} + \frac{4a^3}{2X^2} - \frac{a^4}{3X^3}\right].$

94.5. $\displaystyle\int \frac{x^4 dx}{X^5} = \frac{1}{b^5}\left[\log|X| + \frac{4a}{X} - \frac{6a^2}{2X^2} + \frac{4a^3}{3X^3} - \frac{a^4}{4X^4}\right].$

94.6. $\displaystyle\int \frac{x^4 dx}{X^6} = \frac{1}{b^5}\left[-\frac{1}{X} + \frac{4a}{2X^2} - \frac{6a^2}{3X^3} + \frac{4a^3}{4X^4} - \frac{a^4}{5X^5}\right].$

94.7. $\displaystyle\int \frac{x^4 dx}{X^7} = \frac{1}{b^5}\left[-\frac{1}{2X^2} + \frac{4a}{3X^3} - \frac{6a^2}{4X^4} + \frac{4a^3}{5X^5} - \frac{a^4}{6X^6}\right].$

95. $\displaystyle\int \frac{x^5 dx}{X^n} = \frac{1}{b^6}\left[\frac{-1}{(n-6)X^{n-6}} + \frac{5a}{(n-5)X^{n-5}}\right.$

$$-\frac{10a^2}{(n-4)X^{n-4}} + \frac{10a^3}{(n-3)X^{n-3}}$$

$$\left. -\frac{5a^4}{(n-2)X^{n-2}} + \frac{a^5}{(n-1)X^{n-1}}\right],$$

[except where any one of the exponents of X is 0, see **89**].

95.1. $\displaystyle\int \frac{x^5 dx}{X} = \frac{1}{b^6}\left[\frac{X^5}{5} - \frac{5aX^4}{4} + \frac{10a^2X^3}{3} - \frac{10a^3X^2}{2}\right.$

$$\left. + 5a^4X - a^5\log|X|\right]$$

$$= \frac{x^5}{5b} - \frac{ax^4}{4b^2} + \frac{a^2x^3}{3b^3} - \frac{a^3x^2}{2b^4} + \frac{a^4x}{b^5}$$

$$-\frac{a^5}{b^6}\log|a + bx| + \text{constant}.$$

[Ref. 1, p. 11.]

95.2. $\displaystyle\int \frac{x^5 dx}{X^2} = \frac{1}{b^6}\left[\frac{X^4}{4} - \frac{5aX^3}{3} + \frac{10a^2X^2}{2} - 10aX^3\right.$

$$\left. + 5a^4\log|X| + \frac{a^5}{X}\right].$$

95.3. $\displaystyle\int \frac{x^5 dx}{X^3} = \frac{1}{b^6}\left[\frac{X^3}{3} - \frac{5aX^2}{2} + 10a^2X - 10a^3\log|X|\right.$

$$\left. -\frac{5a^4}{X} + \frac{a^5}{2X^2}\right].$$

95.4. $\displaystyle\int \frac{x^5 dx}{X^4} = \frac{1}{b^6}\left[\frac{X^2}{2} - 5aX + 10a^2 \log |X| + \frac{10a^3}{X}\right.$

$$\left. - \frac{5a^4}{2X^2} + \frac{a^5}{3X^3}\right].$$

95.5. $\displaystyle\int \frac{x^5 dx}{X^5} = \frac{1}{b^6}\left[X - 5a \log |X| - \frac{10a^2}{X} + \frac{10a^3}{2X^2}\right.$

$$\left. - \frac{5a^4}{3X^3} + \frac{a^5}{4X^4}\right].$$

95.6. $\displaystyle\int \frac{x^5 dx}{X^6} = \frac{1}{b^6}\left[\log |X| + \frac{5a}{X} - \frac{10a^2}{2X^2} + \frac{10a^3}{3X^3}\right.$

$$\left. - \frac{5a^4}{4X^4} + \frac{a^5}{5X^5}\right].$$

95.7. $\displaystyle\int \frac{x^5 dx}{X^7} = \frac{1}{b^6}\left[- \frac{1}{X} + \frac{5a}{2X^2} - \frac{10a^2}{3X^3} + \frac{10a^3}{4X^4}\right.$

$$\left. - \frac{5a^4}{5X^5} + \frac{a^5}{6X^6}\right].$$

95.8. $\displaystyle\int \frac{x^5 dx}{X^8} = \frac{1}{b^6}\left[- \frac{1}{2X^2} + \frac{5a}{3X^3} - \frac{10a^2}{4X^4} + \frac{10a^3}{5X^5}\right.$

$$\left. - \frac{5a^4}{6X^6} + \frac{a^5}{7X^7}\right].\quad \text{[Ref. 2, pp. 7–11.]}$$

100. General formula for **101** to **105**:

$$\int \frac{dx}{x^m X^n} = \frac{-1}{a^{m+n-1}} \int \frac{\left(\dfrac{X}{x} - b\right)^{m+n-2}}{\left(\dfrac{X}{x}\right)^n} \, d\left(\frac{X}{x}\right)$$

$$= \frac{-1}{a^{m+n-1}}\left[\sum_{s=0}^{m+n-2} \frac{(m+n-2)! \, X^{m-s-1}(-b)^s}{(m+n-s-2)! \, s! \, (m-s-1)x^{m-s-1}}\right]$$

unless $m - s - 1 = 0$, when the corresponding term in square brackets is

$$\frac{(m+n-2)!}{(m-1)! \, (n-1)!} (-b)^{m-1} \log \left|\frac{X}{x}\right|.$$

101.1. $\displaystyle\int \frac{dx}{xX} = - \frac{1}{a} \log \left|\frac{X}{x}\right|.$

101.2. $\displaystyle\int \frac{dx}{xX^2} = - \frac{1}{a^2}\left[\log \left|\frac{X}{x}\right| + \frac{bx}{X}\right].$

101.3. $\displaystyle\int \frac{dx}{xX^3} = -\frac{1}{a^3}\left[\log\left|\frac{X}{x}\right| + \frac{2bx}{X} - \frac{b^2x^2}{2X^2}\right].$

101.4. $\displaystyle\int \frac{dx}{xX^4} = -\frac{1}{a^4}\left[\log\left|\frac{X}{x}\right| + \frac{3bx}{X} - \frac{3b^2x^2}{2X^2} + \frac{b^3x^3}{3X^3}\right].$

101.5. $\displaystyle\int \frac{dx}{xX^5} = -\frac{1}{a^5}\left[\log\left|\frac{X}{x}\right| + \frac{4bx}{X} - \frac{6b^2x^2}{2X^2} + \frac{4b^3x^3}{3X^3} - \frac{b^4x^4}{4X^4}\right].$

Alternative solutions, which differ by a constant, are:

101.92. $\displaystyle\int \frac{dx}{xX^2} = \frac{1}{aX} - \frac{1}{a^2}\log\left|\frac{X}{x}\right|.$

101.93. $\displaystyle\int \frac{dx}{xX^3} = \frac{1}{2aX^2} + \frac{1}{a^2X} - \frac{1}{a^3}\log\left|\frac{X}{x}\right|.$

101.94. $\displaystyle\int \frac{dx}{xX^4} = \frac{1}{3aX^3} + \frac{1}{2a^2X^2} + \frac{1}{a^3X} - \frac{1}{a^4}\log\left|\frac{X}{x}\right|.$

101.95. $\displaystyle\int \frac{dx}{xX^5} = \frac{1}{4aX^4} + \frac{1}{3a^2X^3} + \frac{1}{2a^3X^2} + \frac{1}{a^4X} - \frac{1}{a^5}\log\left|\frac{X}{x}\right|.$

[Ref. 2, p. 13.]

102.1. $\displaystyle\int \frac{dx}{x^2X} = -\frac{1}{a^2}\left[\frac{X}{x} - b\log\left|\frac{X}{x}\right|\right].$

102.2. $\displaystyle\int \frac{dx}{x^2X^2} = -\frac{1}{a^3}\left[\frac{X}{x} - 2b\log\left|\frac{X}{x}\right| - \frac{b^2x}{X}\right].$

102.3. $\displaystyle\int \frac{dx}{x^2X^3} = -\frac{1}{a^4}\left[\frac{X}{x} - 3b\log\left|\frac{X}{x}\right| - \frac{3b^2x}{X} + \frac{b^3x^2}{2X^2}\right].$

102.4. $\displaystyle\int \frac{dx}{x^2X^4} = -\frac{1}{a^5}\left[\frac{X}{x} - 4b\log\left|\frac{X}{x}\right| - \frac{6b^2x}{X}\right.$
$$\left. + \frac{4b^3x^2}{2X^2} - \frac{b^4x^3}{3X^3}\right].$$

Alternative solutions, which differ by a constant, are:

102.91. $\displaystyle\int \frac{dx}{x^2X} = -\frac{1}{ax} + \frac{b}{a^2}\log\left|\frac{X}{x}\right|.$

102.92. $\displaystyle\int \frac{dx}{x^2X^2} = -b\left[\frac{1}{a^2X} + \frac{1}{a^2bx} - \frac{2}{a^3}\log\left|\frac{X}{x}\right|\right].$

102.93. $\displaystyle \int \frac{dx}{x^2 X^3} = -b\left[\frac{1}{2a^2 X^2} + \frac{2}{a^3 X} + \frac{1}{a^3 bx} - \frac{3}{a^4}\log\left|\frac{X}{x}\right|\right]$,

where $X = a + bx$.

102.94. $\displaystyle \int \frac{dx}{x^2 X^4} = -b\left[\frac{1}{3a^2 X^3} + \frac{2}{2a^3 X^2} + \frac{3}{a^4 X} + \frac{1}{a^4 bx}\right.$

$$\left. - \frac{4}{a^5}\log\left|\frac{X}{x}\right|\right]. \qquad \text{[Ref. 2, p. 14.]}$$

103.1. $\displaystyle \int \frac{dx}{x^3 X} = -\frac{1}{a^3}\left[\frac{X^2}{2x^2} - \frac{2bX}{x} + b^2\log\left|\frac{X}{x}\right|\right]$

$$= -\frac{1}{2ax^2} + \frac{b}{a^2 x} - \frac{b^2}{a^3}\log\left|\frac{X}{x}\right| + \text{constant.}$$

103.2. $\displaystyle \int \frac{dx}{x^3 X^2} = -\frac{1}{a^4}\left[\frac{X^2}{2x^2} - \frac{3bX}{x} + 3b^2\log\left|\frac{X}{x}\right| + \frac{b^3 x}{X}\right].$

103.3. $\displaystyle \int \frac{dx}{x^3 X^3} = -\frac{1}{a^5}\left[\frac{X^2}{2x^2} - \frac{4bX}{x} + 6b^2\log\left|\frac{X}{x}\right|\right.$

$$\left. + \frac{4b^3 x}{X} - \frac{b^4 x^2}{2X^2}\right].$$

104.1. $\displaystyle \int \frac{dx}{x^4 X} = -\frac{1}{a^4}\left[\frac{X^3}{3x^3} - \frac{3bX^2}{2x^2} + \frac{3b^2 X}{x} - b^3\log\left|\frac{X}{x}\right|\right]$

$$= -\frac{1}{3ax^3} + \frac{b}{2a^2 x^2} - \frac{b^2}{a^3 x} + \frac{b^3}{a^4}\log\left|\frac{X}{x}\right| + \text{const.}$$

104.2. $\displaystyle \int \frac{dx}{x^4 X^2} = -\frac{1}{a^5}\left[\frac{X^3}{3x^3} - \frac{4bX^2}{2x^2} + \frac{6b^2 X}{x}\right.$

$$\left. - 4b^3\log\left|\frac{X}{x}\right| - \frac{b^4 x}{X}\right].$$

105.1. $\displaystyle \int \frac{dx}{x^5 X} = -\frac{1}{4ax^4} + \frac{b}{3a^2 x^3} - \frac{b^2}{2a^3 x^2} + \frac{b^3}{a^4 x} - \frac{b^4}{a^5}\log\left|\frac{X}{x}\right|.$

Integrals Involving Linear Factors

110. $\displaystyle \int \frac{(a + x)dx}{(c + x)} = x + (a - c)\log|c + x|.$

110.1. $\displaystyle \int \frac{(a + fx)dx}{(c + gx)} = \frac{fx}{g} + \frac{ag - cf}{g^2}\log|c + gx|.$

111. $\displaystyle\int \frac{dx}{(a+x)(c+x)} = \frac{1}{a-c}\log\left|\frac{c+x}{a+x}\right|,$ $[\![a \neq c]\!].$

If $a = c$, see **90.2.**

111.1. $\displaystyle\int \frac{dx}{(a+fx)(c+gx)} = \frac{1}{ag-cf}\log\left|\frac{c+gx}{a+fx}\right|,$ $[\![ag \neq cf]\!].$

If $ag = cf$, see **90.2.**

111.2. $\displaystyle\int \frac{x\,dx}{(a+x)(c+x)} = \frac{1}{(a-c)}\{a\log|a+x|$

$$- c\log|c+x|\}.$$

112. $\displaystyle\int \frac{dx}{(a+x)(c+x)^2} = \frac{1}{(c-a)(c+x)}$

$$+ \frac{1}{(c-a)^2}\log\left|\frac{a+x}{c+x}\right|.$$

112.1. $\displaystyle\int \frac{x\,dx}{(a+x)(c+x)^2} = \frac{c}{(a-c)(c+x)}$

$$- \frac{a}{(a-c)^2}\log\left|\frac{a+x}{c+x}\right|.$$

112.2. $\displaystyle\int \frac{x^2 dx}{(a+x)(c+x)^2} = \frac{c^2}{(c-a)(c+x)}$

$$+ \frac{a^2}{(c-a)^2}\log|a+x| + \frac{c^2-2ac}{(c-a)^2}\log|c+x|.$$

113. $\displaystyle\int \frac{dx}{(a+x)^2(c+x)^2} = \frac{-1}{(a-c)^2}\left(\frac{1}{a+x} + \frac{1}{c+x}\right)$

$$+ \frac{2}{(a-c)^3}\log\left|\frac{a+x}{c+x}\right|.$$

113.1. $\displaystyle\int \frac{x\,dx}{(a+x)^2(c+x)^2} = \frac{1}{(a-c)^2}\left(\frac{a}{a+x} + \frac{c}{c+x}\right)$

$$+ \frac{a+c}{(a-c)^3}\log\left|\frac{a+x}{c+x}\right|.$$

113.2. $\displaystyle\int \frac{x^2 dx}{(a+x)^2(c+x)^2} = \frac{-1}{(a-c)^2}\left(\frac{a^2}{a+x} + \frac{c^2}{c+x}\right)$

$$+ \frac{2ac}{(a-c)^3}\log\left|\frac{a+x}{c+x}\right|.$$

[Ref. 1, p. 71.]

Integrals Involving $X = a^2 + x^2$

120. $\displaystyle\int \frac{dx}{1 + x^2} = \tan^{-1} x.$

The principal value of $\tan^{-1} x$ is to be taken, that is,

$$-\frac{\pi}{2} < \tan^{-1} x < \frac{\pi}{2}.$$

FIG. 120. Graphs of $1/(1 + x^2)$ and of principal values of $\tan^{-1} x$.

120.01. $\displaystyle\int \frac{dx}{a^2 + b^2 x^2} = \frac{1}{ab} \tan^{-1} \frac{bx}{a}.$

120.1. $\displaystyle\int \frac{dx}{X} = \int \frac{dx}{a^2 + x^2} = \frac{1}{a} \tan^{-1} \frac{x}{a}.$

120.2. $\displaystyle\int \frac{dx}{X^2} = \frac{x}{2a^2 X} + \frac{1}{2a^3} \tan^{-1} \frac{x}{a}.$

120.3. $\displaystyle\int \frac{dx}{X^3} = \frac{x}{4a^2 X^2} + \frac{3x}{8a^4 X} + \frac{3}{8a^5} \tan^{-1} \frac{x}{a}.$

120.4. $\displaystyle\int \frac{dx}{X^4} = \frac{x}{6a^2 X^3} + \frac{5x}{24a^4 X^2} + \frac{5x}{16a^6 X} + \frac{5}{16a^7} \tan^{-1} \frac{x}{a}.$

120.9. $\displaystyle\int \frac{dx}{(a^2 + b^2 x^2)^{n+1}} = \frac{x}{2na^2(a^2 + b^2 x^2)^n}$

$$+ \frac{2n - 1}{2na^2} \int \frac{dx}{(a^2 + b^2 x^2)^n}.$$

[Ref. 2, p. 20.]

121. Integrals of the form

$$\int \frac{x^{2m+1}dx}{(a^2 \pm x^2)^n}$$

by putting $x^2 = z$, become

$$\frac{1}{2} \int \frac{z^m dz}{(a^2 \pm z)^n}$$

for which see **89** to **105** (m positive, negative or zero).

121.1. $\int \frac{x\,dx}{X} = \int \frac{x\,dx}{a^2 + x^2} = \frac{1}{2} \log (a^2 + x^2).$

121.2. $\int \frac{x\,dx}{X^2} = -\frac{1}{2X}.$ **121.3.** $\int \frac{x\,dx}{X^3} = -\frac{1}{4X^2}.$

121.4. $\int \frac{x\,dx}{X^4} = -\frac{1}{6X^3}.$

121.9. $\int \frac{x\,dx}{X^{n+1}} = -\frac{1}{2nX^n},$ $[n \neq 0].$

122.1. $\int \frac{x^2 dx}{X} = x - a \tan^{-1}\frac{x}{a}.$

122.2. $\int \frac{x^2 dx}{X^2} = -\frac{x}{2X} + \frac{1}{2a} \tan^{-1}\frac{x}{a}.$

122.3. $\int \frac{x^2 dx}{X^3} = -\frac{x}{4X^2} + \frac{x}{8a^2 X} + \frac{1}{8a^3} \tan^{-1}\frac{x}{a}.$

122.4. $\int \frac{x^2 dx}{X^4} = -\frac{x}{6X^3} + \frac{x}{24a^2 X^2} + \frac{x}{16a^4 X} + \frac{1}{16a^5} \tan^{-1}\frac{x}{a}.$

122.9. $\int \frac{x^2 dx}{X^{n+1}} = \frac{-x}{2nX^n} + \frac{1}{2n} \int \frac{dx}{X^n}.$

123.1. $\int \frac{x^3 dx}{X} = \frac{x^2}{2} - \frac{a^2}{2} \log X.$

123.2. $\int \frac{x^3 dx}{X^2} = \frac{a^2}{2X} + \frac{1}{2} \log X.$

123.3. $\int \frac{x^3 dx}{X^3} = -\frac{1}{2X} + \frac{a^2}{4X^2}.$

123.4. $\displaystyle\int \frac{x^3 dx}{X^4} = -\frac{1}{4X^2} + \frac{a^2}{6X^3}.$

. . . .

123.9. $\displaystyle\int \frac{x^3 dx}{X^{n+1}} = \frac{-1}{2(n-1)X^{n-1}} + \frac{a^2}{2nX^n},$ $[n > 1].$

124.1. $\displaystyle\int \frac{x^4 dx}{X} = \frac{x^3}{3} - a^2 x + a^3 \tan^{-1}\frac{x}{a}.$

124.2. $\displaystyle\int \frac{x^4 dx}{X^2} = x + \frac{a^2 x}{2X} - \frac{3a}{2}\tan^{-1}\frac{x}{a}.$

124.3. $\displaystyle\int \frac{x^4 dx}{X^3} = \frac{a^2 x}{4X^2} - \frac{5x}{8X} + \frac{3}{8a}\tan^{-1}\frac{x}{a}.$

124.4. $\displaystyle\int \frac{x^4 dx}{X^4} = \frac{a^2 x}{6X^3} - \frac{7x}{24X^2} + \frac{x}{16a^2 X} + \frac{1}{16a^3}\tan^{-1}\frac{x}{a}.$

125.1. $\displaystyle\int \frac{x^5 dx}{X} = \frac{x^4}{4} - \frac{a^2 x^2}{2} + \frac{a^4}{2}\log X.$

125.2. $\displaystyle\int \frac{x^5 dx}{X^2} = \frac{x^2}{2} - \frac{a^4}{2X} - a^2 \log X.$

125.3. $\displaystyle\int \frac{x^5 dx}{X^3} = \frac{a^2}{X} - \frac{a^4}{4X^2} + \frac{1}{2}\log X.$

125.4. $\displaystyle\int \frac{x^5 dx}{X^4} = -\frac{1}{2X} + \frac{a^2}{2X^2} - \frac{a^4}{6X^3}.$

. . . .

125.9. $\displaystyle\int \frac{x^5 dx}{X^{n+1}} = \frac{-1}{2(n-2)X^{n-2}} + \frac{a^2}{(n-1)X^{n-1}} - \frac{a^4}{2nX^n},$

                                         $[n > 2].$

126.1. $\displaystyle\int \frac{x^6 dx}{X} = \frac{x^5}{5} - \frac{a^2 x^3}{3} + a^4 x - a^5 \tan^{-1}\frac{x}{a}.$

127.1. $\displaystyle\int \frac{x^7 dx}{X} = \frac{x^6}{6} - \frac{a^2 x^4}{4} + \frac{a^4 x^2}{2} - \frac{a^6}{2}\log X.$

128.1. $\displaystyle\int \frac{x^8 dx}{X} = \frac{x^7}{7} - \frac{a^2 x^5}{5} + \frac{a^4 x^3}{3} - a^6 x + a^7 \tan^{-1}\frac{x}{a}.$

131.1. $\displaystyle\int \frac{dx}{xX} = \int \frac{dx}{x(a^2 + x^2)} = \frac{1}{2a^2}\log\left(\frac{x^2}{a^2 + x^2}\right).$

Integrals Involving $X = a^2 + x^2$ (continued)

131.2. $\displaystyle\int \frac{dx}{xX^2} = \frac{1}{2a^2X} + \frac{1}{2a^4}\log\frac{x^2}{X}.$

131.3. $\displaystyle\int \frac{dx}{xX^3} = \frac{1}{4a^2X^2} + \frac{1}{2a^4X} + \frac{1}{2a^6}\log\frac{x^2}{X}.$

131.4. $\displaystyle\int \frac{dx}{xX^4} = \frac{1}{6a^2X^3} + \frac{1}{4a^4X^2} + \frac{1}{2a^6X} + \frac{1}{2a^8}\log\frac{x^2}{X}.$

132.1. $\displaystyle\int \frac{dx}{x^2X} = -\frac{1}{a^2x} - \frac{1}{a^3}\tan^{-1}\frac{x}{a}.$

132.2. $\displaystyle\int \frac{dx}{x^2X^2} = -\frac{1}{a^4x} - \frac{x}{2a^4X} - \frac{3}{2a^5}\tan^{-1}\frac{x}{a}.$

132.3. $\displaystyle\int \frac{dx}{x^2X^3} = -\frac{1}{a^6x} - \frac{x}{4a^4X^2} - \frac{7x}{8a^6X} - \frac{15}{8a^7}\tan^{-1}\frac{x}{a}.$

133.1. $\displaystyle\int \frac{dx}{x^3X} = -\frac{1}{2a^2x^2} - \frac{1}{2a^4}\log\frac{x^2}{X}.$

133.2. $\displaystyle\int \frac{dx}{x^3X^2} = -\frac{1}{2a^4x^2} - \frac{1}{2a^4X} - \frac{1}{a^6}\log\frac{x^2}{X}.$

133.3. $\displaystyle\int \frac{dx}{x^3X^3} = -\frac{1}{2a^6x^2} - \frac{1}{a^6X} - \frac{1}{4a^4X^2} - \frac{3}{2a^8}\log\frac{x^2}{X}.$

134.1. $\displaystyle\int \frac{dx}{x^4X} = -\frac{1}{3a^2x^3} + \frac{1}{a^4x} + \frac{1}{a^5}\tan^{-1}\frac{x}{a}.$

134.2. $\displaystyle\int \frac{dx}{x^4X^2} = -\frac{1}{3a^4x^3} + \frac{2}{a^6x} + \frac{x}{2a^6X} + \frac{5}{2a^7}\tan^{-1}\frac{x}{a}.$

135.1. $\displaystyle\int \frac{dx}{x^5X} = -\frac{1}{4a^2x^4} + \frac{1}{2a^4x^2} + \frac{1}{2a^6}\log\frac{x^2}{X}.$

135.2. $\displaystyle\int \frac{dx}{x^5X^2} = -\frac{1}{4a^4x^4} + \frac{1}{a^6x^2} + \frac{1}{2a^6X} + \frac{3}{2a^8}\log\frac{x^2}{X}.$

[See References 1 and 2 for additional integrals of the type of Nos. 120 to 135.]

136. $\displaystyle\int \frac{dx}{(f+gx)(a^2+x^2)} = \frac{1}{(f^2+a^2g^2)}\left[g\log|f+gx| \right.$

$$\left. - \frac{g}{2}\log(a^2+x^2) + \frac{f}{a}\tan^{-1}\frac{x}{a}\right].$$

Integrals Involving $X = a^2 - x^2$

140. $\displaystyle\int \frac{dx}{1 - x^2} = \frac{1}{2} \log \left| \frac{1 + x}{1 - x} \right|.$ [See note under **140.1.**]

The function $1/(1 - x^2)$ and its integral can be plotted for negative values of x. See Fig. 140.

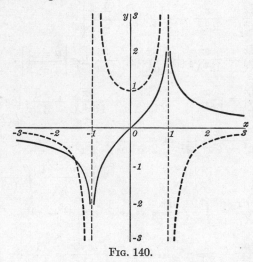

FIG. 140.

Dotted graph, $1/(1 - x^2)$.

Full line graph, $\frac{1}{2} \log \left| \dfrac{1 + x}{1 - x} \right|.$

140.01. $\displaystyle\int \frac{dx}{x^2 - 1} = -\int \frac{dx}{1 - x^2}.$ [See **140.**]

140.02. $\displaystyle\int \frac{dx}{a^2 - b^2 x^2} = \frac{1}{2ab} \log \left| \frac{a + bx}{a - bx} \right|.$

Note that

$$\frac{1}{2ab} \log \frac{a + bx}{a - bx} = \frac{1}{ab} \tanh^{-1} \frac{bx}{a}, \qquad [b^2 x^2 < a^2],$$

and

$$\frac{1}{2ab} \log \frac{bx + a}{bx - a} = \frac{1}{ab} \operatorname{ctnh}^{-1} \frac{bx}{a}, \qquad [b^2 x^2 > a^2].$$

140.1. $\displaystyle\int \frac{dx}{X} = \int \frac{dx}{a^2 - x^2} = \frac{1}{2a} \log \left| \frac{a + x}{a - x} \right|.$

Note:
$$\frac{1}{2a} \log \frac{a+x}{a-x} = \frac{1}{a} \tanh^{-1} \frac{x}{a}, \qquad [x^2 < a^2],$$

$$\frac{1}{2a} \log \frac{x+a}{x-a} = \frac{1}{a} \operatorname{ctnh}^{-1} \frac{x}{a}, \qquad [x^2 > a^2].$$

[Ref. 8, p. 100, (s) and (s').]

140.2. $\displaystyle\int \frac{dx}{X^2} = \frac{x}{2a^2X} + \frac{1}{4a^3} \log \left| \frac{a+x}{a-x} \right|.$

140.3. $\displaystyle\int \frac{dx}{X^3} = \frac{x}{4a^2X^2} + \frac{3x}{8a^4X} + \frac{3}{16a^5} \log \left| \frac{a+x}{a-x} \right|.$

140.4. $\displaystyle\int \frac{dx}{X^4} = \frac{x}{6a^2X^3} + \frac{5x}{24a^4X^2} + \frac{5x}{16a^6X} + \frac{5}{32a^7} \log \left| \frac{a+x}{a-x} \right|.$

140.9. $\displaystyle\int \frac{dx}{(a^2 - b^2x^2)^{n+1}} = \frac{x}{2na^2(a^2 - b^2x^2)^n}$

$$+ \frac{2n-1}{2na^2} \int \frac{dx}{(a^2 - b^2x^2)^n}.$$

141.1. $\displaystyle\int \frac{x\,dx}{X} = \int \frac{x\,dx}{a^2 - x^2} = -\frac{1}{2} \log |a^2 - x^2|.$

141.2. $\displaystyle\int \frac{x\,dx}{X^2} = \frac{1}{2X}.$ **141.3.** $\displaystyle\int \frac{x\,dx}{X^3} = \frac{1}{4X^2}.$

141.4. $\displaystyle\int \frac{x\,dx}{X^4} = \frac{1}{6X^3}.$ **141.9.** $\displaystyle\int \frac{x\,dx}{X^{n+1}} = \frac{1}{2nX^n},$ $[n \neq 0].$

142.1. $\displaystyle\int \frac{x^2dx}{X} = -x + \frac{a}{2} \log \left| \frac{a+x}{a-x} \right|.$

142.2. $\displaystyle\int \frac{x^2dx}{X^2} = \frac{x}{2X} - \frac{1}{4a} \log \left| \frac{a+x}{a-x} \right|.$

142.3. $\displaystyle\int \frac{x^2dx}{X^3} = \frac{x}{4X^2} - \frac{x}{8a^2X} - \frac{1}{16a^3} \log \left| \frac{a+x}{a-x} \right|.$

142.4. $\displaystyle\int \frac{x^2dx}{X^4} = \frac{x}{6X^3} - \frac{x}{24a^2X^2} - \frac{x}{16a^4X} - \frac{1}{32a^5} \log \left| \frac{a+x}{a-x} \right|.$

142.9. $\displaystyle\int \frac{x^2dx}{X^{n+1}} = \frac{x}{2nX^n} - \frac{1}{2n} \int \frac{dx}{X^n}.$

143.1. $\displaystyle\int \frac{x^3dx}{X} = -\frac{x^2}{2} - \frac{a^2}{2} \log |X|.$

143.2. $\displaystyle\int \frac{x^3 dx}{X^2} = \frac{a^2}{2X} + \frac{1}{2} \log |X|.$

143.3. $\displaystyle\int \frac{x^3 dx}{X^3} = \frac{-1}{2X} + \frac{a^2}{4X^2} \cdot$ **143.4.** $\displaystyle\int \frac{x^3 dx}{X^4} = \frac{-1}{4X^2} + \frac{a^2}{6X^3} \cdot$

143.9. $\displaystyle\int \frac{x^3 dx}{X^{n+1}} = \frac{-1}{2(n-1)X^{n-1}} + \frac{a^2}{2nX^n},$ $[n > 1].$

144.1. $\displaystyle\int \frac{x^4 dx}{X} = -\frac{x^3}{3} - a^2 x + \frac{a^3}{2} \log \left| \frac{a+x}{a-x} \right| \cdot$

144.2. $\displaystyle\int \frac{x^4 dx}{X^2} = x + \frac{a^2 x}{2X} - \frac{3a}{4} \log \left| \frac{a+x}{a-x} \right| \cdot$

144.3. $\displaystyle\int \frac{x^4 dx}{X^3} = \frac{a^2 x}{4X^2} - \frac{5x}{8X} + \frac{3}{16a} \log \left| \frac{a+x}{a-x} \right| \cdot$

144.4. $\displaystyle\int \frac{x^4 dx}{X^4} = \frac{a^2 x}{6X^3} - \frac{7x}{24X^2} + \frac{x}{16a^2 X} + \frac{1}{32a^3} \log \left| \frac{a+x}{a-x} \right| \cdot$

145.1. $\displaystyle\int \frac{x^5 dx}{X} = -\frac{x^4}{4} - \frac{a^2 x^2}{2} - \frac{a^4}{2} \log |X|.$

145.2. $\displaystyle\int \frac{x^5 dx}{X^2} = \frac{x^2}{2} + \frac{a^4}{2X} + a^2 \log |X|.$

145.3. $\displaystyle\int \frac{x^5 dx}{X^3} = -\frac{a^2}{X} + \frac{a^4}{4X^2} - \frac{1}{2} \log |X|.$

145.4. $\displaystyle\int \frac{x^5 dx}{X^4} = \frac{1}{2X} - \frac{a^2}{2X^2} + \frac{a^4}{6X^3} \cdot$

145.9. $\displaystyle\int \frac{x^5 dx}{X^{n+1}} = \frac{1}{2(n-2)X^{n-2}} - \frac{a^2}{(n-1)X^{n-1}} + \frac{a^4}{2nX^n},$

$[n > 2].$

146.1. $\displaystyle\int \frac{x^6 dx}{X} = -\frac{x^5}{5} - \frac{a^2 x^3}{3} - a^4 x + \frac{a^5}{2} \log \left| \frac{a+x}{a-x} \right| \cdot$

147.1. $\displaystyle\int \frac{x^7 dx}{X} = -\frac{x^6}{6} - \frac{a^2 x^4}{4} - \frac{a^4 x^2}{2} - \frac{a^6}{2} \log |X|.$

148.1. $\displaystyle\int \frac{x^8 dx}{X} = -\frac{x^7}{7} - \frac{a^2 x^5}{5} - \frac{a^4 x^3}{3} - a^6 x + \frac{a}{2} \log \left| \frac{a+x}{a-x} \right| \cdot$

151.1. $\displaystyle\int \frac{dx}{xX} = \int \frac{dx}{x(a^2 - x^2)} = \frac{1}{2a^2} \log \left| \frac{x^2}{a^2 - x^2} \right|.$

151.2. $\displaystyle\int \frac{dx}{xX^2} = \frac{1}{2a^2X} + \frac{1}{2a^4} \log \left| \frac{x^2}{X} \right|.$

151.3. $\displaystyle\int \frac{dx}{xX^3} = \frac{1}{4a^2X^2} + \frac{1}{2a^4X} + \frac{1}{2a^6} \log \left| \frac{x^2}{X} \right|.$

151.4. $\displaystyle\int \frac{dx}{xX^4} = \frac{1}{6a^2X^3} + \frac{1}{4a^4X^2} + \frac{1}{2a^6X} + \frac{1}{2a^8} \log \left| \frac{x^2}{X} \right|.$

152.1. $\displaystyle\int \frac{dx}{x^2X} = -\frac{1}{a^2x} + \frac{1}{2a^3} \log \left| \frac{a + x}{a - x} \right|.$

152.2. $\displaystyle\int \frac{dx}{x^2X^2} = -\frac{1}{a^4x} + \frac{x}{2a^4X} + \frac{3}{4a^5} \log \left| \frac{a + x}{a - x} \right|.$

152.3. $\displaystyle\int \frac{dx}{x^2X^3} = -\frac{1}{a^6x} + \frac{x}{4a^4X^2} + \frac{7x}{8a^6X} + \frac{15}{16a^7} \log \left| \frac{a + x}{a - x} \right|.$

153.1. $\displaystyle\int \frac{dx}{x^3X} = -\frac{1}{2a^2x^2} + \frac{1}{2a^4} \log \left| \frac{x^2}{X} \right|.$

153.2. $\displaystyle\int \frac{dx}{x^3X^2} = -\frac{1}{2a^4x^2} + \frac{1}{2a^4X} + \frac{1}{a^6} \log \left| \frac{x^2}{X} \right|.$

153.3. $\displaystyle\int \frac{dx}{x^3X^3} = -\frac{1}{2a^6x^2} + \frac{1}{a^6X} + \frac{1}{4a^4X^2} + \frac{3}{2a^8} \log \left| \frac{x^2}{X} \right|.$

154.1. $\displaystyle\int \frac{dx}{x^4X} = -\frac{1}{3a^2x^3} - \frac{1}{a^4x} + \frac{1}{2a^5} \log \left| \frac{a + x}{a - x} \right|.$

154.2. $\displaystyle\int \frac{dx}{x^4X^2} = -\frac{1}{3a^4x^3} - \frac{2}{a^6x} + \frac{x}{2a^6X} + \frac{5}{4a^7} \log \left| \frac{a + x}{a - x} \right|.$

155.1. $\displaystyle\int \frac{dx}{x^5X} = -\frac{1}{4a^2x^4} - \frac{1}{2a^4x^2} + \frac{1}{2a^6} \log \left| \frac{x^2}{X} \right|.$

155.2. $\displaystyle\int \frac{dx}{x^5X^2} = -\frac{1}{4a^4x^4} - \frac{1}{a^6x^2} + \frac{1}{2a^6X} + \frac{3}{2a^8} \log \left| \frac{x^2}{X} \right|.$

[See References 1 and 2 for other integrals of the type of Nos. **140** to **155.**]

156. $\displaystyle\int \frac{dx}{(f + gx)(a^2 - x^2)} = \frac{1}{a^2g^2 - f^2} \left[g \log |f + gx| \right.$

$$\left. - \frac{g}{2} \log |a^2 - x^2| - \frac{f}{2a} \log \left| \frac{a + x}{a - x} \right| \right].$$

Integrals Involving $X = ax^2 + bx + c$

160.01. $\displaystyle\int \frac{dx}{X} = \frac{2}{\sqrt{(4ac - b^2)}} \tan^{-1} \frac{2ax + b}{\sqrt{(4ac - b^2)}},$ $[4ac > b^2],$

$\displaystyle = \frac{1}{\sqrt{(b^2 - 4ac)}} \log \left| \frac{2ax + b - \sqrt{(b^2 - 4ac)}}{2ax + b + \sqrt{(b^2 - 4ac)}} \right|,$

$[b^2 > 4ac],$

$\displaystyle = \frac{1}{a(p - q)} \log \left| \frac{x - p}{x - q} \right|,$ $[b^2 > 4ac],$

where p and q are the roots of $ax^2 + bx + c = 0,$

$\displaystyle = - \frac{2}{\sqrt{(b^2 - 4ac)}} \tanh^{-1} \frac{2ax + b}{\sqrt{(b^2 - 4ac)}},$

$[b^2 > 4ac, \quad (2ax + b)^2 < b^2 - 4ac],$

$\displaystyle = - \frac{2}{\sqrt{(b^2 - 4ac)}} \operatorname{ctnh}^{-1} \frac{2ax + b}{\sqrt{(b^2 - 4ac)}},$

$[b^2 > 4ac, \quad (2ax + b)^2 > b^2 - 4ac],$

$\displaystyle = - \frac{2}{2ax + b},$ $[b^2 = 4ac].$

$[\text{Put } 2ax + b = z.]$

160.02. $\displaystyle\int \frac{dx}{X^2} = \frac{2ax + b}{(4ac - b^2)X} + \frac{2a}{4ac - b^2} \int \frac{dx}{X}.$

$[\text{See } \textbf{160.01.}]$

160.03. $\displaystyle\int \frac{dx}{X^3} = \frac{2ax + b}{2(4ac - b^2)X^2} + \frac{3a(2ax + b)}{(4ac - b^2)^2 X}$

$\displaystyle + \frac{6a^2}{(4ac - b^2)^2} \int \frac{dx}{X}.$ $[\text{See } \textbf{160.01.}]$

160.09. $\displaystyle\int \frac{dx}{X^n} = \frac{2ax + b}{(n - 1)(4ac - b^2)X^{n-1}}$

$\displaystyle + \frac{(2n - 3)2a}{(n - 1)(4ac - b^2)} \int \frac{dx}{X^{n-1}}.$

$[\text{Ref. 1, p. 83.}]$

160.11. $\displaystyle\int \frac{x\,dx}{X} = \frac{1}{2a} \log |X| - \frac{b}{2a} \int \frac{dx}{X}.$ $[\text{See } \textbf{160.01.}]$

Integrals Involving $X = ax^2 + bx + c$ (continued)

160.12. $\displaystyle \int \frac{x\,dx}{X^2} = -\frac{bx + 2c}{(4ac - b^2)X} - \frac{b}{4ac - b^2}\int \frac{dx}{X}.$

[See **160.01.**]

160.19. $\displaystyle \int \frac{x\,dx}{X^n} = -\frac{bx + 2c}{(n-1)(4ac - b^2)X^{n-1}}$

$$-\frac{b(2n-3)}{(n-1)(4ac - b^2)}\int \frac{dx}{X^{n-1}}.$$

160.21. $\displaystyle \int \frac{x^2 dx}{X} = \frac{x}{a} - \frac{b}{2a^2}\log |X| + \frac{b^2 - 2ac}{2a^2}\int \frac{dx}{X}.$

[See **160.01.**]

160.22. $\displaystyle \int \frac{x^2 dx}{X^2} = \frac{(b^2 - 2ac)x + bc}{a(4ac - b^2)X} + \frac{2c}{4ac - b^2}\int \frac{dx}{X}.$

[See **160.01.**]

160.27. $\displaystyle \int \frac{x^m dx}{X} = \frac{x^{m-1}}{(m-1)a} - \frac{c}{a}\int \frac{x^{m-2}dx}{X} - \frac{b}{a}\int \frac{x^{m-1}dx}{X}.$

160.28. $\displaystyle \int \frac{x^m dx}{X^n} = -\frac{x^{m-1}}{(2n-m-1)aX^{n-1}}$

$$+\frac{(m-1)c}{(2n-m-1)a}\int \frac{x^{m-2}dx}{X^n} - \frac{(n-m)b}{(2n-m-1)a}\int \frac{x^{m-1}dx}{X^n},$$

$$[m \neq 2n-1].$$

160.29. When $m = 2n - 1$,

$$\int \frac{x^{2n-1}dx}{X^n} = \frac{1}{a}\int \frac{x^{2n-3}dx}{X^{n-1}} - \frac{c}{a}\int \frac{x^{2n-3}dx}{X^n} - \frac{b}{a}\int \frac{x^{2n-2}dx}{X^n}.$$

[Ref. 4, p. 143.]

161.11. $\displaystyle \int \frac{dx}{xX} = \frac{1}{2c}\log \frac{x^2}{X} - \frac{b}{2c}\int \frac{dx}{X}.$ [See **160.01.**]

161.19. $\displaystyle \int \frac{dx}{xX^n} = \frac{1}{2c(n-1)X^{n-1}} - \frac{b}{2c}\int \frac{dx}{X^n} + \frac{1}{c}\int \frac{dx}{xX^{n-1}}.$

161.21. $\displaystyle \int \frac{dx}{x^2 X} = \frac{b}{2c^2}\log \left| \frac{X}{x^2} \right| - \frac{1}{cx} + \frac{b^2 - 2ac}{2c^2}\int \frac{dx}{X}.$

[See **160.01.**]

161.29. $\displaystyle \int \frac{dx}{x^m X^n} = -\frac{1}{(m-1)cx^{m-1}X^{n-1}}$

$$-\frac{(2n+m-3)a}{(m-1)c}\int \frac{dx}{x^{m-2}X^n} - \frac{(n+m-2)b}{(m-1)c}\int \frac{dx}{x^{m-1}X^n},$$

$$[m > 1].$$

Integrals Involving $a^3 \pm x^3$

165.01. $\displaystyle \int \frac{dx}{a^3 + x^3} = \frac{1}{6a^2} \log \frac{(a + x)^2}{a^2 - ax + x^2} + \frac{1}{a^2\sqrt{3}} \tan^{-1} \frac{2x - a}{a\sqrt{3}}.$

165.02. $\displaystyle \int \frac{dx}{(a^3 + x^3)^2} = \frac{x}{3a^3(a^3 + x^3)} + \frac{2}{3a^3} \int \frac{dx}{a^3 + x^3}.$

165.11. $\displaystyle \int \frac{x\,dx}{a^3 + x^3} = \frac{1}{6a} \log \frac{a^2 - ax + x^2}{(a + x)^2} + \frac{1}{a\sqrt{3}} \tan^{-1} \frac{2x - a}{a\sqrt{3}}.$

165.12. $\displaystyle \int \frac{x\,dx}{(a^3 + x^3)^2} = \frac{x^2}{3a^3(a^3 + x^3)} + \frac{1}{3a^3} \int \frac{x\,dx}{a^3 + x^3}.$

165.21. $\displaystyle \int \frac{x^2 dx}{a^3 + x^3} = \frac{1}{3} \log |a^3 + x^3|.$

165.22. $\displaystyle \int \frac{x^2 dx}{(a^3 + x^3)^2} = - \frac{1}{3(a^3 + x^3)}.$

165.31. $\displaystyle \int \frac{x^3 dx}{a^3 + x^3} = x - a^3 \int \frac{dx}{a^3 + x^3}.$ [See **165.01.**]

165.32. $\displaystyle \int \frac{x^3 dx}{(a^3 + x^3)^2} = \frac{- x}{3(a^3 + x^3)} + \frac{1}{3} \int \frac{dx}{a^3 + x^3}.$

[See **165.01.**]

165.41. $\displaystyle \int \frac{x^4 dx}{a^3 + x^3} = \frac{x^2}{2} - a^3 \int \frac{x\,dx}{a^3 + x^3}.$ [See **165.11.**]

165.42. $\displaystyle \int \frac{x^4 dx}{(a^3 + x^3)^2} = - \frac{x^2}{3(a^3 + x^3)} + \frac{2}{3} \int \frac{x\,dx}{a^3 + x^3}.$

[See **165.11.**]

165.51. $\displaystyle \int \frac{x^5 dx}{a^3 + x^3} = \frac{x^3}{3} - \frac{a^3}{3} \log |a^3 + x^3|.$

165.52. $\displaystyle \int \frac{x^5 dx}{(a^3 + x^3)^2} = \frac{a^3}{3(a^3 + x^3)} + \frac{1}{3} \log |a^3 + x^3|.$

166.11. $\displaystyle \int \frac{dx}{x(a^3 + x^3)} = \frac{1}{3a^3} \log \left| \frac{x^3}{a^3 + x^3} \right|.$

166.12. $\displaystyle \int \frac{dx}{x(a^3 + x^3)^2} = \frac{1}{3a^3(a^3 + x^3)} + \frac{1}{3a^6} \log \left| \frac{x^3}{a^3 + x^3} \right|.$

166.21. $\displaystyle\int \frac{dx}{x^2(a^3+x^3)} = -\frac{1}{a^3x} - \frac{1}{a^3}\int \frac{x\,dx}{a^3+x^3}\cdot$ [See **165.11**.]

166.22. $\displaystyle\int \frac{dx}{x^2(a^3+x^3)^2} = -\frac{1}{a^6x} - \frac{x^2}{3a^6(a^3+x^3)}$
$$-\frac{4}{3a^6}\int \frac{x\,dx}{a^3+x^3}\cdot \quad \text{[See } \mathbf{165.11}.\text{]}$$

166.31. $\displaystyle\int \frac{dx}{x^3(a^3+x^3)} = -\frac{1}{2a^3x^2} - \frac{1}{a^3}\int \frac{dx}{a^3+x^3}\cdot$
$$\text{[See } \mathbf{165.01}.\text{]}$$

166.32. $\displaystyle\int \frac{dx}{x^3(a^3+x^3)^2} = -\frac{1}{2a^6x^2} - \frac{x}{3a^6(a^3+x^3)}$
$$-\frac{5}{3a^6}\int \frac{dx}{a^3+x^3}\cdot \quad \text{[See } \mathbf{165.01}.\text{]}$$

166.41. $\displaystyle\int \frac{dx}{x^4(a^3+x^3)} = -\frac{1}{3a^3x^3} + \frac{1}{3a^6}\log\left|\frac{a^3+x^3}{x^3}\right|\cdot$

166.42. $\displaystyle\int \frac{dx}{x^4(a^3+x^3)^2} = -\frac{1}{3a^6x^3} - \frac{1}{3a^6(a^3+x^3)}$
$$+\frac{2}{3a^9}\log\left|\frac{a^3+x^3}{x^3}\right|\cdot$$

168.01. $\displaystyle\int \frac{dx}{a^3-x^3} = \frac{1}{6a^2}\log\frac{a^2+ax+x^2}{(a-x)^2} + \frac{1}{a^2\sqrt{3}}\tan^{-1}\frac{2x+a}{a\sqrt{3}}\cdot$

168.02. $\displaystyle\int \frac{dx}{(a^3-x^3)^2} = \frac{x}{3a^3(a^3-x^3)} + \frac{2}{3a^3}\int \frac{dx}{a^3-x^3}\cdot$

168.11. $\displaystyle\int \frac{x\,dx}{a^3-x^3} = \frac{1}{6a}\log\frac{a^2+ax+x^2}{(a-x)^2} - \frac{1}{a\sqrt{3}}\tan^{-1}\frac{2x+a}{a\sqrt{3}}\cdot$

168.12. $\displaystyle\int \frac{x\,dx}{(a^3-x^3)^2} = \frac{x^2}{3a^3(a^3-x^3)} + \frac{1}{3a^3}\int \frac{x\,dx}{a^3-x^3}\cdot$

168.21. $\displaystyle\int \frac{x^2dx}{a^3-x^3} = -\frac{1}{3}\log|a^3-x^3|\cdot$

168.22. $\displaystyle\int \frac{x^2dx}{(a^3-x^3)^2} = \frac{1}{3(a^3-x^3)}\cdot$

168.31. $\displaystyle\int \frac{x^3dx}{a^3-x^3} = -x + a^3\int \frac{dx}{a^3-x^3}\cdot$ [See **168.01**.]

168.32. $\displaystyle\int \frac{x^3 dx}{(a^3 - x^3)^2} = \frac{x}{3(a^3 - x^3)} - \frac{1}{3}\int \frac{dx}{a^3 - x^3}$.

[See **168.01.**]

168.41. $\displaystyle\int \frac{x^4 dx}{a^3 - x^3} = -\frac{x^2}{2} + a^3 \int \frac{x\,dx}{a^3 - x^3}$.   [See **168.11.**]

168.42. $\displaystyle\int \frac{x^4 dx}{(a^3 - x^3)^2} = \frac{x^2}{3(a^3 - x^3)} - \frac{2}{3}\int \frac{x\,dx}{a^3 - x^3}$.

[See **168.11.**]

168.51. $\displaystyle\int \frac{x^5 dx}{a^3 - x^3} = -\frac{x^3}{3} - \frac{a^3}{3}\log |a^3 - x^3|$.

168.52. $\displaystyle\int \frac{x^5 dx}{(a^3 - x^3)^2} = \frac{a^3}{3(a^3 - x^3)} + \frac{1}{3}\log |a^3 - x^3|$.

169.11. $\displaystyle\int \frac{dx}{x(a^3 - x^3)} = \frac{1}{3a^3}\log \left|\frac{x^3}{a^3 - x^3}\right|$.

169.12. $\displaystyle\int \frac{dx}{x(a^3 - x^3)^2} = \frac{1}{3a^3(a^3 - x^3)} + \frac{1}{3a^6}\log \left|\frac{x^3}{a^3 - x^3}\right|$.

169.21. $\displaystyle\int \frac{dx}{x^2(a^3 - x^3)} = -\frac{1}{a^3 x} + \frac{1}{a^3}\int \frac{x\,dx}{a^3 - x^3}$.   [See **168.11.**]

169.22. $\displaystyle\int \frac{dx}{x^2(a^3 - x^3)^2} = -\frac{1}{a^6 x} + \frac{x^2}{3a^6(a^3 - x^3)} + \frac{4}{3a^6}\int \frac{x\,dx}{a^3 - x^3}$.

[See **168.11.**]

169.31. $\displaystyle\int \frac{dx}{x^3(a^3 - x^3)} = -\frac{1}{2a^3 x^2} + \frac{1}{a^3}\int \frac{dx}{a^3 - x^3}$.

[See **168.01.**]

169.32. $\displaystyle\int \frac{dx}{x^3(a^3 - x^3)^2} = -\frac{1}{2a^6 x^2} + \frac{x}{3a^6(a^3 - x^3)}$

$$+ \frac{5}{3a^6}\int \frac{dx}{a^3 - x^3}.$$ [See **168.01.**]

169.41. $\displaystyle\int \frac{dx}{x^4(a^3 - x^3)} = -\frac{1}{3a^3 x^3} + \frac{1}{3a^6}\log \left|\frac{x^3}{a^3 - x^3}\right|$.

169.42. $\displaystyle\int \frac{dx}{x^4(a^3 - x^3)^2} = -\frac{1}{3a^6 x^3} + \frac{1}{3a^6(a^3 - x^3)}$

$$+ \frac{2}{3a^9}\log \left|\frac{x^3}{a^3 - x^3}\right|.$$

Integrals Involving $a^4 \pm x^4$

170. $\displaystyle \int \frac{dx}{a^4 + x^4} = \frac{1}{4a^3\sqrt{2}} \log \frac{x^2 + ax\sqrt{2} + a^2}{x^2 - ax\sqrt{2} + a^2}$

$$+ \frac{1}{2a^3\sqrt{2}} \tan^{-1} \frac{ax\sqrt{2}}{a^2 - x^2}.$$

170.1. $\displaystyle \int \frac{x\,dx}{a^4 + x^4} = \frac{1}{2a^2} \tan^{-1} \frac{x^2}{a^2}.$

170.2. $\displaystyle \int \frac{x^2 dx}{a^4 + x^4} = -\frac{1}{4a\sqrt{2}} \log \frac{x^2 + ax\sqrt{2} + a^2}{x^2 - ax\sqrt{2} + a^2}$

$$+ \frac{1}{2a\sqrt{2}} \tan^{-1} \frac{ax\sqrt{2}}{a^2 - x^2}.$$

170.3. $\displaystyle \int \frac{x^3 dx}{a^4 + x^4} = \frac{1}{4} \log (a^4 + x^4).$

171. $\displaystyle \int \frac{dx}{a^4 - x^4} = \frac{1}{4a^3} \log \left| \frac{a + x}{a - x} \right| + \frac{1}{2a^3} \tan^{-1} \frac{x}{a}.$

171.1. $\displaystyle \int \frac{x\,dx}{a^4 - x^4} = \frac{1}{4a^2} \log \left| \frac{a^2 + x^2}{a^2 - x^2} \right|.$

171.2. $\displaystyle \int \frac{x^2 dx}{a^4 - x^4} = \frac{1}{4a} \log \left| \frac{a + x}{a - x} \right| - \frac{1}{2a} \tan^{-1} \frac{x}{a}.$

171.3. $\displaystyle \int \frac{x^3 dx}{a^4 - x^4} = -\frac{1}{4} \log |a^4 - x^4|.$

173. $\displaystyle \int \frac{dx}{x(a + bx^m)} = \frac{1}{am} \log \left| \frac{x^m}{a + bx^m} \right|.$

IRRATIONAL ALGEBRAIC FUNCTIONS

Integrals Involving $x^{1/2}$

180. $\displaystyle\int x^{p/2}dx = \frac{2}{p+2}\, x^{(p+2)/2}.$

180.1. $\displaystyle\int x^{1/2}dx = \int \sqrt{x}\, dx = \frac{2}{3}\, x^{3/2}.$

180.3. $\displaystyle\int x^{3/2}dx = \frac{2}{5}\, x^{5/2}.$ **180.5.** $\displaystyle\int x^{5/2}dx = \frac{2}{7}\, x^{7/2}.$

131. $\displaystyle\int \frac{dx}{x^{p/2}} = -\,\frac{2}{(p-2)x^{(p-2)/2}}.$

181.1. $\displaystyle\int \frac{dx}{x^{1/2}} = \int \frac{dx}{\sqrt{x}} = 2x^{1/2}.$ **181.3.** $\displaystyle\int \frac{dx}{x^{3/2}} = -\,\frac{2}{x^{1/2}}.$

181.5. $\displaystyle\int \frac{dx}{x^{5/2}} = -\,\frac{2}{3x^{3/2}}.$ **181.7.** $\displaystyle\int \frac{dx}{x^{7/2}} = -\,\frac{2}{5x^{5/2}}.$

[NOTE.—Put $x = u^2$, then $dx = 2u\, du$.]

185.11. $\displaystyle\int \frac{x^{1/2}dx}{a^2 + b^2 x} = \frac{2x^{1/2}}{b^2} - \frac{2a}{b^3}\tan^{-1}\frac{bx^{1/2}}{a}.$

185.13. $\displaystyle\int \frac{x^{3/2}dx}{a^2 + b^2 x} = \frac{2}{3}\frac{x^{3/2}}{b^2} - \frac{2a^2 x^{1/2}}{b^4} + \frac{2a^3}{b^5}\tan^{-1}\frac{bx^{1/2}}{a}.$

185.21. $\displaystyle\int \frac{x^{1/2}dx}{(a^2 + b^2 x)^2} = -\,\frac{x^{1/2}}{b^2(a^2 + b^2 x)} + \frac{1}{ab^3}\tan^{-1}\frac{bx^{1/2}}{a}.$

185.23. $\displaystyle\int \frac{x^{3/2}dx}{(a^2 + b^2 x)^2} = \frac{2x^{3/2}}{b^2(a^2 + b^2 x)} + \frac{3a^2 x^{1/2}}{b^4(a^2 + b^2 x)}$
$$- \frac{3a}{b^5}\tan^{-1}\frac{bx^{1/2}}{a}.$$

186.11. $\displaystyle\int \frac{dx}{(a^2 + b^2 x)x^{1/2}} = \frac{2}{ab}\tan^{-1}\frac{bx^{1/2}}{a}.$

186.13. $\displaystyle\int \frac{dx}{(a^2 + b^2 x)x^{3/2}} = -\,\frac{2}{a^2 x^{1/2}} - \frac{2b}{a^3}\tan^{-1}\frac{bx^{1/2}}{a}.$

186.21. $\displaystyle\int \frac{dx}{(a^2 + b^2 x)^2 x^{1/2}} = \frac{x^{1/2}}{a^2(a^2 + b^2 x)} + \frac{1}{a^3 b}\tan^{-1}\frac{bx^{1/2}}{a}.$

186.23. $\displaystyle\int \frac{dx}{(a^2 + b^2x)^2 x^{3/2}} = -\frac{2}{a^2(a^2 + b^2x)x^{1/2}} - \frac{3b^2x^{1/2}}{a^4(a^2 + b^2x)}$
$$-\frac{3b}{a^5}\tan^{-1}\frac{bx^{1/2}}{a}.$$

187.11. $\displaystyle\int \frac{x^{1/2}dx}{a^2 - b^2x} = -\frac{2x^{1/2}}{b^2} + \frac{a}{b^3}\log\left|\frac{a + bx^{1/2}}{a - bx^{1/2}}\right|.$

187.13. $\displaystyle\int \frac{x^{3/2}dx}{a^2 - b^2x} = -\frac{2}{3}\frac{x^{3/2}}{b^2} - \frac{2a^2x^{1/2}}{b^4} + \frac{a^3}{b^5}\log\left|\frac{a + bx^{1/2}}{a - bx^{1/2}}\right|.$

187.21. $\displaystyle\int \frac{x^{1/2}dx}{(a^2 - b^2x)^2} = \frac{x^{1/2}}{b^2(a^2 - b^2x)} - \frac{1}{2ab^3}\log\left|\frac{a + bx^{1/2}}{a - bx^{1/2}}\right|.$

187.23. $\displaystyle\int \frac{x^{3/2}dx}{(a^2 - b^2x)^2} = \frac{3a^2x^{1/2} - 2b^2x^{3/2}}{b^4(a^2 - b^2x)} - \frac{3a}{2b^5}\log\left|\frac{a + bx^{1/2}}{a - bx^{1/2}}\right|.$

188.11. $\displaystyle\int \frac{dx}{(a^2 - b^2x)x^{1/2}} = \frac{1}{ab}\log\left|\frac{a + bx^{1/2}}{a - bx^{1/2}}\right|.$

188.13. $\displaystyle\int \frac{dx}{(a^2 - b^2x)x^{3/2}} = -\frac{2}{a^2x^{1/2}} + \frac{b}{a^3}\log\left|\frac{a + bx^{1/2}}{a - bx^{1/2}}\right|.$

188.21. $\displaystyle\int \frac{dx}{(a^2 - b^2x)^2x^{1/2}} = \frac{x^{1/2}}{a^2(a^2 - b^2x)} + \frac{1}{2a^3b}\log\left|\frac{a + bx^{1/2}}{a - bx^{1/2}}\right|.$

188.23. $\displaystyle\int \frac{dx}{(a^2 - b^2x)^2x^{3/2}} = \frac{-2}{a^2(a^2 - b^2x)x^{1/2}} + \frac{3b^2x^{1/2}}{a^4(a^2 - b^2x)}$
$$+\frac{3b}{2a^5}\log\left|\frac{a + bx^{1/2}}{a - bx^{1/2}}\right|.$$

189.1. $\displaystyle\int \frac{x^{1/2}dx}{a^4 + x^2} = \frac{-1}{2a\sqrt{2}}\log\frac{x + a\sqrt{(2x)} + a^2}{x - a\sqrt{(2x)} + a^2}$
$$+\frac{1}{a\sqrt{2}}\tan^{-1}\frac{a\sqrt{(2x)}}{a^2 - x}.$$

189.2. $\displaystyle\int \frac{dx}{(a^4 + x^2)x^{1/2}} = \frac{1}{2a^3\sqrt{2}}\log\frac{x + a\sqrt{(2x)} + a^2}{x - a\sqrt{(2x)} + a^2}$
$$+\frac{1}{a^3\sqrt{2}}\tan^{-1}\frac{a\sqrt{(2x)}}{a^2 - x}.$$

189.3. $\displaystyle\int \frac{x^{1/2}dx}{a^4 - x^2} = \frac{1}{2a}\log\left|\frac{a + x^{1/2}}{a - x^{1/2}}\right| - \frac{1}{a}\tan^{-1}\frac{x^{1/2}}{a}.$

189.4. $\displaystyle\int \frac{dx}{(a^4 - x^2)x^{1/2}} = \frac{1}{2a^3}\log\left|\frac{a + x^{1/2}}{a - x^{1/2}}\right| + \frac{1}{a^3}\tan^{-1}\frac{x^{1/2}}{a}.$

[Ref. 4, pp. 149–151.]

Integrals Involving $X^{1/2} = (a + bx)^{1/2}$

190. $\displaystyle\int \frac{x^q dx}{X^{p/2}} = \frac{1}{b^{q+1}} \int \frac{(X-a)^q dX}{X^{p/2}},$ $[q > 0].$

Expand the numerator by the binomial theorem, when q is a positive integer.

191. $\displaystyle\int \frac{dx}{X^{p/2}} = \frac{-2}{(p-2)bX^{(p-2)/2}}.$ **191.03.** $\displaystyle\int \frac{dx}{X^{3/2}} = \frac{-2}{bX^{1/2}}.$

191.01. $\displaystyle\int \frac{dx}{X^{1/2}} = \frac{2}{b}X^{1/2}.$ **191.05.** $\displaystyle\int \frac{dx}{X^{5/2}} = \frac{-2}{3bX^{3/2}}.$

191.1. $\displaystyle\int \frac{x\,dx}{X^{p/2}} = \frac{2}{b^2}\left[\frac{-1}{(p-4)X^{(p-4)/2}} + \frac{a}{(p-2)X^{(p-2)/2}}\right].$

191.11. $\displaystyle\int \frac{x\,dx}{X^{1/2}} = \frac{2}{b^2}\left(\frac{X^{3/2}}{3} - aX^{1/2}\right).$

191.13. $\displaystyle\int \frac{x\,dx}{X^{3/2}} = \frac{2}{b^2}\left(X^{1/2} + \frac{a}{X^{1/2}}\right).$

191.15. $\displaystyle\int \frac{x\,dx}{X^{5/2}} = \frac{2}{b^2}\left(\frac{-1}{X^{1/2}} + \frac{a}{3X^{3/2}}\right).$

191.17. $\displaystyle\int \frac{x\,dx}{X^{7/2}} = \frac{2}{b^2}\left(\frac{-1}{3X^{3/2}} + \frac{a}{5X^{5/2}}\right).$

191.2. $\displaystyle\int \frac{x^2 dx}{X^{p/2}} = \frac{2}{b^3}\left[\frac{-1}{(p-6)X^{(p-6)/2}} + \frac{2a}{(p-4)X^{(p-4)/2}}\right.$
$$\left. - \frac{a^2}{(p-2)X^{(p-2)/2}}\right].$$

191.21. $\displaystyle\int \frac{x^2 dx}{X^{1/2}} = \frac{2}{b^3}\left(\frac{X^{5/2}}{5} - \frac{2aX^{3/2}}{3} + a^2X^{1/2}\right).$

191.23. $\displaystyle\int \frac{x^2 dx}{X^{3/2}} = \frac{2}{b^3}\left(\frac{X^{3/2}}{3} - 2aX^{1/2} - \frac{a^2}{X^{1/2}}\right).$

191.25. $\displaystyle\int \frac{x^2 dx}{X^{5/2}} = \frac{2}{b^3}\left(X^{1/2} + \frac{2a}{X^{1/2}} - \frac{a^2}{3X^{3/2}}\right).$

191.27. $\displaystyle\int \frac{x^2 dx}{X^{7/2}} = \frac{2}{b^3}\left(\frac{-1}{X^{1/2}} + \frac{2a}{3X^{3/2}} - \frac{a^2}{5X^{5/2}}\right).$

192.1. $\int \dfrac{dx}{xX^{p/2}} = \dfrac{2}{(p-2)aX^{(p-2)/2}} + \dfrac{1}{a}\int \dfrac{dx}{xX^{(p-2)/2}},$

$$[p > 1]. \qquad [\text{Ref. 2, p. 92.}]$$

192.11. $\int \dfrac{dx}{xX^{1/2}} = \dfrac{1}{a^{1/2}}\log\left|\dfrac{X^{1/2}-a^{1/2}}{X^{1/2}+a^{1/2}}\right|, \qquad [a > 0,\, X > 0],$

$$= -\dfrac{2}{a^{1/2}}\tanh^{-1}\dfrac{X^{1/2}}{a^{1/2}}, \qquad [a > X > 0],$$

$$= -\dfrac{2}{a^{1/2}}\operatorname{ctnh}^{-1}\dfrac{X^{1/2}}{a^{1/2}}, \qquad [X > a > 0],$$

$$= \dfrac{2}{(-a)^{1/2}}\tan^{-1}\dfrac{X^{1/2}}{(-a)^{1/2}}, \qquad [a < 0,\, X > 0].$$

$$[\text{Put } X^{1/2} = z. \quad \text{See Nos. 120.1 and 140.1.}]$$

192.13. $\int \dfrac{dx}{xX^{3/2}} = \dfrac{2}{aX^{1/2}} + \dfrac{1}{a}\int \dfrac{dx}{xX^{1/2}}.$ $\qquad [\text{See 192.11.}]$

192.15. $\int \dfrac{dx}{xX^{5/2}} = \dfrac{2}{3aX^{3/2}} + \dfrac{2}{a^2 X^{1/2}} + \dfrac{1}{a^2}\int \dfrac{dx}{xX^{1/2}}.$

$$[\text{See 192.11.}]$$

192.17. $\int \dfrac{dx}{xX^{7/2}} = \dfrac{2}{5aX^{5/2}} + \dfrac{2}{3a^2 X^{3/2}} + \dfrac{2}{a^3 X^{1/2}} + \dfrac{1}{a^3}\int \dfrac{dx}{xX^{1/2}}.$

$$[\text{See 192.11.}]$$

192.2. $\int \dfrac{dx}{x^2 X^{p/2}} = \dfrac{-1}{axX^{(p-2)/2}} - \dfrac{pb}{2a}\int \dfrac{dx}{xX^{p/2}}.$ $\quad [\text{Ref. 2, p. 94.}]$

192.21. $\int \dfrac{dx}{x^2 X^{1/2}} = \dfrac{-X^{1/2}}{ax} - \dfrac{b}{2a}\int \dfrac{dx}{xX^{1/2}}.$ $\qquad [\text{See 192.11.}]$

192.23. $\int \dfrac{dx}{x^2 X^{3/2}} = \dfrac{-1}{axX^{1/2}} - \dfrac{3b}{a^2 X^{1/2}} - \dfrac{3b}{2a^2}\int \dfrac{dx}{xX^{1/2}}.$

$$[\text{See 192.11.}]$$

192.25. $\int \dfrac{dx}{x^2 X^{5/2}} = \dfrac{-1}{axX^{3/2}} - \dfrac{5b}{3a^2 X^{3/2}} - \dfrac{5b}{a^3 X^{1/2}} - \dfrac{5b}{2a^3}\int \dfrac{dx}{xX^{1/2}}.$

$$[\text{See 192.11.}]$$

192.9. $\int \dfrac{dx}{x^p X^{1/2}} = \dfrac{-X^{1/2}}{(p-1)ax^{p-1}} - \dfrac{(2p-3)b}{(2p-2)a}\int \dfrac{dx}{x^{p-1}X^{1/2}}.$

$$[\text{Ref. 2, p. 94.}]$$

193. $\displaystyle\int X^{p/2}dx = \frac{2X^{(p+2)/2}}{(p+2)b}\cdot$

193.01. $\displaystyle\int X^{1/2}dx = \frac{2X^{3/2}}{3b}\cdot$ **193.03.** $\displaystyle\int X^{3/2}dx = \frac{2X^{5/2}}{5b}\cdot$

193.1. $\displaystyle\int xX^{p/2}dx = \frac{2}{b^2}\left(\frac{X^{(p+4)/2}}{p+4} - \frac{aX^{(p+2)/2}}{p+2}\right)\cdot$

193.11. $\displaystyle\int xX^{1/2}dx = \frac{2}{b^2}\left(\frac{X^{5/2}}{5} - \frac{aX^{3/2}}{3}\right)\cdot$

193.13. $\displaystyle\int xX^{3/2}dx = \frac{2}{b^2}\left(\frac{X^{7/2}}{7} - \frac{aX^{5/2}}{5}\right)\cdot$

193.2. $\displaystyle\int x^2X^{p/2}dx = \frac{2}{b^3}\left(\frac{X^{(p+6)/2}}{p+6} - \frac{2aX^{(p+4)/2}}{p+4} + \frac{a^2X^{(p+2)/2}}{p+2}\right)\cdot$

193.21. $\displaystyle\int x^2X^{1/2}dx = \frac{2}{b^3}\left(\frac{X^{7/2}}{7} - \frac{2aX^{5/2}}{5} + \frac{a^2X^{3/2}}{3}\right)\cdot$

194.1. $\displaystyle\int \frac{X^{p/2}dx}{x} = \frac{2X^{p/2}}{p} + a\int \frac{X^{(p-2)/2}dx}{x}\cdot$ [Ref. 2, p. 91.]

194.11. $\displaystyle\int \frac{X^{1/2}dx}{x} = 2X^{1/2} + a\int \frac{dx}{xX^{1/2}}\cdot$ [See **192.11**].

194.13. $\displaystyle\int \frac{X^{3/2}dx}{x} = \frac{2X^{3/2}}{3} + 2aX^{1/2} + a^2\int \frac{dx}{xX^{1/2}}\cdot$
[See **192.11**.]

194.15. $\displaystyle\int \frac{X^{5/2}dx}{x} = \frac{2X^{5/2}}{5} + \frac{2aX^{3/2}}{3} + 2a^2X^{1/2} + a^3\int \frac{dx}{xX^{1/2}}\cdot$
[See **192.11**.]

194.2. $\displaystyle\int \frac{X^{p/2}dx}{x^2} = -\frac{X^{(p+2)/2}}{ax} + \frac{pb}{2a}\int \frac{X^{p/2}dx}{x}\cdot$

194.21. $\displaystyle\int \frac{X^{1/2}dx}{x^2} = -\frac{X^{1/2}}{x} + \frac{b}{2}\int \frac{dx}{xX^{1/2}}\cdot$ [See **192.11**.]

194.31. $\displaystyle\int \frac{X^{1/2}dx}{x^3} = -\frac{(2a+bx)X^{1/2}}{4ax^2} - \frac{b^2}{8a}\int \frac{dx}{xX^{1/2}}\cdot$
[See **192.11**.]   [Ref. 1, p. 105.]

Integrals Involving $X^{1/2} = (a + bx)^{1/2}$ *and* $U^{1/2} = (f + gx)^{1/2}$
Let $k = ag - bf$

195.01. $\displaystyle\int \frac{dx}{X^{1/2}U^{1/2}} = \frac{2}{\sqrt{(-bg)}} \tan^{-1}\sqrt{\left(\frac{-gX}{bU}\right)}, \quad \begin{bmatrix} b > 0 \\ g < 0 \end{bmatrix},$

$\displaystyle = \frac{-1}{\sqrt{(-bg)}} \sin^{-1}\frac{2bgx + ag + bf}{bf - ag}, \quad \begin{bmatrix} b > 0 \\ g < 0 \end{bmatrix},$

$\displaystyle = \frac{2}{\sqrt{(bg)}} \log \, | \sqrt{(bgX)} + b\sqrt{U}|, \quad\quad [bg > 0].$

195.02. $\displaystyle\int \frac{dx}{X^{1/2}U} = \frac{2}{\sqrt{(-kg)}} \tan^{-1}\frac{gX^{1/2}}{\sqrt{(-kg)}}, \quad\quad [kg < 0],$

$\displaystyle = \frac{1}{\sqrt{(kg)}} \log \left| \frac{gX^{1/2} - \sqrt{(kg)}}{gX^{1/2} + \sqrt{(kg)}} \right|, \quad\quad [kg > 0].$

195.03. $\displaystyle\int \frac{dx}{X^{1/2}U^{3/2}} = -\frac{2X^{1/2}}{kU^{1/2}}.$

195.04. $\displaystyle\int \frac{U^{1/2}dx}{X^{1/2}} = \frac{X^{1/2}U^{1/2}}{b} - \frac{k}{2b}\int \frac{dx}{X^{1/2}U^{1/2}}. \quad$ [See 195.01.]

195.09. $\displaystyle\int \frac{U^n dx}{X^{1/2}} = \frac{2}{(2n+1)b}\left(X^{1/2}U^n - nk\int \frac{U^{n-1}dx}{X^{1/2}} \right).$

196.01. $\displaystyle\int X^{1/2}U^{1/2}dx = \frac{k + 2bU}{4bg} X^{1/2}U^{1/2} - \frac{k^2}{8bg}\int \frac{dx}{X^{1/2}U^{1/2}}.$

[See 195.01.]

196.02. $\displaystyle\int \frac{xdx}{X^{1/2}U^{1/2}} = \frac{X^{1/2}U^{1/2}}{bg} - \frac{ag + bf}{2bg}\int \frac{dx}{X^{1/2}U^{1/2}}.$

[See 195.01.]

196.03. $\displaystyle\int \frac{dx}{X^{1/2}U^n}$

$\displaystyle = -\frac{1}{(n-1)k}\left\{ \frac{X^{1/2}}{U^{n-1}} + \left(n - \frac{3}{2}\right)b\int \frac{dx}{X^{1/2}U^{n-1}} \right\}.$

196.04. $\displaystyle\int X^{1/2}U^n dx = \frac{1}{(2n+3)g}\left(2X^{1/2}U^{n+1} + k\int \frac{U^n dx}{X^{1/2}} \right).$

[See 195.09.]

196.05. $\displaystyle\int \frac{X^{1/2}dx}{U^n} = \frac{1}{(n-1)g}\left(-\frac{X^{1/2}}{U^{n-1}} + \frac{b}{2}\int \frac{dx}{X^{1/2}U^{n-1}} \right).$

197. $\displaystyle\int \frac{f(x^2)dx}{\sqrt{(a + bx^2)}} = \int f\left(\frac{au^2}{1 - bu^2} \right)\frac{du}{(1 - bu^2)}$

where $\quad\quad u = x/\sqrt{(a + bx^2)}.$

Integrals Involving $r = (x^2 + a^2)^{1/2}$

200.01. $\displaystyle \int \frac{dx}{r} = \int \frac{dx}{\sqrt{(x^2 + a^2)}} = \log(x + r).$

Note that

$$\log\left(\frac{x + r}{a}\right) = \sinh^{-1}\frac{x}{a} = \frac{1}{2}\log\left(\frac{r + x}{r - x}\right).$$

The positive values of r and a are to be taken.

Fig. 200.01.   Graphs of $1/\sqrt{(x^2 + 1)}$ and of $\log\{x + \sqrt{(x^2 + 1)}\}$, where x is real.

200.03. $\displaystyle \int \frac{dx}{r^3} = \frac{1}{a^2}\frac{x}{r}.$

200.05. $\displaystyle \int \frac{dx}{r^5} = \frac{1}{a^4}\left[\frac{x}{r} - \frac{1}{3}\frac{x^3}{r^3}\right].$

200.07. $\displaystyle \int \frac{dx}{r^7} = \frac{1}{a^6}\left[\frac{x}{r} - \frac{2}{3}\frac{x^3}{r^3} + \frac{1}{5}\frac{x^5}{r^5}\right].$

200.09. $\displaystyle \int \frac{dx}{r^9} = \frac{1}{a^8}\left[\frac{x}{r} - \frac{3}{3}\frac{x^3}{r^3} + \frac{3}{5}\frac{x^5}{r^5} - \frac{1}{7}\frac{x^7}{r^7}\right].$

200.11. $\displaystyle \int \frac{dx}{r^{11}} = \frac{1}{a^{10}}\left[\frac{x}{r} - \frac{4}{3}\frac{x^3}{r^3} + \frac{6}{5}\frac{x^5}{r^5} - \frac{4}{7}\frac{x^7}{r^7} + \frac{1}{9}\frac{x^9}{r^9}\right].$

200.13. $\displaystyle \int \frac{dx}{r^{13}} = \frac{1}{a^{12}}\left[\frac{x}{r} - \frac{5}{3}\frac{x^3}{r^3} + \frac{10}{5}\frac{x^5}{r^5} - \frac{10}{7}\frac{x^7}{r^7} + \frac{5}{9}\frac{x^9}{r^9} - \frac{1}{11}\frac{x^{11}}{r^{11}}\right].$

200.15. $\displaystyle \int \frac{dx}{r^{15}} = \frac{1}{a^{14}}\left[\frac{x}{r} - \frac{6}{3}\frac{x^3}{r^3} + \frac{15}{5}\frac{x^5}{r^5} - \frac{20}{7}\frac{x^7}{r^7} + \frac{15}{9}\frac{x^9}{r^9}\right.$
$$\left. - \frac{6}{11}\frac{x^{11}}{r^{11}} + \frac{1}{13}\frac{x^{13}}{r^{13}}\right].$$

For **200.03–200.15** let

$$z^2 = \frac{x^2}{x^2 + a^2}; \quad \text{then} \quad dx = \frac{a\,dz}{(1 - z^2)^{3/2}}.$$

201.01. $\int \dfrac{x\,dx}{r} = r.$

201.05. $\int \dfrac{x\,dx}{r^5} = -\dfrac{1}{3r^3}.$

201.03. $\int \dfrac{x\,dx}{r^3} = -\dfrac{1}{r}.$

201.07. $\int \dfrac{x\,dx}{r^7} = -\dfrac{1}{5r^5}.$

201.9. $\int \dfrac{x\,dx}{r^{2p+1}} = -\dfrac{1}{(2p-1)r^{2p-1}}.$

202.01. $\int \dfrac{x^2 dx}{r} = \dfrac{xr}{2} - \dfrac{a^2}{2}\log(x+r).$

[See note under **200.01.**]

202.03. $\int \dfrac{x^2 dx}{r^3} = -\dfrac{x}{r} + \log(x+r).$

202.05. $\int \dfrac{x^2 dx}{r^5} = \dfrac{1}{3a^2}\dfrac{x^3}{r^3}.$

202.07. $\int \dfrac{x^2 dx}{r^7} = \dfrac{1}{a^4}\left[\dfrac{1}{3}\dfrac{x^3}{r^3} - \dfrac{1}{5}\dfrac{x^5}{r^5}\right].$

202.09. $\int \dfrac{x^2 dx}{r^9} = \dfrac{1}{a^6}\left[\dfrac{1}{3}\dfrac{x^3}{r^3} - \dfrac{2}{5}\dfrac{x^5}{r^5} + \dfrac{1}{7}\dfrac{x^7}{r^7}\right].$

202.11. $\int \dfrac{x^2 dx}{r^{11}} = \dfrac{1}{a^8}\left[\dfrac{1}{3}\dfrac{x^3}{r^3} - \dfrac{3}{5}\dfrac{x^5}{r^5} + \dfrac{3}{7}\dfrac{x^7}{r^7} - \dfrac{1}{9}\dfrac{x^9}{r^9}\right].$

202.13. $\int \dfrac{x^2 dx}{r^{13}} = \dfrac{1}{a^{10}}\left[\dfrac{1}{3}\dfrac{x^3}{r^3} - \dfrac{4}{5}\dfrac{x^5}{r^5} + \dfrac{6}{7}\dfrac{x^7}{r^7} - \dfrac{4}{9}\dfrac{x^9}{r^9} + \dfrac{1}{11}\dfrac{x^{11}}{r^{11}}\right].$

202.15. $\int \dfrac{x^2 dx}{r^{15}} = \dfrac{1}{a^{12}}\left[\dfrac{1}{3}\dfrac{x^3}{r^3} - \dfrac{5}{5}\dfrac{x^5}{r^5} + \dfrac{10}{7}\dfrac{x^7}{r^7} - \dfrac{10}{9}\dfrac{x^9}{r^9}\right.$

$$\left. + \dfrac{5}{11}\dfrac{x^{11}}{r^{11}} - \dfrac{1}{13}\dfrac{x^{13}}{r^{13}}\right].$$

203.01. $\int \dfrac{x^3 dx}{r} = \dfrac{r^3}{3} - a^2 r.$

203.03. $\int \dfrac{x^3 dx}{r^3} = r + \dfrac{a^2}{r}.$

203.05. $\int \dfrac{x^3 dx}{r^5} = -\dfrac{1}{r} + \dfrac{a^2}{3r^3}.$

203.07. $\displaystyle\int \frac{x^3 dx}{r^7} = -\frac{1}{3r^3} + \frac{a^2}{5r^5}.$

203.9. $\displaystyle\int \frac{x^3 dx}{r^{2p+1}} = -\frac{1}{(2p-3)r^{2p-3}} + \frac{a^2}{(2p-1)r^{2p-1}}.$

204.01. $\displaystyle\int \frac{x^4 dx}{r} = \frac{x^3 r}{4} - \frac{3}{8}a^2 xr + \frac{3}{8}a^4 \log(x + r).$

[See note under 200.01.]

204.03. $\displaystyle\int \frac{x^4 dx}{r^3} = \frac{xr}{2} + \frac{a^2 x}{r} - \frac{3}{2}a^2 \log(x + r).$

204.05. $\displaystyle\int \frac{x^4 dx}{r^5} = -\frac{x}{r} - \frac{1}{3}\frac{x^3}{r^3} + \log(x + r).$

204.07. $\displaystyle\int \frac{x^4 dx}{r^7} = \frac{1}{5a^2}\frac{x^5}{r^5}.$

204.09. $\displaystyle\int \frac{x^4 dx}{r^9} = \frac{1}{a^4}\left[\frac{1}{5}\frac{x^5}{r^5} - \frac{1}{7}\frac{x^7}{r^7}\right].$

204.11. $\displaystyle\int \frac{x^4 dx}{r^{11}} = \frac{1}{a^6}\left[\frac{1}{5}\frac{x^5}{r^5} - \frac{2}{7}\frac{x^7}{r^7} + \frac{1}{9}\frac{x^9}{r^9}\right].$

204.13. $\displaystyle\int \frac{x^4 dx}{r^{13}} = \frac{1}{a^8}\left[\frac{1}{5}\frac{x^5}{r^5} - \frac{3}{7}\frac{x^7}{r^7} + \frac{3}{9}\frac{x^9}{r^9} - \frac{1}{11}\frac{x^{11}}{r^{11}}\right].$

204.15. $\displaystyle\int \frac{x^4 dx}{r^{15}} = \frac{1}{a^{10}}\left[\frac{1}{5}\frac{x^5}{r^5} - \frac{4}{7}\frac{x^7}{r^7} + \frac{6}{9}\frac{x^9}{r^9} - \frac{4}{11}\frac{x^{11}}{r^{11}} + \frac{1}{13}\frac{x^{13}}{r^{13}}\right].$

205.01. $\displaystyle\int \frac{x^5 dx}{r} = \frac{r^5}{5} - \frac{2}{3}a^2 r^3 + a^4 r.$

205.03. $\displaystyle\int \frac{x^5 dx}{r^3} = \frac{r^3}{3} - 2a^2 r - \frac{a^4}{r}.$

205.05. $\displaystyle\int \frac{x^5 dx}{r^5} = r + \frac{2a^2}{r} - \frac{a^4}{3r^3}.$

205.07. $\displaystyle\int \frac{x^5 dx}{r^7} = -\frac{1}{r} + \frac{2a^2}{3r^3} - \frac{a^4}{5r^5}.$

205.9. $\displaystyle\int \frac{x^5 dx}{r^{2p+1}} = -\frac{1}{(2p-5)r^{2p-5}} + \frac{2a^2}{(2p-3)r^{2p-3}}$
$$-\frac{a^4}{(2p-1)r^{2p-1}}.$$

206.01. $\displaystyle\int \frac{x^6 dx}{r} = \frac{x^5 r}{6} - \frac{5}{24} a^2 x^3 r + \frac{5}{16} a^4 x r - \frac{5}{16} a^6 \log (x + r).$

[See note under **200.01.**]

206.03. $\displaystyle\int \frac{x^6 dx}{r^3} = \frac{x^5}{4r} - \frac{5}{8} \frac{a^2 x^3}{r} - \frac{15}{8} \frac{a^4 x}{r} + \frac{15}{8} a^4 \log (x + r).$

206.05. $\displaystyle\int \frac{x^6 dx}{r^5} = \frac{x^5}{2r^3} + \frac{10}{3} \frac{a^2 x^3}{r^3} + \frac{5}{2} \frac{a^4 x}{r^3} - \frac{5}{2} a^2 \log (x + r).$

206.07. $\displaystyle\int \frac{x^6 dx}{r^7} = -\frac{23}{15} \frac{x^5}{r^5} - \frac{7}{3} \frac{a^2 x^3}{r^5} - \frac{a^4 x}{r^5} + \log (x + r).$

206.09. $\displaystyle\int \frac{x^6 dx}{r^9} = \frac{1}{7a^2} \frac{x^7}{r^7}.$

206.11. $\displaystyle\int \frac{x^6 dx}{r^{11}} = \frac{1}{a^4} \left[\frac{1}{7} \frac{x^7}{r^7} - \frac{1}{9} \frac{x^9}{r^9} \right].$

206.13. $\displaystyle\int \frac{x^6 dx}{r^{13}} = \frac{1}{a^6} \left[\frac{1}{7} \frac{x^7}{r^7} - \frac{2}{9} \frac{x^9}{r^9} + \frac{1}{11} \frac{x^{11}}{r^{11}} \right].$

206.15. $\displaystyle\int \frac{x^6 dx}{r^{15}} = \frac{1}{a^8} \left[\frac{1}{7} \frac{x^7}{r^7} - \frac{3}{9} \frac{x^9}{r^9} + \frac{3}{11} \frac{x^{11}}{r^{11}} - \frac{1}{13} \frac{x^{13}}{r^{13}} \right].$

207.01. $\displaystyle\int \frac{x^7 dx}{r} = \frac{1}{7} r^7 - \frac{3}{5} a^2 r^5 + \frac{3}{3} a^4 r^3 - a^6 r.$

207.03. $\displaystyle\int \frac{x^7 dx}{r^3} = \frac{1}{5} r^5 - \frac{3}{3} a^2 r^3 + 3a^4 r + \frac{a^6}{r}.$

207.05. $\displaystyle\int \frac{x^7 dx}{r^5} = \frac{1}{3} r^3 - 3a^2 r - \frac{3a^4}{r} + \frac{a^6}{3r^3}.$

207.07. $\displaystyle\int \frac{x^7 dx}{r^7} = r + \frac{3a^2}{r} - \frac{3a^4}{3r^3} + \frac{a^6}{5r^5}.$

207.9. $\displaystyle\int \frac{x^7 dx}{r^{2p+1}} = -\frac{1}{(2p-7)r^{2p-7}} + \frac{3a^2}{(2p-5)r^{2p-5}}$
$$- \frac{3a^4}{(2p-3)r^{2p-3}} + \frac{a^6}{(2p-1)r^{2p-1}}.$$

221.01. $\displaystyle\int \frac{dx}{xr} = \int \frac{dx}{x\sqrt{(x^2 + a^2)}} = -\frac{1}{a} \log \left| \frac{a + r}{x} \right|.$

Note that

$$-\frac{1}{a} \log \left| \frac{a + r}{x} \right| = -\frac{1}{a} \operatorname{csch}^{-1} \left| \frac{x}{a} \right| = -\frac{1}{a} \sinh^{-1} \left| \frac{a}{x} \right|$$

$$= -\frac{1}{2a} \log \left(\frac{r + a}{r - a} \right).$$

The positive values of a and r are to be taken.

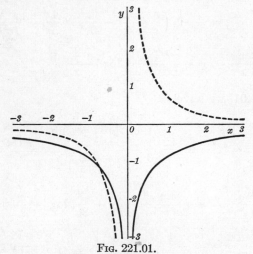

Fig. 221.01.

Dotted graph, $\dfrac{1}{x\sqrt{(x^2 + 1)}}.$

Full line graph, $- \log \left| \dfrac{1 + \sqrt{(x^2 + 1)}}{x} \right|.$

221.03. $\displaystyle\int \frac{dx}{xr^3} = \frac{1}{a^2 r} - \frac{1}{a^3} \log \left| \frac{a + r}{x} \right|.$

221.05. $\displaystyle\int \frac{dx}{xr^5} = \frac{1}{3a^2 r^3} + \frac{1}{a^4 r} - \frac{1}{a^5} \log \left| \frac{a + r}{x} \right|.$

221.07. $\displaystyle\int \frac{dx}{xr^7} = \frac{1}{5a^2 r^5} + \frac{1}{3a^4 r^3} + \frac{1}{a^6 r} - \frac{1}{a^7} \log \left| \frac{a + r}{x} \right|.$

221.09. $\displaystyle\int \frac{dx}{xr^9} = \frac{1}{7a^2 r^7} + \frac{1}{5a^4 r^5} + \frac{1}{3a^6 r^3} + \frac{1}{a^8 r} - \frac{1}{a^9} \log \left| \frac{a + r}{x} \right|.$

222.01. $\int \dfrac{dx}{x^2 r} = -\dfrac{r}{a^2 x}$.

222.03. $\int \dfrac{dx}{x^2 r^3} = -\dfrac{1}{a^4}\left(\dfrac{r}{x} + \dfrac{x}{r}\right)$.

222.05. $\int \dfrac{dx}{x^2 r^5} = -\dfrac{1}{a^6}\left(\dfrac{r}{x} + \dfrac{2x}{r} - \dfrac{x^3}{3r^3}\right)$.

222.07. $\int \dfrac{dx}{x^2 r^7} = -\dfrac{1}{a^8}\left(\dfrac{r}{x} + \dfrac{3x}{r} - \dfrac{3x^3}{3r^3} + \dfrac{x^5}{5r^5}\right)$.

222.09. $\int \dfrac{dx}{x^2 r^9} = -\dfrac{1}{a^{10}}\left(\dfrac{r}{x} + \dfrac{4x}{r} - \dfrac{6x^3}{3r^3} + \dfrac{4x^5}{5r^5} - \dfrac{x^7}{7r^7}\right)$.

223.01. $\int \dfrac{dx}{x^3 r} = -\dfrac{r}{2a^2 x^2} + \dfrac{1}{2a^3}\log\left|\dfrac{a+r}{x}\right|$.

As in **221.01**, we have

$$\log\left|\dfrac{a+r}{x}\right| = \operatorname{csch}^{-1}\left|\dfrac{x}{a}\right| = \sinh^{-1}\left|\dfrac{a}{x}\right|$$
$$= \dfrac{1}{2}\log\left(\dfrac{r+a}{r-a}\right).$$

223.03. $\int \dfrac{dx}{x^3 r^3} = -\dfrac{1}{2a^2 x^2 r} - \dfrac{3}{2a^4 r} + \dfrac{3}{2a^5}\log\left|\dfrac{a+r}{x}\right|$.

223.05. $\int \dfrac{dx}{x^3 r^5} = -\dfrac{1}{2a^2 x^2 r^3} - \dfrac{5}{6a^4 r^3} - \dfrac{5}{2a^6 r} + \dfrac{5}{2a^7}\log\left|\dfrac{a+r}{x}\right|$.

224.01. $\int \dfrac{dx}{x^4 r} = \dfrac{1}{a^4}\left(\dfrac{r}{x} - \dfrac{r^3}{3x^3}\right)$.

224.03. $\int \dfrac{dx}{x^4 r^3} = \dfrac{1}{a^6}\left(\dfrac{x}{r} + \dfrac{2r}{x} - \dfrac{r^3}{3x^3}\right)$.

224.05. $\int \dfrac{dx}{x^4 r^5} = \dfrac{1}{a^8}\left(-\dfrac{x^3}{3r^3} + \dfrac{3x}{r} + \dfrac{3r}{x} - \dfrac{r^3}{3x^3}\right)$.

For **222** and **224**, put

$$z^2 = \dfrac{x^2}{r^2};$$

then

$$dx = \dfrac{a\,dz}{(1-z^2)^{3/2}}.$$

225.01. $\displaystyle\int \frac{dx}{x^5 r} = -\frac{r}{4a^2 x^4} + \frac{3}{8}\frac{r}{a^4 x^2} - \frac{3}{8a^5}\log\left|\frac{a+r}{x}\right|.$

[Ref. 1, p. 121.]

225.03. $\displaystyle\int \frac{dx}{x^5 r^3} = -\frac{1}{4a^2 x^4 r} + \frac{5}{8a^4 x^2 r} + \frac{15}{8a^6 r} - \frac{15}{8a^7}\log\left|\frac{a+r}{x}\right|.$

[Ref. 1, p. 124.]

226.01. $\displaystyle\int \frac{dx}{x^6 r} = \frac{1}{a^6}\left(-\frac{r}{x} + \frac{2r^3}{3x^3} - \frac{r^5}{5x^5}\right).$

226.03. $\displaystyle\int \frac{dx}{x^6 r^3} = \frac{1}{a^8}\left(-\frac{x}{r} - \frac{3r}{x} + \frac{3r^3}{3x^3} - \frac{r^5}{5x^5}\right).$

$\cdot\ \cdot\ \cdot\ \cdot\ \cdot\ \cdot\ \cdot\ \cdot\ \cdot\ \cdot\ \cdot\ \cdot\ \cdot\ \cdot\ \cdot\ \cdot\ \cdot$

230.01. $\displaystyle\int r\,dx = \frac{xr}{2} + \frac{a^2}{2}\log(x+r).$

As in **200.01**, we have

$$\log\left(\frac{x+r}{a}\right) = \sinh^{-1}\frac{x}{a} = \operatorname{csch}^{-1}\frac{a}{x}$$
$$= \frac{1}{2}\log\left(\frac{r+x}{r-x}\right).$$

230.03. $\displaystyle\int r^3 dx = \frac{1}{4}xr^3 + \frac{3}{8}a^2 xr + \frac{3}{8}a^4 \log(x+r).$

230.05. $\displaystyle\int r^5 dx = \frac{1}{6}xr^5 + \frac{5}{24}a^2 xr^3 + \frac{5}{16}a^4 xr + \frac{5}{16}a^6 \log(x+r).$

231.01. $\displaystyle\int xr\,dx = \frac{r^3}{3}.$

231.03. $\displaystyle\int xr^3 dx = \frac{r^5}{5}.$

$\cdot\ \cdot\ \cdot\ \cdot\ \cdot\ \cdot\ \cdot\ \cdot$

231.9. $\displaystyle\int xr^{2p+1}dx = \frac{r^{2p+3}}{2p+3}.$

232.01. $\displaystyle\int x^2 r\,dx = \frac{xr^3}{4} - \frac{a^2 xr}{8} - \frac{a^4}{8}\log(x+r).$

232.03. $\displaystyle\int x^2 r^3 dx = \frac{xr^5}{6} - \frac{a^2 xr^3}{24} - \frac{a^4 xr}{16} - \frac{a^6}{16}\log(x+r).$

233.01. $\int x^3 r \, dx = \dfrac{r^5}{5} - \dfrac{a^2 r^3}{3}.$

233.03. $\int x^3 r^3 dx = \dfrac{r^7}{7} - \dfrac{a^2 r^5}{5}.$

. . . .

233.9. $\int x^3 r^{2p+1} dx = \dfrac{r^{2p+5}}{2p+5} - \dfrac{a^2 r^{2p+3}}{2p+3}.$

234.01. $\int x^4 r \, dx = \dfrac{x^3 r^3}{6} - \dfrac{a^2 x r^3}{8} + \dfrac{a^4 x r}{16} + \dfrac{a^6}{16} \log (x+r).$

As in **200.01** we have

$$\log \left(\frac{x+r}{a} \right) = \sinh^{-1} \frac{x}{a} = \operatorname{csch}^{-1} \frac{a}{x}.$$

$$= \frac{1}{2} \log \left(\frac{r+x}{r-x} \right).$$

234.03. $\int x^4 r^3 dx = \dfrac{x^3 r^5}{8} - \dfrac{a^2 x r^5}{16} + \dfrac{a^4 x r^3}{64} + \dfrac{3}{128} a^6 x r$

$$+ \frac{3}{128} a^8 \log (x+r).$$

235.01. $\int x^5 r \, dx = \dfrac{r^7}{7} - \dfrac{2 a^2 r^5}{5} + \dfrac{a^4 r^3}{3}.$

235.03. $\int x^5 r^3 dx = \dfrac{r^9}{9} - \dfrac{2 a^2 r^7}{7} + \dfrac{a^4 r^5}{5}.$

. . . .

235.9. $\int x^5 r^{2p+1} dx = \dfrac{r^{2p+7}}{2p+7} - \dfrac{2 a^2 r^{2p+5}}{2p+5} + \dfrac{a^4 r^{2p+3}}{2p+3}.$

241.01. $\int \dfrac{r \, dx}{x} = r - a \log \left| \dfrac{a+r}{x} \right|.$

[See note under **221.01.**]

241.03. $\int \dfrac{r^3 dx}{x} = \dfrac{r^3}{3} + a^2 r - a^3 \log \left| \dfrac{a+r}{x} \right|.$

241.05. $\int \dfrac{r^5 dx}{x} = \dfrac{r^5}{5} + \dfrac{a^2 r^3}{3} + a^4 r - a^5 \log \left| \dfrac{a+r}{x} \right|.$

241.07. $\int \dfrac{r^7 dx}{x} = \dfrac{r^7}{7} + \dfrac{a^2 r^5}{5} + \dfrac{a^4 r^3}{3} + a^6 r - a^7 \log \left| \dfrac{a+r}{x} \right|.$

242.01. $\displaystyle\int \frac{r\,dx}{x^2} = -\frac{r}{x} + \log(x + r).$

[See note under 200.01.]

242.03. $\displaystyle\int \frac{r^3 dx}{x^2} = -\frac{r^3}{x} + \frac{3}{2}\,xr + \frac{3}{2}\,a^2 \log(x + r).$

242.05. $\displaystyle\int \frac{r^5 dx}{x^2} = -\frac{r^5}{x} + \frac{5}{4}\,xr^3 + \frac{15}{8}\,a^2 xr + \frac{15}{8}\,a^4 \log(x + r).$

243.01. $\displaystyle\int \frac{r\,dx}{x^3} = -\frac{r}{2x^2} - \frac{1}{2a} \log\left|\frac{a + r}{x}\right|.$

[See note under 221.01.]

243.03. $\displaystyle\int \frac{r^3 dx}{x^3} = -\frac{r^3}{2x^2} + \frac{3}{2}\,r - \frac{3}{2}\,a \log\left|\frac{a + r}{x}\right|.$

243.05. $\displaystyle\int \frac{r^5 dx}{x^3} = -\frac{r^5}{2x^2} + \frac{5}{6}\,r^3 + \frac{5}{2}\,a^2 r - \frac{5}{2}\,a^3 \log\left|\frac{a + r}{x}\right|.$

244.01. $\displaystyle\int \frac{r\,dx}{x^4} = -\frac{r^3}{3a^2 x^3}.$

244.03. $\displaystyle\int \frac{r^3 dx}{x^4} = -\frac{r^3}{3x^3} - \frac{r}{x} + \log(x + r).$

[See note under 200.01.]

244.05. $\displaystyle\int \frac{r^5 dx}{x^4} = -\frac{a^2 r^3}{3x^3} - \frac{2a^2 r}{x} + \frac{xr}{2} + \frac{5}{2}\,a^2 \log(x + r).$

245.01. $\displaystyle\int \frac{r\,dx}{x^5} = -\frac{r}{4x^4} - \frac{r}{8a^2 x^2} + \frac{1}{8a^3} \log\left|\frac{a + r}{x}\right|.$

245.03. $\displaystyle\int \frac{r^3 dx}{x^5} = -\frac{r^3}{4x^4} - \frac{3}{8}\frac{r^3}{a^2 x^2} + \frac{3}{8}\frac{r}{a^2} - \frac{3}{8a} \log\left|\frac{a + r}{x}\right|.$

246.01. $\displaystyle\int \frac{r\,dx}{x^6} = \frac{r^3}{5a^2 x^3}\left(\frac{2}{3a^2} - \frac{1}{x^2}\right).$

246.03. $\displaystyle\int \frac{r^3 dx}{x^6} = -\frac{r^5}{5a^2 x^5}.$

247.01. $\displaystyle\int \frac{r\,dx}{x^7} = -\frac{r}{6x^6} - \frac{r}{24a^2 x^4} + \frac{r}{16a^4 x^2} - \frac{1}{16a^5} \log\left|\frac{a + r}{x}\right|.$

248.01. $\displaystyle\int \frac{r\,dx}{x^8} = \frac{r^3}{7a^2 x^3}\left(-\frac{1}{x^4} + \frac{4}{5a^2 x^2} - \frac{8}{15a^4}\right).$

Integrals Involving $s = (x^2 - a^2)^{1/2}$

260.01. $\displaystyle\int \frac{dx}{s} = \int \frac{dx}{\sqrt{(x^2 - a^2)}} = \log |x + s|$, $\qquad [x^2 > a^2].$

Note that

$$\log \left| \frac{x + s}{a} \right| = \frac{1}{2} \log \left(\frac{x + s}{x - s} \right) = \cosh^{-1} \left| \frac{x}{a} \right|.$$

The positive value of $\cosh^{-1}|x/a|$ is to be taken for positive values of x, and the negative value for negative values of x. The positive value of s is to be taken.

Fig. 260.01.

Dotted graph, $1/\sqrt{(x^2 - 1)}$. Full line graph, $\log |x + \sqrt{(x^2 - 1)}|$.

260.03. $\displaystyle\int \frac{dx}{s^3} = - \frac{1}{a^2} \frac{x}{s}.$

260.05. $\displaystyle\int \frac{dx}{s^5} = \frac{1}{a^4} \left[\frac{x}{s} - \frac{1}{3} \frac{x^3}{s^3} \right].$

260.07. $\displaystyle\int \frac{dx}{s^7} = - \frac{1}{a^6} \left[\frac{x}{s} - \frac{2}{3} \frac{x^3}{s^3} + \frac{1}{5} \frac{x^5}{s^5} \right].$

260.09. $\displaystyle\int \frac{dx}{s^9} = \frac{1}{a^8} \left[\frac{x}{s} - \frac{3}{3} \frac{x^3}{s^3} + \frac{3}{5} \frac{x^5}{s^5} - \frac{1}{7} \frac{x^7}{s^7} \right].$

260.11. $\displaystyle\int \frac{dx}{s^{11}} = - \frac{1}{a^{10}} \left[\frac{x}{s} - \frac{4}{3} \frac{x^3}{s^3} + \frac{6}{5} \frac{x^5}{s^5} - \frac{4}{7} \frac{x^7}{s^7} + \frac{1}{9} \frac{x^9}{s^9} \right].$

260.13. $\displaystyle\int \frac{dx}{s^{13}} = \frac{1}{a^{12}} \left[\frac{x}{s} - \frac{5}{3} \frac{x^3}{s^3} + \frac{10}{5} \frac{x^5}{s^5} - \frac{10}{7} \frac{x^7}{s^7} + \frac{5}{9} \frac{x^9}{s^9} - \frac{1}{11} \frac{x^{11}}{s^{11}} \right].$

260.15. $\displaystyle\int \frac{dx}{s^{15}} = - \frac{1}{a^{14}} \left[\frac{x}{s} - \frac{6}{3} \frac{x^3}{s^3} + \frac{15}{5} \frac{x^5}{s^5} - \frac{20}{7} \frac{x^7}{s^7} + \frac{15}{9} \frac{x^9}{s^9} \right.$

$$\left. - \frac{6}{11} \frac{x^{11}}{s^{11}} + \frac{1}{13} \frac{x^{13}}{s^{13}} \right].$$

For **260.03–260.15**, let

$$z^2 = \frac{x^2}{x^2 - a^2}; \qquad \text{then} \qquad dx = \frac{-a\,dz}{(z^2 - 1)^{3/2}}.$$

261.01. $\displaystyle \int \frac{x\,dx}{s} = s.$ **261.05.** $\displaystyle \int \frac{x\,dx}{s^5} = -\frac{1}{3s^3}.$

261.03. $\displaystyle \int \frac{x\,dx}{s^3} = -\frac{1}{s}.$ **261.07.** $\displaystyle \int \frac{x\,dx}{s^7} = -\frac{1}{5s^5}.$

261.9. $\displaystyle \int \frac{x\,dx}{s^{2p+1}} = -\frac{1}{(2p-1)s^{2p-1}}.$

262.01. $\displaystyle \int \frac{x^2 dx}{s} = \frac{xs}{2} + \frac{a^2}{2} \log |x + s|.$

[See note under 260.01.]

262.03. $\displaystyle \int \frac{x^2 dx}{s^3} = -\frac{x}{s} + \log |x + s|.$

262.05. $\displaystyle \int \frac{x^2 dx}{s^5} = -\frac{1}{3a^2} \frac{x^3}{s^3}.$

262.07. $\displaystyle \int \frac{x^2 dx}{s^7} = \frac{1}{a^4} \left[\frac{1}{3} \frac{x^3}{s^3} - \frac{1}{5} \frac{x^5}{s^5} \right].$

262.09. $\displaystyle \int \frac{x^2 dx}{s^9} = -\frac{1}{a^6} \left[\frac{1}{3} \frac{x^3}{s^3} - \frac{2}{5} \frac{x^5}{s^5} + \frac{1}{7} \frac{x^7}{s^7} \right].$

262.11. $\displaystyle \int \frac{x^2 dx}{s^{11}} = \frac{1}{a^8} \left[\frac{1}{3} \frac{x^3}{s^3} - \frac{3}{5} \frac{x^5}{s^5} + \frac{3}{7} \frac{x^7}{s^7} - \frac{1}{9} \frac{x^9}{s^9} \right].$

262.13. $\displaystyle \int \frac{x^2 dx}{s^{13}} = -\frac{1}{a^{10}} \left[\frac{1}{3} \frac{x^3}{s^3} - \frac{4}{5} \frac{x^5}{s^5} + \frac{6}{7} \frac{x^7}{s^7} - \frac{4}{9} \frac{x^9}{s^9} + \frac{1}{11} \frac{x^{11}}{s^{11}} \right].$

262.15. $\displaystyle \int \frac{x^2 dx}{s^{15}} = \frac{1}{a^{12}} \left[\frac{1}{3} \frac{x^3}{s^3} - \frac{5}{5} \frac{x^5}{s^5} + \frac{10}{7} \frac{x^7}{s^7} - \frac{10}{9} \frac{x^9}{s^9} \right.$
$$\left. + \frac{5}{11} \frac{x^{11}}{s^{11}} - \frac{1}{13} \frac{x^{13}}{s^{13}} \right].$$

263.01. $\displaystyle \int \frac{x^3 dx}{s} = \frac{s^3}{3} + a^2 s.$

263.03. $\displaystyle \int \frac{x^3 dx}{s^3} = s - \frac{a^2}{s}.$

263.05. $\displaystyle \int \frac{x^3 dx}{s^5} = -\frac{1}{s} - \frac{a^2}{3s^3}.$

263.9. $\displaystyle\int \frac{x^3 dx}{s^{2p+1}} = -\frac{1}{(2p-3)s^{2p-3}} - \frac{a^2}{(2p-1)s^{2p-1}}.$

264.01. $\displaystyle\int \frac{x^4 dx}{s} = \frac{x^3 s}{4} + \frac{3}{8} a^2 xs + \frac{3}{8} a^4 \log |x + s|.$

[See note under **260.01.**]

264.03. $\displaystyle\int \frac{x^4 dx}{s^3} = \frac{xs}{2} - \frac{a^2 x}{s} + \frac{3}{2} a^2 \log |x + s|.$

264.05. $\displaystyle\int \frac{x^4 dx}{s^5} = -\frac{x}{s} - \frac{1}{3} \frac{x^3}{s^3} + \log |x + s|.$

264.07. $\displaystyle\int \frac{x^4 dx}{s^7} = -\frac{1}{5a^2} \frac{x^5}{s^5}.$

264.09. $\displaystyle\int \frac{x^4 dx}{s^9} = \frac{1}{a^4} \left[\frac{1}{5} \frac{x^5}{s^5} - \frac{1}{7} \frac{x^7}{s^7} \right].$

264.11. $\displaystyle\int \frac{x^4 dx}{s^{11}} = -\frac{1}{a^6} \left[\frac{1}{5} \frac{x^5}{s^5} - \frac{2}{7} \frac{x^7}{s^7} + \frac{1}{9} \frac{x^9}{s^9} \right].$

264.13. $\displaystyle\int \frac{x^4 dx}{s^{13}} = \frac{1}{a^8} \left[\frac{1}{5} \frac{x^5}{s^5} - \frac{3}{7} \frac{x^7}{s^7} + \frac{3}{9} \frac{x^9}{s^9} - \frac{1}{11} \frac{x^{11}}{s^{11}} \right].$

264.15. $\displaystyle\int \frac{x^4 dx}{s^{15}} = -\frac{1}{a^{10}} \left[\frac{1}{5} \frac{x^5}{s^5} - \frac{4}{7} \frac{x^7}{s^7} + \frac{6}{9} \frac{x^9}{s^9} - \frac{4}{11} \frac{x^{11}}{s^{11}} + \frac{1}{13} \frac{x^{13}}{s^{13}} \right].$

265.01. $\displaystyle\int \frac{x^5 dx}{s} = \frac{s^5}{5} + \frac{2}{3} a^2 s^3 + a^4 s.$

265.03. $\displaystyle\int \frac{x^5 dx}{s^3} = \frac{s^3}{3} + 2a^2 s - \frac{a^4}{s}.$

265.05. $\displaystyle\int \frac{x^5 dx}{s^5} = s - \frac{2a^2}{s} - \frac{a^4}{3s^3}.$

265.07. $\displaystyle\int \frac{x^5 dx}{s^7} = -\frac{1}{s} - \frac{2a^2}{3s^3} - \frac{a^4}{5s^5}.$

265.9. $\displaystyle\int \frac{x^5 dx}{s^{2p+1}} = -\frac{1}{(2p-5)s^{2p-5}} - \frac{2a^2}{(2p-3)s^{2p-3}}$

$$-\frac{a^4}{(2p-1)s^{2p-1}}.$$

266.01. $\int \dfrac{x^6 dx}{s} = \dfrac{x^5 s}{6} + \dfrac{5}{24} a^2 x^3 s + \dfrac{5}{16} a^4 x s + \dfrac{5}{16} a^6 \log |x + s|.$

[See note under 260.01.]

266.03. $\int \dfrac{x^6 dx}{s^3} = \dfrac{x^5}{4s} + \dfrac{5}{8} \dfrac{a^2 x^3}{s} - \dfrac{15}{8} \dfrac{a^4 x}{s} + \dfrac{15}{8} a^4 \log |x + s|.$

266.05. $\int \dfrac{x^6 dx}{s^5} = \dfrac{x^5}{2s^3} - \dfrac{10}{3} \dfrac{a^2 x^3}{s^3} + \dfrac{5}{2} \dfrac{a^4 x}{s^3} + \dfrac{5}{2} a^2 \log |x + s|,$

266.07. $\int \dfrac{x^6 dx}{s^7} = -\dfrac{23}{15} \dfrac{x^5}{s^5} + \dfrac{7}{3} \dfrac{a^2 x^3}{s^5} - \dfrac{a^4 x}{s^5} + \log |x + s|.$

266.09. $\int \dfrac{x^6 dx}{s^9} = -\dfrac{1}{7a^2} \dfrac{x^7}{s^7}.$

266.11. $\int \dfrac{x^6 dx}{s^{11}} = \dfrac{1}{a^4} \left[\dfrac{1}{7} \dfrac{x^7}{s^7} - \dfrac{1}{9} \dfrac{x^9}{s^9} \right].$

266.13. $\int \dfrac{x^6 dx}{s^{13}} = -\dfrac{1}{a^6} \left[\dfrac{1}{7} \dfrac{x^7}{s^7} - \dfrac{2}{9} \dfrac{x^9}{s^9} + \dfrac{1}{11} \dfrac{x^{11}}{s^{11}} \right].$

266.15. $\int \dfrac{x^6 dx}{s^{15}} = \dfrac{1}{a^8} \left[\dfrac{1}{7} \dfrac{x^7}{s^7} - \dfrac{3}{9} \dfrac{x^9}{s^9} + \dfrac{3}{11} \dfrac{x^{11}}{s^{11}} - \dfrac{1}{13} \dfrac{x^{13}}{s^{13}} \right].$

267.01. $\int \dfrac{x^7 dx}{s} = \dfrac{1}{7} s^7 + \dfrac{3}{5} a^2 s^5 + \dfrac{3}{3} a^4 s^3 + a^6 s.$

267.03. $\int \dfrac{x^7 dx}{s^3} = \dfrac{1}{5} s^5 + \dfrac{3}{3} a^2 s^3 + 3a^4 s - \dfrac{a^6}{s}.$

267.05. $\int \dfrac{x^7 dx}{s^5} = \dfrac{1}{3} s^3 + 3a^2 s - \dfrac{3a^4}{s} - \dfrac{a^6}{3s^3}.$

267.07. $\int \dfrac{x^7 dx}{s^7} = s - \dfrac{3a^2}{s} - \dfrac{3a^4}{3s^3} - \dfrac{a^6}{5s^5}.$

267.9. $\int \dfrac{x^7 dx}{s^{2p+1}} = -\dfrac{1}{(2p-7)s^{2p-7}} - \dfrac{3a^2}{(2p-5)s^{2p-5}}$

$$-\dfrac{3a^4}{(2p-3)s^{2p-3}} - \dfrac{a^6}{(2p-1)s^{2p-1}}.$$

281.01. $\int \dfrac{dx}{xs} = \int \dfrac{dx}{x\sqrt{(x^2-a^2)}} = \dfrac{1}{a}\cos^{-1}\left|\dfrac{a}{x}\right| = \dfrac{1}{a}\sec^{-1}\left|\dfrac{x}{a}\right|,$

$$[x^2 > a^2].$$

The positive values of s and a are to be taken. The principal values of $\cos^{-1}|a/x|$ are to be taken, that is, they are to be between 0 and $\pi/2$ since $|a/x|$ is a positive quantity.

Fig. 281.01.

Dotted graph, $\dfrac{1}{x\sqrt{(x^2-1)}}$.

Full line graph, $\cos^{-1}\left|\dfrac{1}{x}\right|$.

281.03. $\int \dfrac{dx}{xs^3} = -\dfrac{1}{a^2s} - \dfrac{1}{a^3}\cos^{-1}\left|\dfrac{a}{x}\right|.$

281.05. $\int \dfrac{dx}{xs^5} = -\dfrac{1}{3a^2s^3} + \dfrac{1}{a^4s} + \dfrac{1}{a^5}\cos^{-1}\left|\dfrac{a}{x}\right|.$

281.07. $\int \dfrac{dx}{xs^7} = -\dfrac{1}{5a^2s^5} + \dfrac{1}{3a^4s^3} - \dfrac{1}{a^6s} - \dfrac{1}{a^7}\cos^{-1}\left|\dfrac{a}{x}\right|.$

281.09. $\int \dfrac{dx}{xs^9} = -\dfrac{1}{7a^2s^7} + \dfrac{1}{5a^4s^5} - \dfrac{1}{3a^6s^3} + \dfrac{1}{a^8s} + \dfrac{1}{a^9}\cos^{-1}\left|\dfrac{a}{x}\right|.$

282.01. $\int \dfrac{dx}{x^2s} = \dfrac{s}{a^2x}.$

282.03. $\int \dfrac{dx}{x^2s^3} = -\dfrac{1}{a^4}\left(\dfrac{s}{x} + \dfrac{x}{s}\right).$

282.05. $\int \dfrac{dx}{x^2s^5} = \dfrac{1}{a^6}\left(\dfrac{s}{x} + \dfrac{2x}{s} - \dfrac{x^3}{3s^3}\right).$

282.07. $\displaystyle \int \frac{dx}{x^2 s^7} = -\frac{1}{a^8}\left(\frac{s}{x} + \frac{3x}{s} - \frac{3x^3}{3s^3} + \frac{x^5}{5s^5}\right).$

282.09. $\displaystyle \int \frac{dx}{x^2 s^9} = \frac{1}{a^{10}}\left(\frac{s}{x} + \frac{4x}{s} - \frac{6x^3}{3s^3} + \frac{4x^5}{5s^5} - \frac{x^7}{7s^7}\right).$

283.01. $\displaystyle \int \frac{dx}{x^3 s} = \frac{s}{2a^2 x^2} + \frac{1}{2a^3}\cos^{-1}\left|\frac{a}{x}\right|.$

[See note under 281.01.]

283.03. $\displaystyle \int \frac{dx}{x^3 s^3} = \frac{1}{2a^2 x^2 s} - \frac{3}{2a^4 s} - \frac{3}{2a^5}\cos^{-1}\left|\frac{a}{x}\right|.$

283.05. $\displaystyle \int \frac{dx}{x^3 s^5} = \frac{1}{2a^2 x^2 s^3} - \frac{5}{6a^4 s^3} + \frac{5}{2a^6 s} + \frac{5}{2a^7}\cos^{-1}\left|\frac{a}{x}\right|.$

284.01. $\displaystyle \int \frac{dx}{x^4 s} = \frac{1}{a^4}\left(\frac{s}{x} - \frac{s^3}{3x^3}\right).$

284.03. $\displaystyle \int \frac{dx}{x^4 s^3} = -\frac{1}{a^6}\left(\frac{x}{s} + \frac{2s}{x} - \frac{s^3}{3x^3}\right).$

284.05. $\displaystyle \int \frac{dx}{x^4 s^5} = \frac{1}{a^8}\left(-\frac{x^3}{3s^3} + \frac{3x}{s} + \frac{3s}{x} - \frac{s^3}{3x^3}\right).$

For **282** and **284**, put

$$z^2 = \frac{x^2}{s^2}; \qquad \text{then} \qquad dx = \frac{-a\,dz}{(z^2 - 1)^{3/2}}.$$

290.01. $\displaystyle \int s\,dx = \frac{xs}{2} - \frac{a^2}{2}\log|x + s|.$ [See note under **260.01.**]

290.03. $\displaystyle \int s^3 dx = \frac{1}{4}xs^3 - \frac{3}{8}a^2 xs + \frac{3}{8}a^4 \log|x + s|.$

290.05. $\displaystyle \int s^5 dx = \frac{1}{6}xs^5 - \frac{5}{24}a^2 xs^3 + \frac{5}{16}a^4 xs - \frac{5}{16}a^6 \log|x + s|.$

291.01. $\displaystyle \int xs\,dx = \frac{s^3}{3}.$ **291.03.** $\displaystyle \int xs^3 dx = \frac{s^5}{5}.$

291.9. $\displaystyle \int xs^{2p+1} dx = \frac{s^{2p+3}}{2p+3}.$

292.01. $\displaystyle \int x^2 s\,dx = \frac{xs^3}{4} + \frac{a^2 xs}{8} - \frac{a^4}{8}\log|x + s|.$

[See note under 260.01.]

292.03. $\displaystyle\int x^2 s^3 dx = \frac{xs^5}{6} + \frac{a^2 xs^3}{24} - \frac{a^4 xs}{16} + \frac{a^6}{16}\log|x+s|.$

293.01. $\displaystyle\int x^3 s\ dx = \frac{s^5}{5} + \frac{a^2 s^3}{3}.$ **293.03.** $\displaystyle\int x^3 s^3 dx = \frac{s^7}{7} + \frac{a^2 s^5}{5}.$

293.9. $\displaystyle\int x^3 s^{2p+1} dx = \frac{s^{2p+5}}{2p+5} + \frac{a^2 s^{2p+3}}{2p+3}.$

294.01. $\displaystyle\int x^4 s\ dx = \frac{x^3 s^3}{6} + \frac{a^2 xs^3}{8} + \frac{a^4 xs}{16} - \frac{a^6}{16}\log|x+s|.$

[See note under 260.01.]

294.03. $\displaystyle\int x^4 s^3 dx = \frac{x^3 s^5}{8} + \frac{a^2 xs^5}{16} + \frac{a^4 xs^3}{64} - \frac{3}{128}a^6 xs$

$$+ \frac{3}{128}a^8 \log|x+s|.$$

295.01. $\displaystyle\int x^5 s\ dx = \frac{s^7}{7} + \frac{2a^2 s^5}{5} + \frac{a^4 s^3}{3}.$

295.03. $\displaystyle\int x^5 s^3 dx = \frac{s^9}{9} + \frac{2a^2 s^7}{7} + \frac{a^4 s^5}{5}.$

295.9. $\displaystyle\int x^5 s^{2p+1} dx = \frac{s^{2p+7}}{2p+7} + \frac{2a^2 s^{2p+5}}{2p+5} + \frac{a^4 s^{2p+3}}{2p+3}.$

301.01. $\displaystyle\int \frac{s\ dx}{x} = s - a\cos^{-1}\left|\frac{a}{x}\right|.$ [See note under **281.01.**]

301.03. $\displaystyle\int \frac{s^3 dx}{x} = \frac{s^3}{3} - a^2 s + a^3 \cos^{-1}\left|\frac{a}{x}\right|.$

301.05. $\displaystyle\int \frac{s^5 dx}{x} = \frac{s^5}{5} - \frac{a^2 s^3}{3} + a^4 s - a^5 \cos^{-1}\left|\frac{a}{x}\right|.$

301.07. $\displaystyle\int \frac{s^7 dx}{x} = \frac{s^7}{7} - \frac{a^2 s^5}{5} + \frac{a^4 s^3}{3} - a^6 s + a^7 \cos^{-1}\left|\frac{a}{x}\right|.$

302.01. $\displaystyle\int \frac{s\ dx}{x^2} = -\frac{s}{x} + \log|x+s|.$

[See note under 260.01.]

302.03. $\displaystyle\int \frac{s^3 dx}{x^2} = -\frac{s^3}{x} + \frac{3}{2}xs - \frac{3}{2}a^2 \log|x+s|.$

302.05. $\int \dfrac{s^5 dx}{x^2} = -\dfrac{s^5}{x} + \dfrac{5}{4} x s^3 - \dfrac{15}{8} a^2 x s + \dfrac{15}{8} a^4 \log |x + s|.$

303.01. $\int \dfrac{s\, dx}{x^3} = -\dfrac{s}{2x^2} + \dfrac{1}{2a} \cos^{-1} \left| \dfrac{a}{x} \right|.$

[See note under 281.01.]

303.03. $\int \dfrac{s^3 dx}{x^3} = -\dfrac{s^3}{2x^2} + \dfrac{3s}{2} - \dfrac{3}{2} a \cos^{-1} \left| \dfrac{a}{x} \right|.$

303.05. $\int \dfrac{s^5 dx}{x^3} = -\dfrac{s^5}{2x^2} + \dfrac{5}{6} s^3 - \dfrac{5}{2} a^2 s + \dfrac{5}{2} a^3 \cos^{-1} \left| \dfrac{a}{x} \right|.$

304.01. $\int \dfrac{s\, dx}{x^4} = \dfrac{s^3}{3a^2 x^3}.$

304.03. $\int \dfrac{s^3 dx}{x^4} = -\dfrac{s^3}{3x^3} - \dfrac{s}{x} + \log |x + s|.$

[See note under 260.01.]

304.05. $\int \dfrac{s^5 dx}{x^4} = \dfrac{a^2 s^3}{3x^3} + \dfrac{2a^2 s}{x} + \dfrac{xs}{2} - \dfrac{5}{2} a^2 \log |x + s|.$

305.01. $\int \dfrac{s\, dx}{x^5} = -\dfrac{s}{4x^4} + \dfrac{s}{8a^2 x^2} + \dfrac{1}{8a^3} \cos^{-1} \left| \dfrac{a}{x} \right|.$

305.03. $\int \dfrac{s^3 dx}{x^5} = -\dfrac{s^3}{4x^4} + \dfrac{3}{8} \dfrac{s^3}{a^2 x^2} - \dfrac{3}{8} \dfrac{s}{a^2} + \dfrac{3}{8a} \cos^{-1} \left| \dfrac{a}{x} \right|.$

306.01. $\int \dfrac{s\, dx}{x^6} = \dfrac{s^3}{5a^2 x^3} \left(\dfrac{1}{x^2} + \dfrac{2}{3a^2} \right).$

306.03. $\int \dfrac{s^3 dx}{x^6} = \dfrac{s^5}{5a^2 x^5}.$

307.01. $\int \dfrac{s\, dx}{x^7} = -\dfrac{s}{6x^6} + \dfrac{s}{24a^2 x^4} + \dfrac{s}{16a^4 x^2} + \dfrac{1}{16a^5} \cos^{-1} \left| \dfrac{a}{x} \right|.$

308.01. $\int \dfrac{s\, dx}{x^8} = \dfrac{s^3}{7a^2 x^3} \left(\dfrac{1}{x^4} + \dfrac{4}{5a^2 x^2} + \dfrac{8}{15a^4} \right).$

Integrals Involving $t = (a^2 - x^2)^{1/2}$

320.01. $\displaystyle\int \frac{dx}{t} = \int \frac{dx}{\sqrt{(a^2 - x^2)}} = \sin^{-1}\frac{x}{a},$ $[x^2 < a^2].$

The principal values of $\sin^{-1}(x/a)$ are to be taken, that is, values between $-\pi/2$ and $\pi/2$. The positive values of t and a are to be taken.

FIG. 320.01.

320.03. $\displaystyle\int \frac{dx}{t^3} = \frac{1}{a^2}\frac{x}{t}.$ **320.05.** $\displaystyle\int \frac{dx}{t^5} = \frac{1}{a^4}\left[\frac{x}{t} + \frac{1}{3}\frac{x^3}{t^3}\right].$

320.07. $\displaystyle\int \frac{dx}{t^7} = \frac{1}{a^6}\left[\frac{x}{t} + \frac{2}{3}\frac{x^3}{t^3} + \frac{1}{5}\frac{x^5}{t^5}\right].$

320.09. $\displaystyle\int \frac{dx}{t^9} = \frac{1}{a^8}\left[\frac{x}{t} + \frac{3}{3}\frac{x^3}{t^3} + \frac{3}{5}\frac{x^5}{t^5} + \frac{1}{7}\frac{x^7}{t^7}\right].$

320.11. $\displaystyle\int \frac{dx}{t^{11}} = \frac{1}{a^{10}}\left[\frac{x}{t} + \frac{4}{3}\frac{x^3}{t^3} + \frac{6}{5}\frac{x^5}{t^5} + \frac{4}{7}\frac{x^7}{t^7} + \frac{1}{9}\frac{x^9}{t^9}\right].$

320.13. $\displaystyle\int \frac{dx}{t^{13}} = \frac{1}{a^{12}}\left[\frac{x}{t} + \frac{5}{3}\frac{x^3}{t^3} + \frac{10}{5}\frac{x^5}{t^5} + \frac{10}{7}\frac{x^7}{t^7} + \frac{5}{9}\frac{x^9}{t^9} + \frac{1}{11}\frac{x^{11}}{t^{11}}\right].$

320.15. $\displaystyle\int \frac{dx}{t^{15}} = \frac{1}{a^{14}}\left[\frac{x}{t} + \frac{6}{3}\frac{x^3}{t^3} + \frac{15}{5}\frac{x^5}{t^5} + \frac{20}{7}\frac{x^7}{t^7} + \frac{15}{9}\frac{x^9}{t^9}\right.$
$$\left. + \frac{6}{11}\frac{x^{11}}{t^{11}} + \frac{1}{13}\frac{x^{13}}{t^{13}}\right].$$

For **320.03–320.15** let

$$z^2 = \frac{x^2}{a^2 - x^2}; \quad \text{then} \quad dx = \frac{a\,dz}{(1 + z^2)^{3/2}}.$$

321.01. $\displaystyle\int \frac{x\,dx}{t} = -\,t.$

321.05. $\displaystyle\int \frac{x\,dx}{t^5} = \frac{1}{3t^3}\cdot$

321.03. $\displaystyle\int \frac{x\,dx}{t^3} = \frac{1}{t}\cdot$

321.07. $\displaystyle\int \frac{x\,dx}{t^7} = \frac{1}{5t^5}\cdot$

321.9. $\displaystyle\int \frac{x\,dx}{t^{2p+1}} = \frac{1}{(2p-1)t^{2p-1}}\cdot$

322.01. $\displaystyle\int \frac{x^2 dx}{t} = -\,\frac{xt}{2} + \frac{a^2}{2}\sin^{-1}\frac{x}{a}\cdot$ [See note under 320.01.]

322.03. $\displaystyle\int \frac{x^2 dx}{t^3} = \frac{x}{t} - \sin^{-1}\frac{x}{a}\cdot$

322.05. $\displaystyle\int \frac{x^2 dx}{t^5} = \frac{1}{3a^2}\frac{x^3}{t^3}\cdot$

322.07. $\displaystyle\int \frac{x^2 dx}{t^7} = \frac{1}{a^4}\left[\frac{1}{3}\frac{x^3}{t^3} + \frac{1}{5}\frac{x^5}{t^5}\right].$

322.09. $\displaystyle\int \frac{x^2 dx}{t^9} = \frac{1}{a^6}\left[\frac{1}{3}\frac{x^3}{t^3} + \frac{2}{5}\frac{x^5}{t^5} + \frac{1}{7}\frac{x^7}{t^7}\right].$

322.11. $\displaystyle\int \frac{x^2 dx}{t^{11}} = \frac{1}{a^8}\left[\frac{1}{3}\frac{x^3}{t^3} + \frac{3}{5}\frac{x^5}{t^5} + \frac{3}{7}\frac{x^7}{t^7} + \frac{1}{9}\frac{x^9}{t^9}\right].$

322.13. $\displaystyle\int \frac{x^2 dx}{t^{13}} = \frac{1}{a^{10}}\left[\frac{1}{3}\frac{x^3}{t^3} + \frac{4}{5}\frac{x^5}{t^5} + \frac{6}{7}\frac{x^7}{t^7} + \frac{4}{9}\frac{x^9}{t^9} + \frac{1}{11}\frac{x^{11}}{t^{11}}\right].$

322.15. $\displaystyle\int \frac{x^2 dx}{t^{15}} = \frac{1}{a^{12}}\left[\frac{1}{3}\frac{x^3}{t^3} + \frac{5}{5}\frac{x^5}{t^5} + \frac{10}{7}\frac{x^7}{t^7} + \frac{10}{9}\frac{x^9}{t^9}\right.$
$$\left. + \frac{5}{11}\frac{x^{11}}{t^{11}} + \frac{1}{13}\frac{x^{13}}{t^{13}}\right].$$

323.01. $\displaystyle\int \frac{x^3 dx}{t} = \frac{t^3}{3} - a^2 t.$

323.03. $\displaystyle\int \frac{x^3 dx}{t^3} = t + \frac{a^2}{t}\cdot$

323.05. $\displaystyle\int \frac{x^3 dx}{t^5} = -\,\frac{1}{t} + \frac{a^2}{3t^3}\cdot$

323.9. $\displaystyle\int \frac{x^3 dx}{t^{2p+1}} = -\,\frac{1}{(2p-3)t^{2p-3}} + \frac{a^2}{(2p-1)t^{2p-1}}\cdot$

324.01. $\displaystyle\int \frac{x^4 dx}{t} = -\,\frac{x^3 t}{4} - \frac{3}{8}a^2 xt + \frac{3}{8}a^4 \sin^{-1}\frac{x}{a}\cdot$

[See note under 320.01.]

324.03. $\displaystyle\int \frac{x^4 dx}{t^3} = \frac{xt}{2} + \frac{a^2 x}{t} - \frac{3}{2}a^2 \sin^{-1}\frac{x}{a}\cdot$

324.05. $\displaystyle\int \frac{x^4 dx}{t^5} = -\frac{x}{t} + \frac{1}{3}\frac{x^3}{t^3} + \sin^{-1}\frac{x}{a}.$

324.07. $\displaystyle\int \frac{x^4 dx}{t^7} = \frac{1}{5a^2}\frac{x^5}{t^5}.$ **324.09.** $\displaystyle\int \frac{x^4 dx}{t^9} = \frac{1}{a^4}\left[\frac{1}{5}\frac{x^5}{t^5} + \frac{1}{7}\frac{x^7}{t^7}\right].$

324.11. $\displaystyle\int \frac{x^4 dx}{t^{11}} = \frac{1}{a^6}\left[\frac{1}{5}\frac{x^5}{t^5} + \frac{2}{7}\frac{x^7}{t^7} + \frac{1}{9}\frac{x^9}{t^9}\right].$

324.13. $\displaystyle\int \frac{x^4 dx}{t^{13}} = \frac{1}{a^8}\left[\frac{1}{5}\frac{x^5}{t^5} + \frac{3}{7}\frac{x^7}{t^7} + \frac{3}{9}\frac{x^9}{t^9} + \frac{1}{11}\frac{x^{11}}{t^{11}}\right].$

324.15. $\displaystyle\int \frac{x^4 dx}{t^{15}} = \frac{1}{a^{10}}\left[\frac{1}{5}\frac{x^5}{t^5} + \frac{4}{7}\frac{x^7}{t^7} + \frac{6}{9}\frac{x^9}{t^9} + \frac{4}{11}\frac{x^{11}}{t^{11}} + \frac{1}{13}\frac{x^{13}}{t^{13}}\right].$

325.01. $\displaystyle\int \frac{x^5 dx}{t} = -\frac{t^5}{5} + \frac{2a^2 t^3}{3} - a^4 t.$

325.03. $\displaystyle\int \frac{x^5 dx}{t^3} = -\frac{t^3}{3} + 2a^2 t + \frac{a^4}{t}.$

325.05. $\displaystyle\int \frac{x^5 dx}{t^5} = -t - \frac{2a^2}{t} + \frac{a^4}{3t^3}.$

325.07. $\displaystyle\int \frac{x^5 dx}{t^7} = \frac{1}{t} - \frac{2a^2}{3t^3} + \frac{a^4}{5t^5}.$

325.9. $\displaystyle\int \frac{x^5 dx}{t^{2p+1}} = \frac{1}{(2p-5)t^{2p-5}} - \frac{2a^2}{(2p-3)t^{2p-3}}$
$$+ \frac{a^4}{(2p-1)t^{2p-1}}.$$

326.01. $\displaystyle\int \frac{x^6 dx}{t} = -\frac{x^5 t}{6} - \frac{5}{24}a^2 x^3 t - \frac{5}{16}a^4 xt + \frac{5}{16}a^6 \sin^{-1}\frac{x}{a}.$
[See note under **320.01.**]

326.03. $\displaystyle\int \frac{x^6 dx}{t^3} = -\frac{x^5}{4t} - \frac{5}{8}\frac{a^2 x^3}{t} + \frac{15}{8}\frac{a^4 x}{t} - \frac{15}{8}a^4 \sin^{-1}\frac{x}{a}.$

326.05. $\displaystyle\int \frac{x^6 dx}{t^5} = -\frac{x^5}{2t^3} + \frac{10}{3}\frac{a^2 x^3}{t^3} - \frac{5}{2}\frac{a^4 x}{t^3} + \frac{5}{2}a^2 \sin^{-1}\frac{x}{a}.$

326.07. $\displaystyle\int \frac{x^6 dx}{t^7} = \frac{23}{15}\frac{x^5}{t^5} - \frac{7}{3}\frac{a^2 x^3}{t^5} + \frac{a^4 x}{t^5} - \sin^{-1}\frac{x}{a}.$

326.09. $\displaystyle\int \frac{x^6 dx}{t^9} = \frac{1}{7a^2}\frac{x^7}{t^7}.$ **326.11.** $\displaystyle\int \frac{x^6 dx}{t^{11}} = \frac{1}{a^4}\left[\frac{1}{7}\frac{x^7}{t^7} + \frac{1}{9}\frac{x^9}{t^9}\right].$

326.13. $\displaystyle\int \frac{x^6 dx}{t^{13}} = \frac{1}{a^6}\left[\frac{1}{7}\frac{x^7}{t^7} + \frac{2}{9}\frac{x^9}{t^9} + \frac{1}{11}\frac{x^{11}}{t^{11}}\right].$

326.15. $\displaystyle\int \frac{x^6 dx}{t^{15}} = \frac{1}{a^8}\left[\frac{1}{7}\frac{x^7}{t^7} + \frac{3}{9}\frac{x^9}{t^9} + \frac{3}{11}\frac{x^{11}}{t^{11}} + \frac{1}{13}\frac{x^{13}}{t^{13}}\right].$

327.01. $\displaystyle\int \frac{x^7 dx}{t} = \frac{1}{7}t^7 - \frac{3}{5}a^2 t^5 + \frac{3}{3}a^4 t^3 - a^6 t.$

327.03. $\displaystyle\int \frac{x^7 dx}{t^3} = \frac{1}{5}t^5 - \frac{3}{3}a^2 t^3 + 3a^4 t + \frac{a^6}{t}.$

327.05. $\displaystyle\int \frac{x^7 dx}{t^5} = \frac{1}{3}t^3 - 3a^2 t - \frac{3a^4}{t} + \frac{a^6}{3t^3}.$

327.07. $\displaystyle\int \frac{x^7 dx}{t^7} = t + \frac{3a^2}{t} - \frac{3a^4}{3t^3} + \frac{a^6}{5t^5}.$

327.9. $\displaystyle\int \frac{x^7 dx}{t^{2p+1}} = -\frac{1}{(2p-7)t^{2p-7}}$

$$+ \frac{3a^2}{(2p-5)t^{2p-5}} - \frac{3a^4}{(2p-3)t^{2p-3}}$$

$$+ \frac{a^6}{(2p-1)t^{2p-1}}.$$

341.01. $\displaystyle\int \frac{dx}{xt} = \int \frac{dx}{x\sqrt{(a^2 - x^2)}}$

$$= -\frac{1}{a}\log\left|\frac{a+t}{x}\right|,$$

$$[x^2 < a^2].$$

Note that

$$-\frac{1}{a}\log\left|\frac{a+t}{x}\right| = -\frac{1}{a}\operatorname{sech}^{-1}\left|\frac{x}{a}\right|$$

$$= -\frac{1}{a}\cosh^{-1}\left|\frac{a}{x}\right|$$

$$= -\frac{1}{2a}\log\left(\frac{a+t}{a-t}\right).$$

The positive values of $\operatorname{sech}^{-1}|x/a|$, $\cosh^{-1}|a/x|$, a and t are to be taken.

Fig. 341.01

Dotted graph,

$$\frac{1}{x\sqrt{(1 - x^2)}}.$$

Full line graph,

$$-\log\left|\frac{1 + \sqrt{(1 - x^2)}}{x}\right|.$$

341.03. $\displaystyle\int \frac{dx}{xt^3} = \frac{1}{a^2t} - \frac{1}{a^3}\log\left|\frac{a+t}{x}\right|.$

341.05. $\displaystyle\int \frac{dx}{xt^5} = \frac{1}{3a^2t^3} + \frac{1}{a^4t} - \frac{1}{a^5}\log\left|\frac{a+t}{x}\right|.$

341.07. $\displaystyle\int \frac{dx}{xt^7} = \frac{1}{5a^2t^5} + \frac{1}{3a^4t^3} + \frac{1}{a^6t} - \frac{1}{a^7}\log\left|\frac{a+t}{x}\right|.$

341.09. $\displaystyle\int \frac{dx}{xt^9} = \frac{1}{7a^2t^7} + \frac{1}{5a^4t^5} + \frac{1}{3a^6t^3} + \frac{1}{a^8t} - \frac{1}{a^9}\log\left|\frac{a+t}{x}\right|.$

342.01. $\displaystyle\int \frac{dx}{x^2t} = -\frac{t}{a^2x}.$ **342.03.** $\displaystyle\int \frac{dx}{x^2t^3} = \frac{1}{a^4}\left(-\frac{t}{x} + \frac{x}{t}\right).$

342.05. $\displaystyle\int \frac{dx}{x^2t^5} = \frac{1}{a^6}\left(-\frac{t}{x} + \frac{2x}{t} + \frac{x^3}{3t^3}\right).$

342.07. $\displaystyle\int \frac{dx}{x^2t^7} = \frac{1}{a^8}\left(-\frac{t}{x} + \frac{3x}{t} + \frac{3x^3}{3t^3} + \frac{x^5}{5t^5}\right).$

342.09. $\displaystyle\int \frac{dx}{x^2t^9} = \frac{1}{a^{10}}\left(-\frac{t}{x} + \frac{4x}{t} + \frac{6x^3}{3t^3} + \frac{4x^5}{5t^5} + \frac{x^7}{7t^7}\right).$

343.01. $\displaystyle\int \frac{dx}{x^3t} = -\frac{t}{2a^2x^2} - \frac{1}{2a^3}\log\left|\frac{a+t}{x}\right|.$ [See **341.01.**]

343.03. $\displaystyle\int \frac{dx}{x^3t^3} = -\frac{1}{2a^2x^2t} + \frac{3}{2a^4t} - \frac{3}{2a^5}\log\left|\frac{a+t}{x}\right|.$

343.05. $\displaystyle\int \frac{dx}{x^3t^5} = -\frac{1}{2a^2x^2t^3} + \frac{5}{6a^4t^3} + \frac{5}{2a^6t} - \frac{5}{2a^7}\log\left|\frac{a+t}{x}\right|.$

344.01. $\displaystyle\int \frac{dx}{x^4t} = -\frac{1}{a^4}\left(\frac{t}{x} + \frac{t^3}{3x^3}\right).$

344.03. $\displaystyle\int \frac{dx}{x^4t^3} = -\frac{1}{a^6}\left(-\frac{x}{t} + \frac{2t}{x} + \frac{t^3}{3x^3}\right).$

344.05. $\displaystyle\int \frac{dx}{x^4t^5} = -\frac{1}{a^8}\left(-\frac{x^3}{3t^3} - \frac{3x}{t} + \frac{3t}{x} + \frac{t^3}{3x^3}\right).$

For **342** and **344**, put $z^2 = \dfrac{x^2}{t^2}$;   then $dx = \dfrac{a\,dz}{(1+z^2)^{3/2}}.$

345.01. $\displaystyle\int \frac{dx}{x^5t} = -\left[\frac{t}{4a^2x^4} + \frac{3}{8}\frac{t}{a^4x^2} + \frac{3}{8a^5}\log\left|\frac{a+t}{x}\right|\right].$

345.03. $\displaystyle \int \frac{dx}{x^5 t^3} = -\left[\frac{1}{4a^2 x^4 t} + \frac{5}{8a^4 x^2 t} - \frac{15}{8a^6 t} + \frac{15}{8a^7}\log\left|\frac{a+t}{x}\right|\right].$

346.01. $\displaystyle \int \frac{dx}{x^6 t} = -\frac{1}{a^6}\left(\frac{t}{x} + \frac{2t^3}{3x^3} + \frac{t^5}{5x^5}\right).$

346.03. $\displaystyle \int \frac{dx}{x^6 t^3} = -\frac{1}{a^8}\left(-\frac{x}{t} + \frac{3t}{x} + \frac{3t^3}{3x^3} + \frac{t^5}{5x^5}\right).$

350.01. $\displaystyle \int t\, dx = \frac{xt}{2} + \frac{a^2}{2}\sin^{-1}\frac{x}{a}.$ [See note under **320.01.**]

350.03. $\displaystyle \int t^3 dx = \frac{xt^3}{4} + \frac{3}{8}a^2 xt + \frac{3}{8}a^4 \sin^{-1}\frac{x}{a}.$

350.05. $\displaystyle \int t^5 dx = \frac{xt^5}{6} + \frac{5}{24}a^2 xt^3 + \frac{5}{16}a^4 xt + \frac{5}{16}a^6 \sin^{-1}\frac{x}{a}.$

351.01. $\displaystyle \int xt\, dx = -\frac{t^3}{3}.$ **351.03.** $\displaystyle \int xt^3 dx = -\frac{t^5}{5}.$

351.9. $\displaystyle \int xt^{2p+1} dx = -\frac{t^{2p+3}}{2p+3}.$

352.01. $\displaystyle \int x^2 t\, dx = -\frac{xt^3}{4} + \frac{a^2 xt}{8} + \frac{a^4}{8}\sin^{-1}\frac{x}{a}.$

[See note under **320.01.**]

352.03. $\displaystyle \int x^2 t^3 dx = -\frac{xt^5}{6} + \frac{a^2 xt^3}{24} + \frac{a^4 xt}{16} + \frac{a^6}{16}\sin^{-1}\frac{x}{a}.$

353.01. $\displaystyle \int x^3 t\, dx = \frac{t^5}{5} - \frac{a^2 t^3}{3}.$ **353.03.** $\displaystyle \int x^3 t^3 dx = \frac{t^7}{7} - \frac{a^2 t^5}{5}.$

353.9. $\displaystyle \int x^3 t^{2p+1} dx = \frac{t^{2p+5}}{2p+5} - \frac{a^2 t^{2p+3}}{2p+3}.$

354.01. $\displaystyle \int x^4 t\, dx = -\frac{x^3 t^3}{6} - \frac{a^2 xt^3}{8} + \frac{a^4 xt}{16} + \frac{a^6}{16}\sin^{-1}\frac{x}{a}.$

[See note under **320.01.**]

354.03. $\displaystyle \int x^4 t^3 dx = -\frac{x^3 t^5}{8} - \frac{a^2 xt^5}{16} + \frac{a^4 xt^3}{64} + \frac{3}{128}a^6 xt$

$$+ \frac{3}{128}a^8 \sin^{-1}\frac{x}{a}.$$

355.01. $\displaystyle \int x^5 t \, dx = -\frac{t^7}{7} + \frac{2a^2t^5}{5} - \frac{a^4t^3}{3}$.

355.03. $\displaystyle \int x^5 t^3 dx = -\frac{t^9}{9} + \frac{2a^2t^7}{7} - \frac{a^4t^5}{5}$.

355.9. $\displaystyle \int x^5 t^{2p+1} dx = -\frac{t^{2p+7}}{2p+7} + \frac{2a^2t^{2p+5}}{2p+5} - \frac{a^4t^{2p+3}}{2p+3}$.

361.01. $\displaystyle \int \frac{t \, dx}{x} = t - a \log \left| \frac{a+t}{x} \right|$.   [See note under **341.01.**]

361.03. $\displaystyle \int \frac{t^3 dx}{x} = \frac{t^3}{3} + a^2 t - a^3 \log \left| \frac{a+t}{x} \right|$.

361.05. $\displaystyle \int \frac{t^5 dx}{x} = \frac{t^5}{5} + \frac{a^2 t^3}{3} + a^4 t - a^5 \log \left| \frac{a+t}{x} \right|$.

361.07. $\displaystyle \int \frac{t^7 dx}{x} = \frac{t^7}{7} + \frac{a^2 t^5}{5} + \frac{a^4 t^3}{3} + a^6 t - a^7 \log \left| \frac{a+t}{x} \right|$.

362.01. $\displaystyle \int \frac{t \, dx}{x^2} = -\frac{t}{x} - \sin^{-1} \frac{x}{a}$.   [See note under **320.01.**]

362.03. $\displaystyle \int \frac{t^3 dx}{x^2} = -\frac{t^3}{x} - \frac{3}{2} xt - \frac{3}{2} a^2 \sin^{-1} \frac{x}{a}$.

362.05. $\displaystyle \int \frac{t^5 dx}{x^2} = -\frac{t^5}{x} - \frac{5}{4} xt^3 - \frac{15}{8} a^2 xt - \frac{15}{8} a^4 \sin^{-1} \frac{x}{a}$.

363.01. $\displaystyle \int \frac{t \, dx}{x^3} = -\frac{t}{2x^2} + \frac{1}{2a} \log \left| \frac{a+t}{x} \right|$.

[See note under **341.01.**]

363.03. $\displaystyle \int \frac{t^3 dx}{x^3} = -\frac{t^3}{2x^2} - \frac{3t}{2} + \frac{3a}{2} \log \left| \frac{a+t}{x} \right|$.

363.05. $\displaystyle \int \frac{t^5 dx}{x^3} = -\frac{t^5}{2x^2} - \frac{5}{6} t^3 - \frac{5}{2} a^2 t + \frac{5}{2} a^3 \log \left| \frac{a+t}{x} \right|$.

364.01. $\displaystyle \int \frac{t \, dx}{x^4} = -\frac{t^3}{3a^2x^3}$.

364.03. $\displaystyle \int \frac{t^3 dx}{x^4} = -\frac{t^3}{3x^3} + \frac{t}{x} + \sin^{-1} \frac{x}{a}$.

[See note under **320.01.**]

364.05. $\displaystyle\int \frac{t^5 dx}{x^4} = -\frac{a^2 t^3}{3x^3} + \frac{2a^2 t}{x} + \frac{xt}{2} + \frac{5}{2}\, a^2 \sin^{-1}\frac{x}{a}.$

365.01. $\displaystyle\int \frac{t\, dx}{x^5} = -\frac{t}{4x^4} + \frac{t}{8a^2 x^2} + \frac{1}{8a^3} \log\left|\frac{a+t}{x}\right|.$

365.03. $\displaystyle\int \frac{t^3 dx}{x^5} = -\frac{t^3}{4x^4} + \frac{3}{8}\frac{t^3}{a^2 x^2} + \frac{3}{8}\frac{t}{a^2} - \frac{3}{8a} \log\left|\frac{a+t}{x}\right|.$

366.01. $\displaystyle\int \frac{t\, dx}{x^6} = -\frac{t^3}{5a^2 x^3}\left(\frac{1}{x^2} + \frac{2}{3a^2}\right).$

366.03. $\displaystyle\int \frac{t^3 dx}{x^6} = -\frac{t^5}{5a^2 x^5}.$

367.01. $\displaystyle\int \frac{t\, dx}{x^7} = -\frac{t}{6x^6} + \frac{t}{24a^2 x^4} + \frac{t}{16a^4 x^2} + \frac{1}{16a^5} \log\left|\frac{a+t}{x}\right|.$

368.01. $\displaystyle\int \frac{t\, dx}{x^8} = -\frac{t^3}{7a^2 x^3}\left(\frac{1}{x^4} + \frac{4}{5a^2 x^2} + \frac{8}{15a^4}\right).$

Integrals of Binomial Differentials
Reduction Formulas

370. $\displaystyle\int x^m(ax^n + b)^p\, dx$

$$= \frac{1}{m + np + 1}\left[x^{m+1} u^p + npb \int x^m u^{p-1}\, dx \right].$$

371. $\displaystyle\int x^m(ax^n + b)^p\, dx = \frac{1}{bn(p+1)}\Big[-x^{m+1}u^{p+1}$

$$+ (m + n + np + 1)\int x^m u^{p+1}\, dx \Big].$$

372. $\displaystyle\int x^m(ax^n + b)^p\, dx = \frac{1}{(m+1)b}\Big[x^{m+1}u^{p+1}$

$$- a(m + n + np + 1)\int x^{m+n}u^p\, dx \Big].$$

373. $\displaystyle\int x^m(ax^n + b)^p\, dx = \frac{1}{a(m + np + 1)}\Big[x^{m-n+1}u^{p+1}$

$$- (m - n + 1)b \int x^{m-n}u^p\, dx \Big].$$

Here $u = ax^n + b$, and a, b, p, m, and n may be any numbers for which no denominator vanishes.

Integrals Involving $X^{1/2} = (ax^2 + bx + c)^{1/2}$

380.001. $\displaystyle\int \frac{dx}{X^{1/2}} = \frac{1}{a^{1/2}} \log |2(aX)^{1/2} + 2ax + b|, \qquad [a > 0],$

$$= \frac{1}{a^{1/2}} \sinh^{-1} \frac{2ax + b}{(4ac - b^2)^{1/2}}, \qquad \begin{bmatrix} a > 0, \\ 4ac > b^2 \end{bmatrix},$$

$$= \frac{1}{a^{1/2}} \log |2ax + b|, \qquad \begin{bmatrix} a > 0, \\ b^2 = 4ac \end{bmatrix},$$

$$= \frac{-1}{(-a)^{1/2}} \sin^{-1} \frac{(2ax + b)}{(b^2 - 4ac)^{1/2}},$$

$$\begin{bmatrix} a < 0, \quad b^2 > 4ac, \\ |2ax + b| < (b^2 - 4ac)^{1/2} \end{bmatrix}.$$

The principal values of $\sin^{-1}$, between $-\pi/2$ and $\pi/2$, are to be taken.

380.003. $\displaystyle\int \frac{dx}{X^{3/2}} = \frac{4ax + 2b}{(4ac - b^2)X^{1/2}}.$

380.005. $\displaystyle\int \frac{dx}{X^{5/2}} = \frac{4ax + 2b}{3(4ac - b^2)X^{1/2}} \left(\frac{1}{X} + \frac{8a}{4ac - b^2} \right).$

380.009. $\displaystyle\int \frac{dx}{X^{(2n+1)/2}} = \frac{4ax + 2b}{(2n - 1)(4ac - b^2)X^{(2n-1)/2}}$

$$+ \frac{8a(n - 1)}{(2n - 1)(4ac - b^2)} \int \frac{dx}{X^{(2n-1)/2}}.$$

380.011. $\displaystyle\int \frac{x\,dx}{X^{1/2}} = \frac{X^{1/2}}{a} - \frac{b}{2a} \int \frac{dx}{X^{1/2}}.$ [See **380.001.**]

380.013. $\displaystyle\int \frac{x\,dx}{X^{3/2}} = -\frac{2bx + 4c}{(4ac - b^2)X^{1/2}}.$

380.019. $\displaystyle\int \frac{x\,dx}{X^{(2n+1)/2}} = -\frac{1}{(2n - 1)aX^{(2n-1)/2}} - \frac{b}{2a} \int \frac{dx}{X^{(2n+1)/2}}.$

380.021. $\displaystyle\int \frac{x^2\,dx}{X^{1/2}} = \left(\frac{x}{2a} - \frac{3b}{4a^2} \right) X^{1/2} + \frac{3b^2 - 4ac}{8a^2} \int \frac{dx}{X^{1/2}}.$

[See **380.001.**]

380.111. $\displaystyle \int \frac{dx}{xX^{1/2}} = -\frac{1}{c^{1/2}} \log \left| \frac{2(cX)^{1/2}}{x} + \frac{2c}{x} + b \right|$, $[c > 0]$,

$\displaystyle = -\frac{1}{c^{1/2}} \sinh^{-1} \frac{bx + 2c}{x(4ac - b^2)^{1/2}}$, $\begin{bmatrix} c > 0, \\ 4ac > b^2 \end{bmatrix}$,

$\displaystyle = -\frac{1}{c^{1/2}} \log \left| \frac{bx + 2c}{x} \right|$, $\begin{bmatrix} c > 0, \\ b^2 = 4ac \end{bmatrix}$,

$\displaystyle = \frac{1}{(-c)^{1/2}} \sin^{-1} \frac{bx + 2c}{x(b^2 - 4ac)^{1/2}}$, $\begin{bmatrix} c < 0, \\ b^2 > 4ac \end{bmatrix}$.

380.119. $\displaystyle \int \frac{dx}{xX^{(2n+1)/2}} = \frac{1}{(2n-1)cX^{(2n-1)/2}}$

$\displaystyle + \frac{1}{c} \int \frac{dx}{xX^{(2n-1)/2}} - \frac{b}{2c} \int \frac{dx}{X^{(2n+1)/2}}$.

380.121. $\displaystyle \int \frac{dx}{x^2 X^{1/2}} = -\frac{X^{1/2}}{cx} - \frac{b}{2c} \int \frac{dx}{xX^{1/2}}$.   [See **380.111.**]

380.201. $\displaystyle \int X^{1/2} dx = \frac{2ax + b}{4a} X^{1/2} + \frac{4ac - b^2}{8a} \int \frac{dx}{X^{1/2}}$.

[See **380.001.**]

380.209. $\displaystyle \int X^{(2n+1)/2} dx = \frac{(2ax + b)X^{(2n+1)/2}}{4a(n+1)}$

$\displaystyle + \frac{(4ac - b^2)(2n + 1)}{8a(n+1)} \int X^{(2n-1)/2} dx$.

380.211. $\displaystyle \int xX^{1/2} dx = \frac{X^{3/2}}{3a} - \frac{b(2ax + b)}{8a^2} X^{1/2}$

$\displaystyle - \frac{b(4ac - b^2)}{16a^2} \int \frac{dx}{X^{1/2}}$.   [See **380.001.**]

380.219. $\displaystyle \int xX^{(2n+1)/2} dx = \frac{X^{(2n+3)/2}}{(2n+3)a} - \frac{b}{2a} \int X^{(2n+1)/2} dx$.

380.311. $\displaystyle \int \frac{X^{1/2} dx}{x} = X^{1/2} + \frac{b}{2} \int \frac{dx}{X^{1/2}} + c \int \frac{dx}{xX^{1/2}}$.

[See **380.001** and **380.111.**]

380.319. $\displaystyle\int \frac{X^{(2n+1)/2}dx}{x} = \frac{X^{(2n+1)/2}}{2n+1}$

$$+ \frac{b}{2}\int X^{(2n-1)/2}dx + c \int \frac{X^{(2n-1)/2}dx}{x}.$$

380.321. $\displaystyle\int \frac{X^{1/2}dx}{x^2} = - \frac{X^{1/2}}{x} + a \int \frac{dx}{X^{1/2}} + \frac{b}{2}\int \frac{dx}{xX^{1/2}},$

where $X = ax^2 + bx + c$.   [See **380.001** and **380.111**.]

383.1. $\displaystyle\int \frac{dx}{x(ax^2+bx)^{1/2}} = - \frac{2}{bx}(ax^2+bx)^{1/2}.$

383.2. $\displaystyle\int \frac{dx}{(2ax-x^2)^{1/2}} = \sin^{-1}\frac{x-a}{a}.$

383.3. $\displaystyle\int \frac{xdx}{(2ax-x^2)^{1/2}} = -(2ax-x^2)^{1/2} + a\sin^{-1}\left(\frac{x-a}{a}\right).$

383.4. $\displaystyle\int (2ax-x^2)^{1/2}dx$

$$= \frac{x-a}{2}(2ax-x^2)^{1/2} + \frac{a^2}{2}\sin^{-1}\frac{x-a}{a}.$$

384.1. $\displaystyle\int \frac{dx}{x(x^n+a^2)^{1/2}} = -\frac{2}{na}\log\left|\frac{a+(x^n+a^2)^{1/2}}{x^{n/2}}\right|.$

384.2. $\displaystyle\int \frac{dx}{x(x^n-a^2)^{1/2}} = \frac{2}{na}\cos^{-1}\left|\frac{a}{x^{n/2}}\right|$

[See note under **281.01**.]

384.3. $\displaystyle\int \frac{x^{1/2}dx}{(a^3-x^3)^{1/2}} = \frac{2}{3}\sin^{-1}\left(\frac{x}{a}\right)^{3/2}.$

387. $\displaystyle\int \frac{dx}{(ax^2+b)\sqrt{(fx^2+g)}}$

$$= \frac{1}{\sqrt{b}\sqrt{(ag-bf)}}\tan^{-1}\frac{x\sqrt{(ag-bf)}}{\sqrt{b}\sqrt{(fx^2+g)}}, \quad [ag > bf],$$

$$= \frac{1}{2\sqrt{b}\sqrt{(bf-ag)}}\log\frac{\sqrt{b}\sqrt{(fx^2+g)}+x\sqrt{(bf-ag)}}{\sqrt{b}\sqrt{(fx^2+g)}-x\sqrt{(bf-ag)}},$$

$$[bf > ag].$$

TRIGONOMETRIC FUNCTIONS

400.01. $\sin^2 A + \cos^2 A = 1.$

400.02. $\sin A = \sqrt{(1 - \cos^2 A)}.$

400.03. $\cos A = \sqrt{(1 - \sin^2 A)}.$

400.04. $\tan A = \sin A / \cos A.$

400.05. $\operatorname{ctn} A = \cos A / \sin A = 1/\tan A.$

400.06. $\sec A = 1/\cos A.$

400.07. $\csc A = 1/\sin A.$

400.08. $\sin(-A) = -\sin A.$

400.09. $\cos(-A) = \cos A.$

400.10. $\tan(-A) = -\tan A.$

400.11. $\sec^2 A - \tan^2 A = 1.$

400.12. $\sec A = \sqrt{(1 + \tan^2 A)}.$

400.13. $\tan A = \sqrt{(\sec^2 A - 1)}.$

400.14. $\csc^2 A - \operatorname{ctn}^2 A = 1.$

400.15. $\csc A = \sqrt{(1 + \operatorname{ctn}^2 A)}.$

400.16. $\operatorname{ctn} A = \sqrt{(\csc^2 A - 1)}.$

400.17. $\operatorname{vers} A = 1 - \cos A.$

Note that for real values of A the sign of the above radicals depends on the quadrant in which the angle A lies.

401.01. $\sin(A + B) = \sin A \cos B + \cos A \sin B.$

401.02. $\sin(A - B) = \sin A \cos B - \cos A \sin B.$

401.03. $\cos(A + B) = \cos A \cos B - \sin A \sin B.$

401.04. $\cos(A - B) = \cos A \cos B + \sin A \sin B.$

401.05. $2 \sin A \cos B = \sin(A + B) + \sin(A - B).$

401.06. $2 \cos A \cos B = \cos(A + B) + \cos(A - B).$

401.07. $2 \sin A \sin B = \cos(A - B) - \cos(A + B).$

401.08. $\sin A + \sin B = 2 \sin \frac{1}{2}(A + B) \cos \frac{1}{2}(A - B).$

401.09. $\sin A - \sin B = 2 \sin \frac{1}{2}(A - B) \cos \frac{1}{2}(A + B).$

401.10. $\cos A + \cos B = 2 \cos \frac{1}{2}(A + B) \cos \frac{1}{2}(A - B).$

401.11. $\cos A - \cos B = 2 \sin \frac{1}{2}(A + B) \sin \frac{1}{2}(B - A).$

401.12. $\sin^2 A - \sin^2 B = \sin(A + B) \sin(A - B).$

401.13. $\cos^2 A - \cos^2 B = \sin(A + B) \sin(B - A).$

401.14. $\cos^2 A - \sin^2 B = \cos(A + B) \cos(A - B)$
$$= \cos^2 B - \sin^2 A.$$

401.15. $\sec^2 A + \csc^2 A = \sec^2 A \csc^2 A = \dfrac{1}{\sin^2 A \cos^2 A}.$

401.2. $p \cos A + q \sin A = r \sin (A + \theta),$

where

$$r = \sqrt{(p^2 + q^2)}, \quad \sin \theta = p/r, \quad \cos \theta = q/r$$

or

$$p \cos A + q \sin A = r \cos (A - \varphi),$$

where

$$r = \sqrt{(p^2 + q^2)}, \quad \cos \varphi = p/r, \quad \sin \varphi = q/r.$$

Note that p and q may be positive or negative.

402.01. $\sin (A + B + C)$
$$= \sin A \cos B \cos C + \cos A \sin B \cos C$$
$$+ \cos A \cos B \sin C - \sin A \sin B \sin C.$$

402.02. $\cos (A + B + C)$
$$= \cos A \cos B \cos C - \sin A \sin B \cos C$$
$$- \sin A \cos B \sin C - \cos A \sin B \sin C.$$

402.03. $4 \sin A \sin B \sin C$
$$= \sin (A + B - C) + \sin (B + C - A)$$
$$+ \sin (C + A - B) - \sin (A + B + C).$$

402.04. $4 \sin A \cos B \cos C$
$$= \sin (A + B - C) - \sin (B + C - A)$$
$$+ \sin (C + A - B) + \sin (A + B + C).$$

402.05. $4 \sin A \sin B \cos C$
$$= - \cos (A + B - C) + \cos (B + C - A)$$
$$+ \cos (C + A - B) - \cos (A + B + C).$$

402.06. $4 \cos A \cos B \cos C$
$$= \cos (A + B - C) + \cos (B + C - A)$$
$$+ \cos (C + A - B) + \cos (A + B + C).$$

403.02. $\sin 2A = 2 \sin A \cos A = \dfrac{2 \tan A}{1 + \tan^2 A}.$

403.03. $\sin 3A = 3 \sin A - 4 \sin^3 A.$

403.04. $\sin 4A = \cos A (4 \sin A - 8 \sin^3 A).$

403.05. $\sin 5A = 5 \sin A - 20 \sin^3 A + 16 \sin^5 A.$

403.06. $\sin 6A = \cos A (6 \sin A - 32 \sin^3 A + 32 \sin^5 A).$

403.07. $\sin 7A = 7 \sin A - 56 \sin^3 A + 112 \sin^5 A - 64 \sin^7 A.$

403.10. When n is an even, positive integer,

$$\sin nA = (-1)^{(n/2)+1} \cos A \left[2^{n-1} \sin^{n-1} A - \frac{(n-2)}{1!} 2^{n-3} \sin^{n-3} A \right.$$

$$+ \frac{(n-3)(n-4)}{2!} 2^{n-5} \sin^{n-5} A$$

$$\left. - \frac{(n-4)(n-5)(n-6)}{3!} 2^{n-7} \sin^{n-7} A + \cdots \right],$$

the series terminating where a coefficient = 0.

403.11. An alternative series, giving the same results for numerical values of n, is

$$\sin nA = n \cos A \left[\sin A - \frac{(n^2 - 2^2)}{3!} \sin^3 A \right.$$

$$+ \frac{(n^2 - 2^2)(n^2 - 4^2)}{5!} \sin^5 A$$

$$\left. - \frac{(n^2 - 2^2)(n^2 - 4^2)(n^2 - 6^2)}{7!} \sin^7 A + \cdots \right],$$

$$[n \text{ even and} > 0]. \quad [\text{Ref. 34, p. 181.}]$$

403.12. When n is an odd integer > 1

$$\sin nA = (-1)^{(n-1)/2} \left[2^{n-1} \sin^n A - \frac{n}{1!} 2^{n-3} \sin^{n-2} A \right.$$

$$+ \frac{n(n-3)}{2!} 2^{n-5} \sin^{n-4} A - \frac{n(n-4)(n-5)}{3!} 2^{n-7} \sin^{n-6} A$$

$$\left. + \frac{n(n-5)(n-6)(n-7)}{4!} 2^{n-9} \sin^{n-8} A - \cdots \right],$$

the series terminating where a coefficient = 0.

403.13. An alternative series is

$$\sin nA = n \sin A - \frac{n(n^2 - 1^2)}{3!} \sin^3 A$$

$$+ \frac{n(n^2 - 1^2)(n^2 - 3^2)}{5!} \sin^5 A - \cdots,$$

$$[n \text{ odd and} > 0]. \quad [\text{Ref. 34, p. 180.}]$$

403.22. $\cos 2A = \cos^2 A - \sin^2 A = 2 \cos^2 A - 1 = 1 - 2 \sin^2 A$

$$= \frac{1 - \tan^2 A}{1 + \tan^2 A} = \frac{\operatorname{ctn} A - \tan A}{\operatorname{ctn} A + \tan A}.$$

403.23. $\cos 3A = 4 \cos^3 A - 3 \cos A.$

403.24. $\cos 4A = 8 \cos^4 A - 8 \cos^2 A + 1.$

403.25. $\cos 5A = 16 \cos^5 A - 20 \cos^3 A + 5 \cos A.$

403.26. $\cos 6A = 32 \cos^6 A - 48 \cos^4 A + 18 \cos^2 A - 1.$

403.27. $\cos 7A = 64 \cos^7 A - 112 \cos^5 A + 56 \cos^3 A - 7 \cos A.$

403.3. $\cos nA = 2^{n-1} \cos^n A - \dfrac{n}{1!} 2^{n-3} \cos^{n-2} A$

$+ \dfrac{n(n-3)}{2!} 2^{n-5} \cos^{n-4} A - \dfrac{n(n-4)(n-5)}{3!} 2^{n-7} \cos^{n-6} A$

$+ \dfrac{n(n-5)(n-6)(n-7)}{4!} 2^{n-9} \cos^{n-8} A - \cdots,$

terminating where a coefficient $= 0$, [n an integer > 2].
[Ref. 4, pp. 409, 416 and 417, and Ref. 34, p. 177.]

403.4. $\sin \frac{1}{2}A = \sqrt{\{\frac{1}{2}(1 - \cos A)\}}.$

403.5. $\cos \frac{1}{2}A = \sqrt{\{\frac{1}{2}(1 + \cos A)\}}.$

404.12. $\sin^2 A = \frac{1}{2}(-\cos 2A + 1).$

404.13. $\sin^3 A = \frac{1}{4}(-\sin 3A + 3 \sin A).$

404.14. $\sin^4 A = \frac{1}{8}(\cos 4A - 4 \cos 2A + \frac{6}{2}).$

404.15. $\sin^5 A = \frac{1}{16}(\sin 5A - 5 \sin 3A + 10 \sin A).$

404.16. $\sin^6 A = \frac{1}{32}(-\cos 6A + 6 \cos 4A - 15 \cos 2A + \frac{20}{2}).$

404.17. $\sin^7 A = \frac{1}{64}(-\sin 7A + 7 \sin 5A - 21 \sin 3A$
$+ 35 \sin A).$

404.22. $\cos^2 A = \frac{1}{2}(\cos 2A + 1).$

404.23. $\cos^3 A = \frac{1}{4}(\cos 3A + 3 \cos A).$

404.24. $\cos^4 A = \frac{1}{8}(\cos 4A + 4 \cos 2A + \frac{6}{2}).$

404.25. $\cos^5 A = \frac{1}{16}(\cos 5A + 5 \cos 3A + 10 \cos A).$

404.26. $\cos^6 A = \frac{1}{32}(\cos 6A + 6 \cos 4A + 15 \cos 2A + \frac{20}{2}).$

404.27. $\cos^7 A = \frac{1}{64}(\cos 7A + 7 \cos 5A + 21 \cos 3A$
$+ 35 \cos A).$

[No. **404** can be extended by inspection by using binomial coefficients.]

405.01. $\tan (A + B) = \dfrac{\tan A + \tan B}{1 - \tan A \tan B} = \dfrac{\operatorname{ctn} A + \operatorname{ctn} B}{\operatorname{ctn} A \operatorname{ctn} B - 1}.$

405.02. $\tan (A - B) = \dfrac{\tan A - \tan B}{1 + \tan A \tan B} = \dfrac{\operatorname{ctn} B - \operatorname{ctn} A}{\operatorname{ctn} A \operatorname{ctn} B + 1}.$

405.03. $\quad \operatorname{ctn} (A + B) = \dfrac{\operatorname{ctn} A \operatorname{ctn} B - 1}{\operatorname{ctn} A + \operatorname{ctn} B} = \dfrac{1 - \tan A \tan B}{\tan A + \tan B}.$

405.04. $\quad \operatorname{ctn} (A - B) = \dfrac{\operatorname{ctn} A \operatorname{ctn} B + 1}{\operatorname{ctn} B - \operatorname{ctn} A} = \dfrac{1 + \tan A \tan B}{\tan A - \tan B}.$

405.05. $\quad \tan A + \tan B = \dfrac{\sin (A + B)}{\cos A \cos B}.$

405.06. $\quad \tan A - \tan B = \dfrac{\sin (A - B)}{\cos A \cos B}.$

405.07. $\quad \operatorname{ctn} A + \operatorname{ctn} B = \dfrac{\sin (A + B)}{\sin A \sin B}.$

405.08. $\quad \operatorname{ctn} A - \operatorname{ctn} B = \dfrac{\sin (B - A)}{\sin A \sin B}.$

405.09. $\quad \tan A + \operatorname{ctn} B = \dfrac{\cos (A - B)}{\cos A \sin B}.$

405.10. $\quad \operatorname{ctn} A - \tan B = \dfrac{\cos (A + B)}{\sin A \cos B}.$

406.02. $\quad \tan 2A = \dfrac{2 \tan A}{1 - \tan^2 A} = \dfrac{2 \operatorname{ctn} A}{\operatorname{ctn}^2 A - 1} = \dfrac{2}{\operatorname{ctn} A - \tan A}.$

406.03. $\quad \tan 3A = \dfrac{3 \tan A - \tan^3 A}{1 - 3 \tan^2 A}.$

406.04. $\quad \tan 4A = \dfrac{4 \tan A - 4 \tan^3 A}{1 - 6 \tan^2 A + \tan^4 A}.$

406.12. $\quad \operatorname{ctn} 2A = \dfrac{\operatorname{ctn}^2 A - 1}{2 \operatorname{ctn} A} = \dfrac{1 - \tan^2 A}{2 \tan A} = \dfrac{\operatorname{ctn} A - \tan A}{2}.$

406.13. $\quad \operatorname{ctn} 3A = \dfrac{\operatorname{ctn}^3 A - 3 \operatorname{ctn} A}{3 \operatorname{ctn}^2 A - 1}.$

406.14. $\quad \operatorname{ctn} 4A = \dfrac{\operatorname{ctn}^4 A - 6 \operatorname{ctn}^2 A + 1}{4 \operatorname{ctn}^3 A - 4 \operatorname{ctn} A}.$

406.2. $\quad \tan \tfrac{1}{2}A = \dfrac{1 - \cos A}{\sin A} = \dfrac{\sin A}{1 + \cos A} = \sqrt{\left(\dfrac{1 - \cos A}{1 + \cos A} \right)}.$

406.3. $\quad \operatorname{ctn} \tfrac{1}{2}A = \dfrac{\sin A}{1 - \cos A} = \dfrac{1 + \cos A}{\sin A} = \sqrt{\left(\dfrac{1 + \cos A}{1 - \cos A} \right)}.$

407. $\sin 0°$ $= 0$ $= \cos 90°.$

$$\sin 15° = \sin \frac{\pi}{12} = \frac{\sqrt{3}-1}{2\sqrt{2}} \qquad = \cos 75°.$$

$$\sin 18° = \sin \frac{\pi}{10} = \frac{\sqrt{5}-1}{4} \qquad = \cos 72°.$$

$$\sin 30° = \sin \frac{\pi}{6} = \frac{1}{2} \qquad = \cos 60°.$$

$$\sin 36° = \sin \frac{\pi}{5} = \frac{\sqrt{(5-\sqrt{5})}}{2\sqrt{2}} = \cos 54°.$$

$$\sin 45° = \sin \frac{\pi}{4} = \frac{1}{\sqrt{2}} \qquad = \cos 45°.$$

$$\sin 54° = \sin \frac{3\pi}{10} = \frac{\sqrt{5}+1}{4} \qquad = \cos 36°.$$

$$\sin 60° = \sin \frac{\pi}{3} = \frac{\sqrt{3}}{2} \qquad = \cos 30°.$$

$$\sin 72° = \sin \frac{2\pi}{5} = \frac{\sqrt{(5+\sqrt{5})}}{2\sqrt{2}} = \cos 18°.$$

$$\sin 75° = \sin \frac{5\pi}{12} = \frac{\sqrt{3}+1}{2\sqrt{2}} \qquad = \cos 15°.$$

$$\sin 90° = \sin \frac{\pi}{2} = 1 \qquad = \cos 0.$$

[Ref. 4, pp. 406–407.]

$$\sin 120° = \sin \frac{2\pi}{3} = \frac{\sqrt{3}}{2}. \qquad\qquad \sin 240° = \sin \frac{4\pi}{3} = -\frac{\sqrt{3}}{2}.$$

$$\cos 120° = \cos \frac{2\pi}{3} = -\frac{1}{2}. \qquad\qquad \cos 240° = \cos \frac{4\pi}{3} = -\frac{1}{2}.$$

$$\sin 180° = \sin \pi = 0. \qquad\qquad \sin 270° = \sin \frac{3\pi}{2} = -1.$$

$$\cos 180° = \cos \pi = -1. \qquad\qquad \cos 270° = \cos \frac{3\pi}{2} = 0.$$

408.01. $\sin x = \dfrac{1}{2i}\left(e^{ix} - e^{-ix}\right)$, where $i = +\sqrt{(-1)}$.

Note that in electrical work the letter j is often used instead of i.

408.02. $\cos x = \dfrac{1}{2}(e^{ix} + e^{-ix}).$

408.03. $\tan x = -i\left(\dfrac{e^{ix} - e^{-ix}}{e^{ix} + e^{-ix}}\right) = -i\left(\dfrac{e^{2ix} - 1}{e^{2ix} + 1}\right).$

408.04. $e^{ix} = \cos x + i \sin x,$ [EULER'S FORMULA].

408.05. $e^{z+ix} = e^{z}(\cos x + i \sin x).$

408.06. $a^{z+ix} = a^{z}[\cos (x \log a) + i \sin (x \log a)].$

408.07. $(\cos x + i \sin x)^{n} = e^{inx} = \cos nx + i \sin nx,$

[DE MOIVRE'S FORMULA].

408.08. $(\cos x + i \sin x)^{-n} = \cos nx - i \sin nx.$

408.09. $(\cos x + i \sin x)^{-1} = \cos x - i \sin x.$

408.10. $\sin (ix) = i \sinh x.$ **408.13.** $\operatorname{ctn} (ix) = -i \operatorname{ctnh} x.$

408.11. $\cos (ix) = \cosh x.$ **408.14.** $\sec (ix) = \operatorname{sech} x.$

408.12. $\tan (ix) = i \tanh x.$ **408.15.** $\csc (ix) = -i \operatorname{csch} x.$

408.16. $\sin (x \pm iy) = \sin x \cosh y \pm i \cos x \sinh y.$

408.17. $\cos (x \pm iy) = \cos x \cosh y \mp i \sin x \sinh y.$

408.18. $\tan (x \pm iy) = \dfrac{\sin 2x \pm i \sinh 2y}{\cos 2x + \cosh 2y}.$

408.19. $\operatorname{ctn} (x \pm iy) = \dfrac{\sin 2x \mp i \sinh 2y}{\cosh 2y - \cos 2x}.$

409.01. $ce^{ix} = ce^{i(x+2k\pi)},$ where k is an integer or 0,

$= c(\cos x + i \sin x) = c\underline{/x}.$ [Ref. 37, p. 51.]

409.02. $1 = e^{0+2k\pi i} = \cos 0 + i \sin 0.$ Note that

$\cos 2k\pi = \cos 2\pi = \cos 0 = 1.$

409.03. $-1 = e^{0+(2k+1)\pi i} = \cos \pi + i \sin \pi.$ Note that

$\log (-1) = (2k + 1)\pi i.$

409.04. $\sqrt{1} = e^{2k\pi i/2}.$ This has two different values, depending on whether k is even or odd. They are, respectively,

$e^{2r\pi i} = \cos 0 + i \sin 0 = 1,$ $e^{(2r+1)\pi i} = \cos \pi + i \sin \pi = -1,$

where r is an integer or 0.

409.05. $\sqrt{(-1)} = e^{(2r+1)\pi i/2}.$ This square root has two different values, depending on whether r is even or odd; they are, respectively,

$\cos \dfrac{\pi}{2} + i \sin \dfrac{\pi}{2} = i,$ $\cos \dfrac{3\pi}{2} + i \sin \dfrac{3\pi}{2} = -i.$

409.06. $\sqrt[3]{1} = e^{2k\pi i/3}$. This has three different values:

$$e^{2r\pi i} = \cos 0 + i \sin 0 = 1,$$

$$e^{(2r\pi + 2\pi/3)i} = \cos \frac{2\pi}{3} + i \sin \frac{2\pi}{3} = -\frac{1}{2} + i\frac{\sqrt{3}}{2} = \omega,$$

$$e^{(2r\pi + 4\pi/3)i} = \cos \frac{4\pi}{3} + i \sin \frac{4\pi}{3} = -\frac{1}{2} - i\frac{\sqrt{3}}{2} = \omega^2.$$

409.07. $\sqrt[4]{1} = e^{2k\pi i/4}$; this has four different values:

$$e^{2r\pi i} = \cos 0 + i \sin 0 = 1,$$

$$e^{(2r\pi + 2\pi/4)i} = \cos \frac{\pi}{2} + i \sin \frac{\pi}{2} = i,$$

$$e^{(2r\pi + 4\pi/4)i} = \cos \pi + i \sin \pi = -1,$$

$$e^{(2r\pi + 6\pi/4)i} = \cos \frac{3\pi}{2} + i \sin \frac{3\pi}{2} = -i. \quad \text{[See \textbf{409.04} and \textbf{.05}.]}$$

409.08. $\sqrt{i} = e^{(4s+1)\pi i/4}$, from **409.05**, putting $r = 2s$.
This has 2 values:

$$e^{\pi i/4} = \cos \frac{\pi}{4} + i \sin \frac{\pi}{4} = \frac{1}{\sqrt{2}} + \frac{i}{\sqrt{2}}, \qquad (s \text{ even}),$$

$$e^{5\pi i/4} = \cos \frac{5\pi}{4} + i \sin \frac{5\pi}{4} = -\left(\frac{1}{\sqrt{2}} + \frac{i}{\sqrt{2}}\right), \qquad (s \text{ odd}).$$

409.09. $\sqrt[n]{1} = e^{2k\pi i/n} = \cos \frac{2k\pi}{n} + i \sin \frac{2k\pi}{n}.$

There are n different values, corresponding to different values of k. The equation $\omega^n = 1$ has n different roots:

$$\omega_0 = \cos 0 + i \sin 0 = 1, \qquad \omega_1 = \cos \frac{2\pi}{n} + i \sin \frac{2\pi}{n},$$

$$\omega_2 = \cos 2\left(\frac{2\pi}{n}\right) + i \sin 2\left(\frac{2\pi}{n}\right), \cdots \omega_k = \cos k\frac{2\pi}{n} + i \sin k\frac{2\pi}{n},$$

$$\omega_{n-1} = \cos (n-1)\frac{2\pi}{n} + i \sin (n-1)\frac{2\pi}{n}.$$

Note that, by **408.07**,

$$\omega_2 = \omega_1^2, \quad \omega_3 = \omega_1^3, \quad \omega_k = \omega_1^k, \quad \omega_0 = \omega_1^n.$$

409.10.   All the nth roots of a quantity may be obtained from any root by multiplying this root by the n roots of unity given in **409.09**.   [Ref. 10, pp. 21–22.]

410. Formulas for Plane Triangles. Let a, b, and c be the sides opposite the angles A, B, and C.

410.01. $a^2 = b^2 + c^2 - 2bc \cos A.$

410.02. $\dfrac{a}{\sin A} = \dfrac{b}{\sin B} = \dfrac{c}{\sin C}.$

410.03. $a = b \cos C + c \cos B.$

410.04. $A + B + C = \pi \text{ radians} = 180°.$

410.05. $\sin \dfrac{A}{2} = \sqrt{\left(\dfrac{(s-b)(s-c)}{bc}\right)}$, where $s = \frac{1}{2}(a + b + c)$.

410.06. $\cos \dfrac{A}{2} = \sqrt{\left(\dfrac{s(s-a)}{bc}\right)}.$

410.07. $\tan \dfrac{A}{2} = \sqrt{\left(\dfrac{(s-b)(s-c)}{s(s-a)}\right)}.$

410.08. $\tan \dfrac{A-B}{2} = \dfrac{a-b}{a+b} \operatorname{ctn} \dfrac{C}{2}.$

410.09. To find c from a, b and C, when using logarithmic trigonometric tables, let

$$\tan \theta = \frac{a+b}{a-b} \tan \frac{C}{2}; \qquad \text{then} \qquad c = (a - b) \cos \frac{C}{2} \sec \theta.$$

410.10. The area of a triangle is

$$\frac{1}{2} ab \sin C = \sqrt{\{s(s-a)(s-b)(s-c)\}} = \frac{a^2}{2} \frac{\sin B \sin C}{\sin A}.$$

410.11. If $C = 90°$, $c^2 = a^2 + b^2$. To find $c \equiv \sqrt{(a^2 + b^2)}$ when using logarithmic tables, let $\tan \theta = b/a$; then $c = a \sec \theta$.

This is useful also in other types of work. See also Table 1000.

410.12. In a plane triangle,

$$\log a = \log b - \left(\frac{c}{b} \cos A + \frac{c^2}{2b^2} \cos 2A + \cdots\right.$$
$$\left. + \frac{c^n}{nb^n} \cos nA + \cdots\right), \qquad [c < b],$$
$$= \log c - \left(\frac{b}{c} \cos A + \frac{b^2}{2c^2} \cos 2A + \cdots\right.$$
$$\left. + \frac{b^n}{nc^n} \cos nA + \cdots\right), \qquad [b < c].$$

[See **418.**]

Trigonometric Series

415.01. $\sin x = x - \dfrac{x^3}{3!} + \dfrac{x^5}{5!} - \dfrac{x^7}{7!} + \cdots,$ $[x^2 < \infty].$

415.02. $\cos x = 1 - \dfrac{x^2}{2!} + \dfrac{x^4}{4!} - \dfrac{x^6}{6!} + \cdots,$ $[x^2 < \infty].$

415.03. $\tan x = x + \dfrac{x^3}{3} + \dfrac{2}{15} x^5 + \dfrac{17}{315} x^7 + \dfrac{62}{2835} x^9 + \cdots$

$$\cdots + \dfrac{2^{2n}(2^{2n} - 1)B_n}{(2n)!} x^{2n-1} + \cdots, \qquad \left[x^2 < \dfrac{\pi^2}{4} \right].$$
[See **45.**]

415.04. $\operatorname{ctn} x = \dfrac{1}{x} - \dfrac{x}{3} - \dfrac{x^3}{45} - \dfrac{2x^5}{945} - \dfrac{x^7}{4725} - \cdots$

$$\cdots - \dfrac{2^{2n}B_n}{(2n)!} x^{2n-1} - \cdots, \qquad [x^2 < \pi^2].$$
[See **45.**]

415.05. $\sec x = 1 + \dfrac{x^2}{2} + \dfrac{5}{24} x^4 + \dfrac{61}{720} x^6 + \dfrac{277}{8064} x^8 + \cdots$

$$\cdots + \dfrac{E_n x^{2n}}{(2n)!} + \cdots, \qquad \left[x^2 < \dfrac{\pi^2}{4} \right].$$
[See **45.**]

415.06. $\csc x = \dfrac{1}{x} + \dfrac{x}{6} + \dfrac{7}{360} x^3 + \dfrac{31}{15,120} x^5 + \dfrac{127}{604,800} x^7 + \cdots$

$$\cdots + \dfrac{2(2^{2n-1} - 1)}{(2n)!} B_n x^{2n-1} + \cdots, \qquad [x^2 < \pi^2].$$
[See **45.**]

415.07. $\sin (\theta + x) = \sin \theta + x \cos \theta - \dfrac{x^2 \sin \theta}{2!}$

$$- \dfrac{x^3 \cos \theta}{3!} + \dfrac{x^4 \sin \theta}{4!} + \cdots.$$

415.08. $\cos (\theta + x) = \cos \theta - x \sin \theta - \dfrac{x^2 \cos \theta}{2!}$

$$+ \dfrac{x^3 \sin \theta}{3!} + \dfrac{x^4 \cos \theta}{4!} - \cdots.$$

416.01. $\dfrac{\pi}{4} = \sin x + \dfrac{\sin 3x}{3} + \dfrac{\sin 5x}{5} + \dfrac{\sin 7x}{7} + \cdots,$

$$[0 < x < \pi, \text{ exclusive}].$$

416.02. c, a constant, $= \dfrac{4c}{\pi} \left(\sin x + \dfrac{\sin 3x}{3} + \dfrac{\sin 5x}{5} \right.$

$$\left. + \dfrac{\sin 7x}{7} + \cdots \right), \qquad [0 < x < \pi, \text{ exclusive}].$$

416.03. $c = \dfrac{4c}{\pi} \left(\sin \dfrac{\pi x}{a} + \dfrac{1}{3} \sin \dfrac{3\pi x}{a} + \dfrac{1}{5} \sin \dfrac{5\pi x}{a} \right.$

$$\left. + \dfrac{1}{7} \sin \dfrac{7\pi x}{a} + \cdots \right), \qquad [0 < x < a, \text{ exclusive}].$$

416.04. $\dfrac{\pi}{4} = \cos x - \dfrac{\cos 3x}{3} + \dfrac{\cos 5x}{5} - \dfrac{\cos 7x}{7} + \cdots,$

$$\left[-\dfrac{\pi}{2} < x < \dfrac{\pi}{2}, \text{ exclusive} \right].$$

416.05. c, a constant, $= \dfrac{4c}{\pi} \left(\cos x - \dfrac{\cos 3x}{3} + \dfrac{\cos 5x}{5} \right.$

$$\left. - \dfrac{\cos 7x}{7} + \cdots \right), \qquad \left[-\dfrac{\pi}{2} < x < \dfrac{\pi}{2}, \text{ exclusive} \right].$$

416.06. $c = \dfrac{4c}{\pi} \left(\cos \dfrac{\pi x}{a} - \dfrac{1}{3} \cos \dfrac{3\pi x}{a} + \dfrac{1}{5} \cos \dfrac{5\pi x}{a} \right.$

$$\left. - \dfrac{1}{7} \cos \dfrac{7\pi x}{a} + \cdots \right), \qquad \left[-\dfrac{a}{2} < x < \dfrac{a}{2}, \text{ exclusive} \right].$$

416.07. $x = 2 \left(\sin x - \dfrac{\sin 2x}{2} + \dfrac{\sin 3x}{3} - \dfrac{\sin 4x}{4} + \cdots \right),$

$$[-\pi < x < \pi, \text{ exclusive}].$$

416.08. $x = \pi - 2 \left(\sin x + \dfrac{\sin 2x}{2} + \dfrac{\sin 3x}{3} + \dfrac{\sin 4x}{4} + \cdots \right),$

$$[0 < x < 2\pi, \text{ exclusive}].$$

416.09. $x = \dfrac{4}{\pi} \left(\sin x - \dfrac{\sin 3x}{3^2} + \dfrac{\sin 5x}{5^2} - \dfrac{\sin 7x}{7^2} + \cdots \right),$

$$\left[-\dfrac{\pi}{2} < x < \dfrac{\pi}{2}, \text{ inclusive} \right].$$

416.10. $x = \dfrac{\pi}{2} - \dfrac{4}{\pi} \left(\cos x + \dfrac{\cos 3x}{3^2} + \dfrac{\cos 5x}{5^2} + \dfrac{\cos 7x}{7^2} + \cdots \right),$

$$[0 < x < \pi, \text{ inclusive}].$$

416.11. $x^2 = \dfrac{\pi^2}{3} - 4\left(\cos x - \dfrac{\cos 2x}{2^2} + \dfrac{\cos 3x}{3^2}\right.$

$$\left. - \dfrac{\cos 4x}{4^2} + \cdots\right),$$

$[-\pi < x < \pi, \text{ inclusive}]$.

416.12. $x^2 = \dfrac{\pi^2}{4} - \dfrac{8}{\pi}\left(\cos x - \dfrac{\cos 3x}{3^3} + \dfrac{\cos 5x}{5^3} - \dfrac{\cos 7x}{7^3} + \cdots\right)$.

416.13. $x^3 - \pi^2 x = -12\left(\sin x - \dfrac{\sin 2x}{2^3} + \dfrac{\sin 3x}{3^3}\right.$

$$\left. - \dfrac{\sin 4x}{4^3} + \cdots\right).$$

416.14. $\sin x = \dfrac{4}{\pi}\left(\dfrac{1}{2} - \dfrac{\cos 2x}{1\cdot 3} - \dfrac{\cos 4x}{3\cdot 5} - \dfrac{\cos 6x}{5\cdot 7} - \cdots\right)$.

416.15. $\cos x = \dfrac{8}{\pi}\left\{\dfrac{\sin 2x}{1\cdot 3} + \dfrac{2}{3\cdot 5}\sin 4x + \dfrac{3}{5\cdot 7}\sin 6x + \cdots\right.$

$$\left. \cdots + \dfrac{n}{(2n-1)(2n+1)}\sin 2nx + \cdots\right\},$$

$[0 < x < \pi, \text{ exclusive}]$.

416.16. $\sin ax = \dfrac{2\sin a\pi}{\pi}\left\{\dfrac{\sin x}{1^2 - a^2} - \dfrac{2\sin 2x}{2^2 - a^2}\right.$

$$\left. + \dfrac{3\sin 3x}{3^2 - a^2} - \cdots\right\}$$

where a is not an integer, $[0 < x, \text{ inclusive}; x < \pi, \text{ exclusive}]$.

416.17. $\cos ax = \dfrac{2a\sin a\pi}{\pi}\left\{\dfrac{1}{2a^2} + \dfrac{\cos x}{1^2 - a^2} - \dfrac{\cos 2x}{2^2 - a^2}\right.$

$$\left. + \dfrac{\cos 3x}{3^2 - a^2} + \cdots\right\}, \qquad [0 < x < \pi, \text{ inclusive}],$$

where a is not an integer. [Ref. 7, pp. 301–309.]

416.18. $\sec x = 2(\cos x - \cos 3x + \cos 5x - \cos 7x + \cdots)$.

416.19. $\sec^2 x = 2^2(\cos 2x - 2\cos 4x + 3\cos 6x$

$$- 4\cos 8x + \cdots).$$

416.20. $\sec^3 x = 2^3 \left(\cos 3x - \dfrac{3}{1!} \cos 5x + \dfrac{3\cdot 4}{2!} \cos 7x \right.$

$$\left. - \dfrac{3\cdot 4\cdot 5}{3!} \cos 9x + \cdots \right).$$

416.21. $\csc x = 2(\sin x + \sin 3x + \sin 5x + \sin 7x + \cdots).$

416.22. $\csc^2 x = -2^2(\cos 2x + 2\cos 4x + 3\cos 6x$

$$+ 4\cos 8x + \cdots).$$

416.23. $\csc^3 x = -2^3 \left(\sin 3x + \dfrac{3}{1!} \sin 5x + \dfrac{3\cdot 4}{2!} \sin 7x \right.$

$$\left. + \dfrac{3\cdot 4\cdot 5}{3!} \sin 9x + \cdots \right).$$

[Ref. 4, pp. 414 and 421.]

417.1. $\dfrac{1}{1 - 2a\cos\theta + a^2} = 1 + \dfrac{1}{\sin\theta} (a\sin 2\theta + a^2\sin 3\theta$

$$+ a^3\sin 4\theta + \cdots), \qquad [a^2 < 1].$$

[Ref. 29, p. 87.]

417.2. $\dfrac{1 - a^2}{1 - 2a\cos\theta + a^2} = 1 + 2(a\cos\theta + a^2\cos 2\theta$

$$+ a^3\cos 3\theta + \cdots), \qquad [a^2 < 1].$$

417.3. $\dfrac{1 - a\cos\theta}{1 - 2a\cos\theta + a^2} = 1 + a\cos\theta + a^2\cos 2\theta$

$$+ a^3\cos 3\theta + \cdots, \qquad [a^2 < 1].$$

417.4. $\dfrac{\sin\theta}{1 - 2a\cos\theta + a^2} = \sin\theta + a\sin 2\theta + a^2\sin 3\theta + \cdots,$

$$[a^2 < 1].$$

418. $\log(1 - 2a\cos\theta + a^2)$

$$= -2\left(a\cos\theta + \dfrac{a^2}{2}\cos 2\theta + \dfrac{a^3}{3}\cos 3\theta + \cdots \right),$$

$$[a^2 < 1],$$

$$= 2\log|a| - 2\left(\dfrac{\cos\theta}{a} + \dfrac{\cos 2\theta}{2a^2} + \dfrac{\cos 3\theta}{3a^3} + \cdots \right),$$

$$[a^2 > 1]. \qquad [\text{Ref. 7, Art. 292.}]$$

419.1. $e^{ax} \sin bx = \dfrac{rx \sin \theta}{1!} + \dfrac{r^2x^2 \sin 2\theta}{2!} + \dfrac{r^3x^3 \sin 3\theta}{3!} + \cdots,$

where $r = \sqrt{(a^2 + b^2)}$, $a = r \cos \theta$,   and $b = r \sin \theta$.

419.2. $e^{ax} \cos bx = 1 + \dfrac{rx \cos \theta}{1!} + \dfrac{r^2x^2 \cos 2\theta}{2!}$

$$+ \dfrac{r^3x^3 \cos 3\theta}{3!} + \cdots,$$

where r and θ are as in **419.1.**

420.1. $\sin \alpha + \sin 2\alpha + \sin 3\alpha + \cdots + \sin n\alpha$

$$= \dfrac{\sin \dfrac{n+1}{2} \alpha \sin \dfrac{n\alpha}{2}}{\sin \dfrac{\alpha}{2}}.$$

420.2. $\cos \alpha + \cos 2\alpha + \cos 3\alpha + \cdots + \cos n\alpha$

$$= \dfrac{\cos \dfrac{n+1}{2} \alpha \sin \dfrac{n\alpha}{2}}{\sin \dfrac{\alpha}{2}}.$$

420.3. $\sin \alpha + \sin (\alpha + \delta) + \sin (\alpha + 2\delta) + \cdots$

$$+ \sin \{\alpha + (n-1)\delta\} = \dfrac{\sin \left(\alpha + \dfrac{n-1}{2}\delta\right) \sin \dfrac{n\delta}{2}}{\sin \dfrac{\delta}{2}}.$$

420.4. $\cos \alpha + \cos (\alpha + \delta) + \cos (\alpha + 2\delta) + \cdots$

$$+ \cos \{\alpha + (n-1)\delta\} = \dfrac{\cos \left(\alpha + \dfrac{n-1}{2}\delta\right) \sin \dfrac{n\delta}{2}}{\sin \dfrac{\delta}{2}}.$$

[Ref. 29, Chap. V.]

421.   If $\sin \theta = x \sin (\theta + \alpha)$,

$\theta + r\pi = x \sin \alpha + \frac{1}{2}x^2 \sin 2\alpha + \frac{1}{3}x^3 \sin 3\alpha + \cdots,$ $[x^2 < 1]$,
where r is an integer.   [Ref. 29, Art. 78.]

422.1. $\sin \theta = \theta \left(1 - \dfrac{\theta^2}{\pi^2}\right)\left(1 - \dfrac{\theta^2}{2^2\pi^2}\right)\left(1 - \dfrac{\theta^2}{3^2\pi^2}\right) \cdots, \ [\theta^2 < \infty]$.

422.2. $\cos \theta = \left(1 - \dfrac{4\theta^2}{\pi^2}\right)\left(1 - \dfrac{4\theta^2}{3^2\pi^2}\right)\left(1 - \dfrac{4\theta^2}{5^2\pi^2}\right) \cdots, \ [\theta^2 < \infty]$.

TRIGONOMETRIC FUNCTIONS—DERIVATIVES

427.1. $\dfrac{d \sin x}{dx} = \cos x.$ **427.4.** $\dfrac{d \operatorname{ctn} x}{dx} = -\csc^2 x.$

427.2. $\dfrac{d \cos x}{dx} = -\sin x.$ **427.5.** $\dfrac{d \sec x}{dx} = \sec x \tan x.$

427.3. $\dfrac{d \tan x}{dx} = \sec^2 x.$ **427.6.** $\dfrac{d \csc x}{dx} = -\csc x \operatorname{ctn} x.$

TRIGONOMETRIC FUNCTIONS—INTEGRALS

In integrating from one point to another, a process of curve plotting is frequently of assistance. Some of the curves, such as the tan curve, have more than one branch. In general, integration should not be carried out from a point on one branch to a point on another branch.

	$u =$	du	$\sin x$	$\cos x$	$\tan x$	x	dx
(1)	$\sin x$	$\cos x\, dx$	u	$\sqrt{1-u^2}$	$\dfrac{u}{\sqrt{1-u^2}}$	$\sin^{-1} u$	$\dfrac{du}{\sqrt{1-u^2}}$
(2)	$\cos x$	$-\sin x\, dx$	$\sqrt{1-u^2}$	u	$\dfrac{\sqrt{1-u^2}}{u}$	$\cos^{-1} u$	$-\dfrac{du}{\sqrt{1-u^2}}$
(3)	$\tan x$	$\sec^2 x\, dx$	$\dfrac{u}{\sqrt{1+u^2}}$	$\dfrac{1}{\sqrt{1+u^2}}$	u	$\tan^{-1} u$	$\dfrac{du}{1+u^2}$
(4)	$\sec x$	$\sec x \tan x\, dx$	$\dfrac{\sqrt{u^2-1}}{u}$	$\dfrac{1}{u}$	$\sqrt{u^2-1}$	$\sec^{-1} u$	$\dfrac{du}{u\sqrt{u^2-1}}$
(5)	$\tan \dfrac{x}{2}$	$\dfrac{1}{2}\sec^2 \dfrac{x}{2}\, dx$	$\dfrac{2u}{1+u^2}$	$\dfrac{1-u^2}{1+u^2}$	$\dfrac{2u}{1-u^2}$	$2\tan^{-1} u$	$\dfrac{2\,du}{1+u^2}$

429. Substitutions:*

Replace $\operatorname{ctn} x$, $\sec x$, $\csc x$ by $1/\tan x$, $1/\cos x$, $1/\sin x$, respectively.

Notes.   (a) $\int F(\sin x) \cos x\, dx$,—use (1).

(b) $\int F(\cos x) \sin x\, dx$,—use (2).

(c) $\int F(\tan x) \sec^2 x\, dx$,—use (3).

(d) *Inspection* of this table shows *desirable substitutions* from trigonometric to algebraic, and conversely. Thus, if only $\tan x$, $\sin^2 x$, $\cos^2 x$ appear, use (3).

* From *Macmillan Mathematical Tables.*

Integrals Involving sin x

430.10. $\displaystyle\int \sin x\, dx = -\cos x.$

430.101. $\displaystyle\int \sin (a + bx)dx = -\frac{1}{b}\cos (a + bx).$

430.102. $\displaystyle\int \sin \frac{x}{a}\, dx = -a\cos \frac{x}{a}.$

430.11. $\displaystyle\int x \sin x\, dx = \sin x - x \cos x.$

430.12. $\displaystyle\int x^2 \sin x\, dx = 2x \sin x - (x^2 - 2)\cos x.$

430.13. $\displaystyle\int x^3 \sin x\, dx = (3x^2 - 6)\sin x - (x^3 - 6x)\cos x.$

430.14. $\displaystyle\int x^4 \sin x\, dx = (4x^3 - 24x)\sin x$
$$- (x^4 - 12x^2 + 24)\cos x.$$

430.15. $\displaystyle\int x^5 \sin x\, dx = (5x^4 - 60x^2 + 120)\sin x$
$$- (x^5 - 20x^3 + 120x)\cos x.$$

430.16. $\displaystyle\int x^6 \sin x\, dx = (6x^5 - 120x^3 + 720x)\sin x$
$$- (x^6 - 30x^4 + 360x^2 - 720)\cos x.$$

430.19. $\displaystyle\int x^m \sin x\, dx = -x^m \cos x + m\int x^{m-1}\cos x\, dx.$

[See **440.**]   [Ref. 2, p. 137.]

430.20. $\displaystyle\int \sin^2 x\, dx = \frac{x}{2} - \frac{\sin 2x}{4} = \frac{x}{2} - \frac{\sin x \cos x}{2}.$

430.21. $\displaystyle\int x \sin^2 x\, dx = \frac{x^2}{4} - \frac{x \sin 2x}{4} - \frac{\cos 2x}{8}.$

430.22. $\displaystyle\int x^2 \sin^2 x\, dx = \frac{x^3}{6} - \left(\frac{x^2}{4} - \frac{1}{8}\right)\sin 2x - \frac{x \cos 2x}{4}.$

430.23. $\displaystyle\int x^3 \sin^2 x\, dx = \frac{x^4}{8} - \left(\frac{x^3}{4} - \frac{3x}{8}\right)\sin 2x$
$$- \left(\frac{3x^2}{8} - \frac{3}{16}\right)\cos 2x.$$

430.30. $\displaystyle\int \sin^3 x\, dx = \frac{\cos^3 x}{3} - \cos x.$

430.31. $\displaystyle\int x \sin^3 x\, dx = \frac{x\cos 3x}{12} - \frac{\sin 3x}{36} - \frac{3}{4}x\cos x + \frac{3}{4}\sin x.$
$$\text{[Expand } \sin^3 x \text{ by } \mathbf{404.13.]}$$

430.40. $\displaystyle\int \sin^4 x\, dx = \frac{3x}{8} - \frac{\sin 2x}{4} + \frac{\sin 4x}{32}.$

430.50. $\displaystyle\int \sin^5 x\, dx = -\frac{5\cos x}{8} + \frac{5\cos 3x}{48} - \frac{\cos 5x}{80}.$

430.60. $\displaystyle\int \sin^6 x\, dx = \frac{5x}{16} - \frac{15\sin 2x}{64} + \frac{3\sin 4x}{64} - \frac{\sin 6x}{192}.$

430.70. $\displaystyle\int \sin^7 x\, dx = -\frac{35\cos x}{64} + \frac{7\cos 3x}{64}$
$$- \frac{7\cos 5x}{320} + \frac{\cos 7x}{448}.$$

$$\text{[Ref. 1, p. 239.   Integrate expressions in } \mathbf{404.]}$$

431.11. $\displaystyle\int \frac{\sin x\, dx}{x} = \operatorname{Si}(x) = x - \frac{x^3}{3\cdot 3!} + \frac{x^5}{5\cdot 5!} - \frac{x^7}{7\cdot 7!} + \cdots.$

For table of numerical values, see Ref. 4, pp. 291 and 295 and Ref. 55 f.

431.12. $\displaystyle\int \frac{\sin x\, dx}{x^2} = -\frac{\sin x}{x} + \int \frac{\cos x\, dx}{x}.$ [See **441.11.**]

431.13. $\displaystyle\int \frac{\sin x\, dx}{x^3} = -\frac{\sin x}{2x^2} - \frac{\cos x}{2x} - \frac{1}{2}\int \frac{\sin x\, dx}{x}.$
$$\text{[See } \mathbf{431.11.]}$$

431.14. $\displaystyle\int \frac{\sin x\, dx}{x^4} = -\frac{\sin x}{3x^3} - \frac{\cos x}{6x^2} + \frac{\sin x}{6x} - \frac{1}{6}\int \frac{\cos x\, dx}{x}.$
$$\text{[See } \mathbf{441.11.]}$$

431.19. $\displaystyle\int \frac{\sin x\, dx}{x^m} = -\frac{\sin x}{(m-1)x^{m-1}} + \frac{1}{m-1}\int \frac{\cos x\, dx}{x^{m-1}}.$
$$\text{[Ref. 2, p. 138.]}$$

431.21. $\displaystyle\int \frac{\sin^2 x \, dx}{x} = \frac{1}{2} \log |x| - \frac{1}{2} \int \frac{\cos 2x \, d(2x)}{2x}.$

[See **441.11.**]

431.31. $\displaystyle\int \frac{\sin^3 x \, dx}{x} = \frac{3}{4} \int \frac{\sin x \, dx}{x} - \frac{1}{4} \int \frac{\sin 3x \, d(3x)}{3x}.$

[See **431.11.**]

431.9. $\displaystyle\int \frac{\sin^n x \, dx}{x^m}.$ Expand $\sin^n x$ by **404** and integrate each term by **431.1** and **441.1.**

432.10. $\displaystyle\int \frac{dx}{\sin x} = \int \csc x \, dx = \log \left| \tan \frac{x}{2} \right|$

$\displaystyle\qquad\qquad = -\frac{1}{2} \log \frac{1 + \cos x}{1 - \cos x} = \log |\csc x - \operatorname{ctn} x|$

$\displaystyle\qquad\qquad = \lambda \left(x - \frac{\pi}{2} \right),$ (Lambda function).

[See **603.6.**]

FIG. 432.10. Dotted graph, csc x. Full line graph, $\log \left| \tan \dfrac{x}{2} \right|$.

432.11. $\displaystyle\int \frac{x \, dx}{\sin x} = x + \frac{x^3}{3 \cdot 3!} + \frac{7x^5}{3 \cdot 5 \cdot 5!} + \frac{31x^7}{3 \cdot 7 \cdot 7!} + \frac{127x^9}{3 \cdot 5 \cdot 9!}$

$\displaystyle\qquad\qquad + \cdots + \frac{2(2^{2n-1} - 1)}{(2n + 1)!} B_n x^{2n+1} + \cdots.$

[See **45.**]

432.12. $\displaystyle\int \frac{x^2 \, dx}{\sin x} = \frac{x^2}{2} + \frac{x^4}{4 \cdot 3!} + \frac{7x^6}{3 \cdot 6 \cdot 5!} + \frac{31x^8}{3 \cdot 8 \cdot 7!} + \frac{127x^{10}}{5 \cdot 5 \cdot 6 \cdot 8!}$

$\displaystyle\qquad\qquad + \cdots + \frac{2(2^{2n-1} - 1)}{(2n + 2)(2n)!} B_n x^{2n+2} + \cdots.$

[See **45.**]

432.19. $\displaystyle\int \frac{x^m dx}{\sin x}$.   Expand $\dfrac{1}{\sin x}$ by **415.06**, multiply by x^m and integrate, $[m > 0]$.

432.20. $\displaystyle\int \frac{dx}{\sin^2 x} = \int \csc^2 x \, dx = -\operatorname{ctn} x.$

432.21. $\displaystyle\int \frac{x \, dx}{\sin^2 x} = -x \operatorname{ctn} x + \log |\sin x|.$

432.29. $\displaystyle\int \frac{x^m dx}{\sin^2 x}$.   Expand $\dfrac{1}{\sin^2 x}$ by **416.22**, $[m > 1]$.

432.30. $\displaystyle\int \frac{dx}{\sin^3 x} = -\frac{\cos x}{2 \sin^2 x} + \frac{1}{2} \log \left| \tan \frac{x}{2} \right|.$

432.31. $\displaystyle\int \frac{x \, dx}{\sin^3 x} = -\frac{x \cos x}{2 \sin^2 x} - \frac{1}{2 \sin x} + \frac{1}{2} \int \frac{x \, dx}{\sin x}.$
           [See **432.11**.]

432.40. $\displaystyle\int \frac{dx}{\sin^4 x} = -\frac{\cos x}{3 \sin^3 x} - \frac{2}{3} \operatorname{ctn} x = -\operatorname{ctn} x - \frac{\operatorname{ctn}^3 x}{3}.$

432.41. $\displaystyle\int \frac{x \, dx}{\sin^4 x} = -\frac{x \cos x}{3 \sin^3 x} - \frac{1}{6 \sin^2 x} - \frac{2}{3} x \operatorname{ctn} x$
$$+ \frac{2}{3} \log |\sin x|.$$

432.50. $\displaystyle\int \frac{dx}{\sin^5 x} = -\frac{\cos x}{4 \sin^4 x} - \frac{3}{8} \frac{\cos x}{\sin^2 x} + \frac{3}{8} \log \left| \tan \frac{x}{2} \right|.$

432.60. $\displaystyle\int \frac{dx}{\sin^6 x} = -\frac{\cos x}{5 \sin^5 x} - \frac{4}{15} \frac{\cos x}{\sin^3 x} - \frac{8}{15} \operatorname{ctn} x.$

432.90. $\displaystyle\int \frac{dx}{\sin^n x} = \int \csc^n x \, dx$
$$= -\frac{\cos x}{(n-1) \sin^{n-1} x} + \frac{n-2}{n-1} \int \frac{dx}{\sin^{n-2} x},$$
$$[n > 1].$$

432.91. $\displaystyle\int \frac{x \, dx}{\sin^n x} = -\frac{x \cos x}{(n-1) \sin^{n-1} x} - \frac{1}{(n-1)(n-2) \sin^{n-2} x}$
$$+ \frac{n-2}{n-1} \int \frac{x \, dx}{\sin^{n-2} x}, \qquad [n > 2].$$

433.01. $\displaystyle\int \frac{dx}{1 + \sin x} = -\tan\left(\frac{\pi}{4} - \frac{x}{2}\right) \cdot$

433.02. $\displaystyle\int \frac{dx}{1 - \sin x} = \tan\left(\frac{\pi}{4} + \frac{x}{2}\right) \cdot$

433.03. $\displaystyle\int \frac{x\,dx}{1 + \sin x} = -x\tan\left(\frac{\pi}{4} - \frac{x}{2}\right)$
$$+ 2\log\left|\cos\left(\frac{\pi}{4} - \frac{x}{2}\right)\right| \cdot$$

433.04. $\displaystyle\int \frac{x\,dx}{1 - \sin x} = x\,\text{ctn}\left(\frac{\pi}{4} - \frac{x}{2}\right) + 2\log\left|\sin\left(\frac{\pi}{4} - \frac{x}{2}\right)\right| \cdot$

433.05. $\displaystyle\int \frac{\sin x\,dx}{1 + \sin x} = x + \tan\left(\frac{\pi}{4} - \frac{x}{2}\right) \cdot$

433.06. $\displaystyle\int \frac{\sin x\,dx}{1 - \sin x} = -x + \tan\left(\frac{\pi}{4} + \frac{x}{2}\right) \cdot$

433.07. $\displaystyle\int \frac{dx}{\sin x(1 + \sin x)} = \tan\left(\frac{\pi}{4} - \frac{x}{2}\right) + \log\left|\tan\frac{x}{2}\right| \cdot$

433.08. $\displaystyle\int \frac{dx}{\sin x(1 - \sin x)} = \tan\left(\frac{\pi}{4} + \frac{x}{2}\right) + \log\left|\tan\frac{x}{2}\right| \cdot$

434.01. $\displaystyle\int \frac{dx}{(1 + \sin x)^2} = -\frac{1}{2}\tan\left(\frac{\pi}{4} - \frac{x}{2}\right) - \frac{1}{6}\tan^3\left(\frac{\pi}{4} - \frac{x}{2}\right) \cdot$

434.02. $\displaystyle\int \frac{dx}{(1 - \sin x)^2} = \frac{1}{2}\,\text{ctn}\left(\frac{\pi}{4} - \frac{x}{2}\right) + \frac{1}{6}\,\text{ctn}^3\left(\frac{\pi}{4} - \frac{x}{2}\right) \cdot$

434.03. $\displaystyle\int \frac{\sin x\,dx}{(1 + \sin x)^2} = -\frac{1}{2}\tan\left(\frac{\pi}{4} - \frac{x}{2}\right) + \frac{1}{6}\tan^3\left(\frac{\pi}{4} - \frac{x}{2}\right) \cdot$

434.04. $\displaystyle\int \frac{\sin x\,dx}{(1 - \sin x)^2} = -\frac{1}{2}\,\text{ctn}\left(\frac{\pi}{4} - \frac{x}{2}\right) + \frac{1}{6}\,\text{ctn}^3\left(\frac{\pi}{4} - \frac{x}{2}\right) \cdot$

434.05. $\displaystyle\int \frac{dx}{1 + \sin^2 x} = \frac{1}{2\sqrt{2}}\sin^{-1}\left(\frac{3\sin^2 x - 1}{\sin^2 x + 1}\right) \cdot$ [See **436.6.**]

434.06. $\displaystyle\int \frac{dx}{1 - \sin^2 x} = \int \frac{dx}{\cos^2 x} = \tan x.$ [See **442.20.**]

435. $\displaystyle\int \sin mx \sin nx\,dx = \frac{\sin(m-n)x}{2(m-n)} - \frac{\sin(m+n)x}{2(m+n)},$
$$[m^2 \neq n^2. \ \text{If } m^2 = n^2, \text{ see } \mathbf{430.20}]$$

436.00. $\int \dfrac{dx}{a + b \sin x}$

$$= \frac{2}{\sqrt{(a^2 - b^2)}} \tan^{-1} \frac{a \tan (x/2) + b}{\sqrt{(a^2 - b^2)}}, \quad [a^2 > b^2],$$

$$= \frac{1}{\sqrt{(b^2 - a^2)}} \log \left| \frac{a \tan (x/2) + b - \sqrt{(b^2 - a^2)}}{a \tan (x/2) + b + \sqrt{(b^2 - a^2)}} \right|,$$
$$[b^2 > a^2],$$

$$= \frac{-2}{\sqrt{(b^2 - a^2)}} \tanh^{-1} \frac{a \tan (x/2) + b}{\sqrt{(b^2 - a^2)}},$$
$$[b^2 > a^2, \quad |a \tan (x/2) + b| < \sqrt{(b^2 - a^2)}]$$

$$= \frac{-2}{\sqrt{(b^2 - a^2)}} \operatorname{ctnh}^{-1} \frac{a \tan (x/2) + b}{\sqrt{(b^2 - a^2)}},$$
$$[b^2 > a^2, \quad |a \tan (x/2) + b| > \sqrt{(b^2 - a^2)}].$$

[See **160.01**.　Also Ref. 7, p. 16 and Ref. 5, No. 298.]

The integration should not be carried out from a point on one branch of the curve to a point on another branch. The function becomes infinite at $x = \sin^{-1}(-a/b)$, which can occur when $|x| < \pi$.

436.01. $\int \dfrac{\sin x\, dx}{a + b \sin x} = \dfrac{x}{b} - \dfrac{a}{b} \int \dfrac{dx}{a + b \sin x}.$

436.02. $\int \dfrac{dx}{\sin x (a + b \sin x)} = \dfrac{1}{a} \log \left| \tan \dfrac{x}{2} \right| - \dfrac{b}{a} \int \dfrac{dx}{a + b \sin x}.$

436.03. $\int \dfrac{dx}{(a + b \sin x)^2} = \dfrac{b \cos x}{(a^2 - b^2)(a + b \sin x)}$

$$+ \frac{a}{a^2 - b^2} \int \frac{dx}{a + b \sin x}.$$

436.04. $\int \dfrac{\sin x\, dx}{(a + b \sin x)^2} = \dfrac{a \cos x}{(b^2 - a^2)(a + b \sin x)}$

$$+ \frac{b}{b^2 - a^2} \int \frac{dx}{a + b \sin x}.$$

[For **436.01** to **436.04**, see **436.00**.]

436.5. $\int \dfrac{dx}{a^2 + b^2 \sin^2 x} = \dfrac{1}{a \sqrt{(a^2 + b^2)}} \tan^{-1} \dfrac{\sqrt{(a^2 + b^2)} \tan x}{a},$
$$[a > 0].$$

436.6.   When $a = b = 1$,

$$\int \frac{dx}{1 + \sin^2 x} = \frac{1}{\sqrt{2}} \tan^{-1} (\sqrt{2} \tan x).$$

See also the alternative solution in **434.05**, which differs by a constant.

436.7. $\displaystyle\int \frac{dx}{a^2 - b^2 \sin^2 x}$

$$= \frac{1}{a\sqrt{(a^2 - b^2)}} \tan^{-1} \frac{\sqrt{(a^2 - b^2)} \tan x}{a},$$

$$[a^2 > b^2, \quad a > 0],$$

$$= \frac{1}{2a\sqrt{(b^2 - a^2)}} \log \left| \frac{\sqrt{(b^2 - a^2)} \tan x + a}{\sqrt{(b^2 - a^2)} \tan x - a} \right|,$$

$$[b^2 > a^2, \quad a > 0].$$

$$\text{If } b^2 = a^2, \text{ see } \textbf{434.06.}$$

437.1. $\displaystyle\int \frac{\sin x \, dx}{\sqrt{(1 + m^2 \sin^2 x)}} = -\frac{1}{m} \sin^{-1} \frac{m \cos x}{\sqrt{(1 + m^2)}}.$

437.2. $\displaystyle\int \frac{\sin x \, dx}{\sqrt{(1 - m^2 \sin^2 x)}}$

$$= -\frac{1}{m} \log \{ m \cos x + \sqrt{(1 - m^2 \sin^2 x)} \}.$$

437.3. $\displaystyle\int (\sin x) \sqrt{(1 + m^2 \sin^2 x)} \, dx$

$$= -\frac{\cos x}{2} \sqrt{(1 + m^2 \sin^2 x)} - \frac{1 + m^2}{2m} \sin^{-1} \frac{m \cos x}{\sqrt{(1 + m^2)}}.$$

437.4. $\displaystyle\int (\sin x) \sqrt{(1 - m^2 \sin^2 x)} \, dx$

$$= -\frac{\cos x}{2} \sqrt{(1 - m^2 \sin^2 x)}$$

$$-\frac{1 - m^2}{2m} \log \{ m \cos x + \sqrt{(1 - m^2 \sin^2 x)} \}.$$

Integrals Involving cos x

440.10. $\int \cos x \, dx = \sin x.$

440.101. $\int \cos (a + bx) dx = \frac{1}{b} \sin (a + bx).$

440.102. $\int \cos \frac{x}{a} \, dx = a \sin \frac{x}{a}.$

440.11. $\int x \cos x \, dx = \cos x + x \sin x.$

440.12. $\int x^2 \cos x \, dx = 2x \cos x + (x^2 - 2) \sin x.$

440.13. $\int x^3 \cos x \, dx = (3x^2 - 6) \cos x + (x^3 - 6x) \sin x.$

440.14. $\int x^4 \cos x \, dx = (4x^3 - 24x) \cos x + (x^4 - 12x^2 + 24) \sin x.$

440.15. $\int x^5 \cos x \, dx = (5x^4 - 60x^2 + 120) \cos x + (x^5 - 20x^3 + 120x) \sin x.$

440.16. $\int x^6 \cos x \, dx = (6x^5 - 120x^3 + 720x) \cos x + (x^6 - 30x^4 + 360x^2 - 720) \sin x.$

440.19. $\int x^m \cos x \, dx = x^m \sin x - m \int x^{m-1} \sin x \, dx.$

[See **430.**]   [Ref. 2, p. 137.]

440.20. $\int \cos^2 x \, dx = \frac{x}{2} + \frac{\sin 2x}{4} = \frac{x}{2} + \frac{\sin x \cos x}{2}.$

440.21. $\int x \cos^2 x \, dx = \frac{x^2}{4} + \frac{x \sin 2x}{4} + \frac{\cos 2x}{8}.$

440.22. $\int x^2 \cos^2 x \, dx = \frac{x^3}{6} + \left(\frac{x^2}{4} - \frac{1}{8}\right) \sin 2x + \frac{x \cos 2x}{4}.$

440.23. $\int x^3 \cos^2 x \, dx = \frac{x^4}{8} + \left(\frac{x^3}{4} - \frac{3x}{8}\right) \sin 2x + \left(\frac{3x^2}{8} - \frac{3}{16}\right) \cos 2x.$

440.30. $\displaystyle\int \cos^3 x \, dx = \sin x - \frac{\sin^3 x}{3}.$

440.31. $\displaystyle\int x \cos^3 x \, dx = \frac{x \sin 3x}{12} + \frac{\cos 3x}{36} + \frac{3}{4} x \sin x + \frac{3}{4} \cos x.$

[Expand $\cos^3 x$ by **404.23.**]

440.40. $\displaystyle\int \cos^4 x \, dx = \frac{3x}{8} + \frac{\sin 2x}{4} + \frac{\sin 4x}{32}.$

440.50. $\displaystyle\int \cos^5 x \, dx = \frac{5 \sin x}{8} + \frac{5 \sin 3x}{48} + \frac{\sin 5x}{80}.$

440.60. $\displaystyle\int \cos^6 x \, dx = \frac{5x}{16} + \frac{15 \sin 2x}{64} + \frac{3 \sin 4x}{64} + \frac{\sin 6x}{192}.$

440.70. $\displaystyle\int \cos^7 x \, dx = \frac{35 \sin x}{64} + \frac{7 \sin 3x}{64} + \frac{7 \sin 5x}{320} + \frac{\sin 7x}{448}.$

[Ref. 1, p. 240. Integrate expressions in **404.**]

441.11. $\displaystyle\int \frac{\cos x \, dx}{x} = \log |x| - \frac{x^2}{2 \cdot 2!} + \frac{x^4}{4 \cdot 4!} - \frac{x^6}{6 \cdot 6!} + \cdots.$

For table of numerical values, see Ref. 4, pp. 291 and 294 and Ref. 55 f.

441.12. $\displaystyle\int \frac{\cos x \, dx}{x^2} = -\frac{\cos x}{x} - \int \frac{\sin x \, dx}{x}.$ [See **431.11.**]

441.13. $\displaystyle\int \frac{\cos x \, dx}{x^3} = -\frac{\cos x}{2x^2} + \frac{\sin x}{2x} - \frac{1}{2} \int \frac{\cos x \, dx}{x}.$

[See **441.11.**]

441.14. $\displaystyle\int \frac{\cos x \, dx}{x^4} = -\frac{\cos x}{3x^3} + \frac{\sin x}{6x^2} + \frac{\cos x}{6x} + \frac{1}{6} \int \frac{\sin x \, dx}{x}.$

[See **431.11.**]

441.19. $\displaystyle\int \frac{\cos x \, dx}{x^m} = -\frac{\cos x}{(m-1)x^{m-1}} - \frac{1}{m-1} \int \frac{\sin x \, dx}{x^{m-1}}.$

441.21. $\displaystyle\int \frac{\cos^2 x \, dx}{x} = \frac{1}{2} \log |x| + \frac{1}{2} \int \frac{\cos 2x \, d(2x)}{2x}.$

[See **441.11.**]

441.31. $\displaystyle\int \frac{\cos^3 x \, dx}{x} = \frac{3}{4} \int \frac{\cos x \, dx}{x} + \frac{1}{4} \int \frac{\cos 3x \, d(3x)}{3x}.$

[See **441.11.**]

441.9. $\displaystyle\int \frac{\cos^n x\, dx}{x^m}.$

Expand $\cos^n x$ by **404** and integrate each term by **441.1**.

442.10. $\displaystyle\int \frac{dx}{\cos x} = \int \sec x\, dx = \log\left| \tan\left(\frac{\pi}{4} + \frac{x}{2}\right)\right|$

$$= \log|\sec x + \tan x| = \frac{1}{2}\log\frac{1 + \sin x}{1 - \sin x}$$

$$= \lambda(x), \quad (\text{Lambda Function}). \qquad [\text{See } \mathbf{640}.]$$

442.11. $\displaystyle\int \frac{x\, dx}{\cos x} = \frac{x^2}{2} + \frac{x^4}{4\cdot 2!} + \frac{5x^6}{6\cdot 4!} + \frac{61x^8}{8\cdot 6!} + \frac{1385x^{10}}{10\cdot 8!} + \cdots$

$$\cdots + \frac{E_n x^{2n+2}}{(2n+2)(2n)!} + \cdots. \qquad [\text{See } \mathbf{45}.]$$

442.12. $\displaystyle\int \frac{x^2 dx}{\cos x} = \frac{x^3}{3} + \frac{x^5}{5\cdot 2!} + \frac{5x^7}{7\cdot 4!} + \frac{61x^9}{9\cdot 6!} + \frac{1385x^{11}}{11\cdot 8!} + \cdots$

$$\cdots + \frac{E_{n-1} x^{2n+1}}{(2n+1)(2n-2)!} + \cdots. \qquad [\text{See } \mathbf{45}.]$$

442.19. $\displaystyle\int \frac{x^m dx}{\cos x}.$ Expand $\dfrac{1}{\cos x}$ by **415.05**, multiply by x^m and integrate, $[m \neq 0].$

442.20. $\displaystyle\int \frac{dx}{\cos^2 x} = \int \sec^2 x\, dx = \tan x.$

442.21. $\displaystyle\int \frac{x\, dx}{\cos^2 x} = x \tan x + \log|\cos x|.$

442.29. $\displaystyle\int \frac{x^m dx}{\cos^2 x}.$ Expand $\dfrac{1}{\cos^2 x}$ by **416.19**, $[m > 1].$

442.30. $\displaystyle\int \frac{dx}{\cos^3 x} = \frac{\sin x}{2\cos^2 x} + \frac{1}{2}\log\left|\tan\left(\frac{\pi}{4} + \frac{x}{2}\right)\right|.$

442.31. $\displaystyle\int \frac{x\, dx}{\cos^3 x} = \frac{x \sin x}{2\cos^2 x} - \frac{1}{2\cos x} + \frac{1}{2}\int \frac{x\, dx}{\cos x}.$

$$[\text{See } \mathbf{442.11}.]$$

442.40. $\displaystyle\int \frac{dx}{\cos^4 x} = \frac{\sin x}{3\cos^3 x} + \frac{2}{3}\tan x = \tan x + \frac{\tan^3 x}{3}.$

442.41. $\displaystyle\int \frac{x\,dx}{\cos^4 x} = \frac{x\sin x}{3\cos^3 x} - \frac{1}{6\cos^2 x} + \frac{2}{3}\,x\tan x$

$$+ \frac{2}{3}\log|\cos x|.$$

442.50. $\displaystyle\int \frac{dx}{\cos^5 x} = \frac{\sin x}{4\cos^4 x} + \frac{3}{8}\frac{\sin x}{\cos^2 x} + \frac{3}{8}\log\left|\tan\left(\frac{\pi}{4} + \frac{x}{2}\right)\right|.$

442.60. $\displaystyle\int \frac{dx}{\cos^6 x} = \frac{\sin x}{5\cos^5 x} + \frac{4}{15}\frac{\sin x}{\cos^3 x} + \frac{8}{15}\tan x.$

442.90. $\displaystyle\int \frac{dx}{\cos^n x} = \int \sec^n x\,dx$

$$= \frac{\sin x}{(n-1)\cos^{n-1} x} + \frac{n-2}{n-1}\int \frac{dx}{\cos^{n-2} x}, \qquad [n > 1].$$

442.91. $\displaystyle\int \frac{x\,dx}{\cos^n x} = \frac{x\sin x}{(n-1)\cos^{n-1} x} - \frac{1}{(n-1)(n-2)\cos^{n-2} x}$

$$+ \frac{n-2}{n-1}\int \frac{x\,dx}{\cos^{n-2} x}, \qquad [n > 2].$$

443.01. $\displaystyle\int \frac{dx}{1+\cos x} = \tan\frac{x}{2}.$

443.02. $\displaystyle\int \frac{dx}{1-\cos x} = -\ctn\frac{x}{2}.$

443.03. $\displaystyle\int \frac{x\,dx}{1+\cos x} = x\tan\frac{x}{2} + 2\log\left|\cos\frac{x}{2}\right|.$

443.04. $\displaystyle\int \frac{x\,dx}{1-\cos x} = -x\ctn\frac{x}{2} + 2\log\left|\sin\frac{x}{2}\right|.$

443.05. $\displaystyle\int \frac{\cos x\,dx}{1+\cos x} = x - \tan\frac{x}{2}.$

443.06. $\displaystyle\int \frac{\cos x\,dx}{1-\cos x} = -x - \ctn\frac{x}{2}.$

443.07. $\displaystyle\int \frac{dx}{\cos x(1+\cos x)} = \log\left|\tan\left(\frac{\pi}{4} + \frac{x}{2}\right)\right| - \tan\frac{x}{2}.$

443.08. $\displaystyle\int \frac{dx}{\cos x(1-\cos x)} = \log\left|\tan\left(\frac{\pi}{4} + \frac{x}{2}\right)\right| - \ctn\frac{x}{2}.$

444.01. $\displaystyle\int \frac{dx}{(1 + \cos x)^2} = \frac{1}{2} \tan \frac{x}{2} + \frac{1}{6} \tan^3 \frac{x}{2}.$

444.02. $\displaystyle\int \frac{dx}{(1 - \cos x)^2} = -\frac{1}{2} \operatorname{ctn} \frac{x}{2} - \frac{1}{6} \operatorname{ctn}^3 \frac{x}{2}.$

444.03. $\displaystyle\int \frac{\cos x\ dx}{(1 + \cos x)^2} = \frac{1}{2} \tan \frac{x}{2} - \frac{1}{6} \tan^3 \frac{x}{2}.$

444.04. $\displaystyle\int \frac{\cos x\ dx}{(1 - \cos x)^2} = \frac{1}{2} \operatorname{ctn} \frac{x}{2} - \frac{1}{6} \operatorname{ctn}^3 \frac{x}{2}.$

444.05. $\displaystyle\int \frac{dx}{1 + \cos^2 x} = \frac{1}{2\sqrt{2}} \sin^{-1} \left(\frac{1 - 3 \cos^2 x}{1 + \cos^2 x} \right).$ [See **446.6.**]

444.06. $\displaystyle\int \frac{dx}{1 - \cos^2 x} = \int \frac{dx}{\sin^2 x} = - \operatorname{ctn} x.$ [See **432.20.**]

445. $\displaystyle\int \cos mx \cos nx\ dx = \frac{\sin (m - n)x}{2(m - n)} + \frac{\sin (m + n)x}{2(m + n)},$

$$[m^2 \neq n^2. \text{ If } m^2 = n^2, \text{ see } \mathbf{440.20}].$$

446.00. $\displaystyle\int \frac{dx}{a + b \cos x}$

$$= \frac{2}{\sqrt{(a^2 - b^2)}} \tan^{-1} \frac{(a - b) \tan (x/2)}{\sqrt{(a^2 - b^2)}}, \qquad [a^2 > b^2],$$

$$= \frac{1}{\sqrt{(b^2 - a^2)}} \log \left| \frac{(b - a) \tan (x/2) + \sqrt{(b^2 - a^2)}}{(b - a) \tan (x/2) - \sqrt{(b^2 - a^2)}} \right|,$$

$$[b^2 > a^2],$$

$$= \frac{2}{\sqrt{(b^2 - a^2)}} \tanh^{-1} \frac{(b - a) \tan (x/2)}{\sqrt{(b^2 - a^2)}},$$

$$[b^2 > a^2, \ |(b - a) \tan (x/2)| < \sqrt{(b^2 - a^2)}],$$

$$= \frac{2}{\sqrt{(b^2 - a^2)}} \operatorname{ctnh}^{-1} \frac{(b - a) \tan (x/2)}{\sqrt{(b^2 - a^2)}},$$

$$[b^2 > a^2, \ |(b - a) \tan (x/2)| > \sqrt{(b^2 - a^2)}].$$

$$[\text{See Ref. 7, p. 15, and Ref. 5, No. 300.}]$$

The integration should not be carried out from a point on one branch of the curve to a point on another branch. The function becomes infinite at $x = \cos^{-1} (- a/b)$ which can occur when $|x| < \pi.$

446.01. $\displaystyle\int \frac{\cos x\ dx}{a + b \cos x} = \frac{x}{b} - \frac{a}{b} \int \frac{dx}{a + b \cos x}.$

446.02. $\displaystyle\int \frac{dx}{\cos x(a + b\cos x)} = \frac{1}{a}\log\left|\tan\left(\frac{x}{2} + \frac{\pi}{4}\right)\right|$

$$-\frac{b}{a}\int \frac{dx}{a + b\cos x}.$$

446.03. $\displaystyle\int \frac{dx}{(a + b\cos x)^2} = \frac{b\sin x}{(b^2 - a^2)(a + b\cos x)}$

$$-\frac{a}{b^2 - a^2}\int \frac{dx}{a + b\cos x}.$$

446.04. $\displaystyle\int \frac{\cos x\,dx}{(a + b\cos x)^2} = \frac{a\sin x}{(a^2 - b^2)(a + b\cos x)}$

$$-\frac{b}{a^2 - b^2}\int \frac{dx}{a + b\cos x}.$$

[For **446.01** to **446.04**, see **446.00**.]

446.2. $\displaystyle\int \frac{dx}{a^2 + b^2 - 2ab\cos x}$

$$= \frac{2}{|a^2 - b^2|}\tan^{-1}\left[\left|\frac{a + b}{a - b}\right|\tan\frac{x}{2}\right], \quad [a \neq b].$$

[Ref. 38, p. 52.]   [See **446.00**.]

446.5. $\displaystyle\int \frac{dx}{a^2 + b^2\cos^2 x} = \frac{1}{a\sqrt{(a^2 + b^2)}}\tan^{-1}\frac{a\tan x}{\sqrt{(a^2 + b^2)}},$

$$[a > 0].$$

446.6. When $a = b = 1$,

$$\int \frac{dx}{1 + \cos^2 x} = \frac{1}{\sqrt{2}}\tan^{-1}\left(\frac{\tan x}{\sqrt{2}}\right).$$

See also the alternative solution in **444.05**, which differs by a constant.

446.7. $\displaystyle\int \frac{dx}{a^2 - b^2\cos^2 x}$

$$= \frac{1}{a\sqrt{(a^2 - b^2)}}\tan^{-1}\frac{a\tan x}{\sqrt{(a^2 - b^2)}},$$

$$[a^2 > b^2, \quad a > 0],$$

$$= \frac{1}{2a\sqrt{(b^2 - a^2)}}\log\left|\frac{a\tan x - \sqrt{(b^2 - a^2)}}{a\tan x + \sqrt{(b^2 - a^2)}}\right|,$$

$$[b^2 > a^2, \quad a > 0].$$

If $b^2 = a^2$, see **444.06**.

Integrals Involving sin x and cos x

450.11. $\displaystyle\int \sin x \cos x \, dx = \frac{\sin^2 x}{2} = -\frac{\cos^2 x}{2} + \text{constant}$

$$= -\frac{\cos 2x}{4} + \text{constant.}$$

450.12. $\displaystyle\int \sin x \cos^2 x \, dx = -\frac{\cos^3 x}{3} \cdot$

450.13. $\displaystyle\int \sin x \cos^3 x \, dx = -\frac{\cos^4 x}{4} \cdot$

450.19. $\displaystyle\int \sin x \cos^n x \, dx = -\frac{\cos^{n+1} x}{n+1} \cdot$

450.21. $\displaystyle\int \sin^2 x \cos x \, dx = \frac{\sin^3 x}{3} \cdot$

450.22. $\displaystyle\int \sin^2 x \cos^2 x \, dx = \frac{1}{8}\left(x - \frac{\sin 4x}{4}\right) \cdot$

450.23. $\displaystyle\int \sin^2 x \cos^3 x \, dx = \frac{\sin^3 x \cos^2 x}{5} + \frac{2}{15} \sin^3 x.$

450.31. $\displaystyle\int \sin^3 x \cos x \, dx = \frac{\sin^4 x}{4} \cdot$

450.81. $\displaystyle\int \sin^m x \cos x \, dx = \frac{\sin^{m+1} x}{m+1}, \qquad [m \neq -1].$

$$[\text{If } m = -1, \text{ see } \mathbf{453.11.}]$$

450.9. $\displaystyle\int \sin^m x \cos^n x \, dx$

$$= \frac{\sin^{m+1} x \cos^{n-1} x}{m+n} + \frac{n-1}{m+n} \int \sin^m x \cos^{n-2} x \, dx$$

$$= -\frac{\sin^{m-1} x \cos^{n+1} x}{m+n} + \frac{m-1}{m+n} \int \sin^{m-2} x \cos^n x \, dx,$$

$[m \neq -n, \text{ see } \mathbf{480.9}].$ \qquad [See also **461.**]

451.11. $\displaystyle\int \frac{dx}{\sin x \cos x} = \log|\tan x|.$

451.12. $\displaystyle\int \frac{dx}{\sin x \cos^2 x} = \frac{1}{\cos x} + \log\left|\tan \frac{x}{2}\right| \cdot$

451.13. $\displaystyle\int \frac{dx}{\sin x \cos^3 x} = \frac{1}{2 \cos^2 x} + \log |\tan x|.$

451.14. $\displaystyle\int \frac{dx}{\sin x \cos^4 x} = \frac{1}{3 \cos^3 x} + \frac{1}{\cos x} + \log \left| \tan \frac{x}{2} \right|.$

451.15. $\displaystyle\int \frac{dx}{\sin x \cos^5 x} = \frac{1}{4 \cos^4 x} + \frac{1}{2 \cos^2 x} + \log |\tan x|.$

451.19. $\displaystyle\int \frac{dx}{\sin x \cos^n x} = \frac{1}{(n-1) \cos^{n-1} x} + \int \frac{dx}{\sin x \cos^{n-2} x},$
$$[n \neq 1].$$

451.21. $\displaystyle\int \frac{dx}{\sin^2 x \cos x} = -\frac{1}{\sin x} + \log \left| \tan \left(\frac{\pi}{4} + \frac{x}{2} \right) \right|.$

451.22. $\displaystyle\int \frac{dx}{\sin^2 x \cos^2 x} = -2 \operatorname{ctn} 2x.$

451.23. $\displaystyle\int \frac{dx}{\sin^2 x \cos^3 x} = \frac{\sin x}{2 \cos^2 x} - \frac{1}{\sin x}$
$$+ \frac{3}{2} \log \left| \tan \left(\frac{\pi}{4} + \frac{x}{2} \right) \right|.$$

451.24. $\displaystyle\int \frac{dx}{\sin^2 x \cos^4 x} = \frac{1}{3 \sin x \cos^3 x} - \frac{8}{3} \operatorname{ctn} 2x.$

451.31. $\displaystyle\int \frac{dx}{\sin^3 x \cos x} = -\frac{1}{2 \sin^2 x} + \log |\tan x|.$

451.32. $\displaystyle\int \frac{dx}{\sin^3 x \cos^2 x} = \frac{1}{\cos x} - \frac{\cos x}{2 \sin^2 x} + \frac{3}{2} \log \left| \tan \frac{x}{2} \right|.$

451.33. $\displaystyle\int \frac{dx}{\sin^3 x \cos^3 x} = -\frac{2 \cos 2x}{\sin^2 2x} + 2 \log |\tan x|.$

451.41. $\displaystyle\int \frac{dx}{\sin^4 x \cos x} = \frac{3 \cos^2 x - 4}{3 \sin^3 x} + \log \left| \tan \left(\frac{\pi}{4} + \frac{x}{2} \right) \right|.$
$$[\text{Ref. 1, pp. 260–263.}]$$

451.91. $\displaystyle\int \frac{dx}{\sin^m x \cos x} = -\frac{1}{(m-1) \sin^{m-1} x}$
$$+ \int \frac{dx}{\sin^{m-2} x \cos x}, \quad [m \neq 1].$$

451.92. $\displaystyle\int \frac{dx}{\sin^n x \cos^n x} = 2^{n-1} \int \frac{d(2x)}{\sin^n (2x)}$.   [See **432.**]

451.93. $\displaystyle\int \frac{dx}{\sin^m x \cos^n x}$

$$= \frac{1}{(n-1)\sin^{m-1} x \cos^{n-1} x} + \frac{m+n-2}{n-1} \int \frac{dx}{\sin^m x \cos^{n-2} x} ,$$
$$[n > 1],$$
$$= -\frac{1}{(m-1)\sin^{m-1} x \cos^{n-1} x} + \frac{m+n-2}{m-1} \int \frac{dx}{\sin^{m-2} x \cos^n x} ,$$
$$[m > 1].$$

452.11. $\displaystyle\int \frac{\sin x \, dx}{\cos x} = \int \tan x \, dx = - \log |\cos x|$
$$= \log |\sec x|.\qquad\text{[See } \mathbf{480.1.]}$$

452.12. $\displaystyle\int \frac{\sin x \, dx}{\cos^2 x} = \frac{1}{\cos x} = \sec x.$

452.13. $\displaystyle\int \frac{\sin x \, dx}{\cos^3 x} = \frac{1}{2 \cos^2 x} = \frac{1}{2} \tan^2 x + \text{constant.}$

452.14. $\displaystyle\int \frac{\sin x \, dx}{\cos^4 x} = \frac{1}{3 \cos^3 x}$.

452.19. $\displaystyle\int \frac{\sin x \, dx}{\cos^n x} = \frac{1}{(n-1) \cos^{n-1} x} ,\qquad [n \neq 1].$

452.21. $\displaystyle\int \frac{\sin^2 x \, dx}{\cos x} = - \sin x + \log \left| \tan \left(\frac{\pi}{4} + \frac{x}{2} \right) \right|$.

452.22. $\displaystyle\int \frac{\sin^2 x \, dx}{\cos^2 x} = \int \tan^2 x \, dx = \tan x - x.$ [See **480.2.**]

452.23. $\displaystyle\int \frac{\sin^2 x \, dx}{\cos^3 x} = \frac{\sin x}{2 \cos^2 x} - \frac{1}{2} \log \left| \tan \left(\frac{\pi}{4} + \frac{x}{2} \right) \right|$.

452.24. $\displaystyle\int \frac{\sin^2 x \, dx}{\cos^4 x} = \frac{1}{3} \tan^3 x.$

452.29. $\displaystyle\int \frac{\sin^2 x \, dx}{\cos^n x} = \frac{\sin x}{(n-1) \cos^{n-1} x} - \frac{1}{n-1} \int \frac{dx}{\cos^{n-2} x} ,$
$$[n \neq 1].$$

452.31. $\displaystyle\int \frac{\sin^3 x \, dx}{\cos x} = - \frac{\sin^2 x}{2} - \log |\cos x|.$

452.32. $\displaystyle\int \frac{\sin^3 x\, dx}{\cos^2 x} = \cos x + \sec x.$

452.33. $\displaystyle\int \frac{\sin^3 x\, dx}{\cos^3 x} = \int \tan^3 x\, dx = \frac{1}{2}\tan^2 x + \log|\cos x|.$

[See **480.3.**]

452.34. $\displaystyle\int \frac{\sin^3 x\, dx}{\cos^4 x} = \frac{1}{3\cos^3 x} - \frac{1}{\cos x}.$

452.35. $\displaystyle\int \frac{\sin^3 x\, dx}{\cos^5 x} = \frac{1}{4}\tan^4 x = \frac{1}{4\cos^4 x} - \frac{1}{2\cos^2 x} + \text{constant.}$

452.39. $\displaystyle\int \frac{\sin^3 x\, dx}{\cos^n x} = \frac{1}{(n-1)\cos^{n-1} x} - \frac{1}{(n-3)\cos^{n-3} x},$

$[n \neq 1 \text{ or } 3].$

452.41. $\displaystyle\int \frac{\sin^4 x\, dx}{\cos x} = -\frac{\sin^3 x}{3} - \sin x + \log\left|\tan\left(\frac{\pi}{4} + \frac{x}{2}\right)\right|.$

452.7. $\displaystyle\int \frac{\sin^{n-2} x\, dx}{\cos^n x} = \frac{\tan^{n-1} x}{n-1},$ $\qquad\qquad [n \neq 1].$

452.8. $\displaystyle\int \frac{\sin^n x\, dx}{\cos^n x} = \int \tan^n x\, dx = \frac{\tan^{n-1} x}{n-1} - \int \tan^{n-2} x\, dx,$

$[n \neq 1. \text{ See } \mathbf{480.9}].$

452.9. $\displaystyle\int \frac{\sin^m x\, dx}{\cos^n x}$

$$= \frac{\sin^{m+1} x}{(n-1)\cos^{n-1} x} - \frac{m-n+2}{n-1}\int \frac{\sin^m x\, dx}{\cos^{n-2} x},$$

$[n \neq 1],$

$$= -\frac{\sin^{m-1} x}{(m-n)\cos^{n-1} x} + \frac{m-1}{m-n}\int \frac{\sin^{m-2} x\, dx}{\cos^n x},$$

$[m \neq n],$

$$= \frac{\sin^{m-1} x}{(n-1)\cos^{n-1} x} - \frac{m-1}{n-1}\int \frac{\sin^{m-2} x\, dx}{\cos^{n-2} x},$$

$[n \neq 1].$

453.11. $\displaystyle\int \frac{\cos x\, dx}{\sin x} = \int \operatorname{ctn} x\, dx = \log|\sin x|.$ [See **490.1.**]

453.12. $\displaystyle\int \frac{\cos x\, dx}{\sin^2 x} = -\frac{1}{\sin x} = -\csc x.$

453.13. $\displaystyle\int \frac{\cos x\, dx}{\sin^3 x} = -\frac{1}{2\sin^2 x} = -\frac{\operatorname{ctn}^2 x}{2} + \text{constant.}$

453.14. $\displaystyle\int \frac{\cos x\, dx}{\sin^4 x} = -\frac{1}{3\sin^3 x}.$

453.19. $\displaystyle\int \frac{\cos x\, dx}{\sin^n x} = -\frac{1}{(n-1)\sin^{n-1} x},$ $[n \neq 1].$

453.21. $\displaystyle\int \frac{\cos^2 x\, dx}{\sin x} = \cos x + \log\left|\tan \frac{x}{2}\right|.$

453.22. $\displaystyle\int \frac{\cos^2 x\, dx}{\sin^2 x} = \int \operatorname{ctn}^2 x\, dx = -\operatorname{ctn} x - x.$
$[\text{See } \mathbf{490.2.}]$

453.23. $\displaystyle\int \frac{\cos^2 x\, dx}{\sin^3 x} = -\frac{\cos x}{2\sin^2 x} - \frac{1}{2}\log\left|\tan \frac{x}{2}\right|.$

453.24. $\displaystyle\int \frac{\cos^2 x\, dx}{\sin^4 x} = -\frac{1}{3}\operatorname{ctn}^3 x.$

453.29. $\displaystyle\int \frac{\cos^2 x\, dx}{\sin^n x} = -\frac{\cos x}{(n-1)\sin^{n-1} x} - \frac{1}{n-1}\int \frac{dx}{\sin^{n-2} x},$
$[n \neq 1].$

453.31. $\displaystyle\int \frac{\cos^3 x\, dx}{\sin x} = \frac{\cos^2 x}{2} + \log|\sin x|.$

453.32. $\displaystyle\int \frac{\cos^3 x\, dx}{\sin^2 x} = -\sin x - \csc x.$

453.33. $\displaystyle\int \frac{\cos^3 x\, dx}{\sin^3 x} = \int \operatorname{ctn}^3 x\, dx = -\frac{\operatorname{ctn}^2 x}{2} - \log|\sin x|.$
$[\text{See } \mathbf{490.3.}]$

453.34. $\displaystyle\int \frac{\cos^3 x\, dx}{\sin^4 x} = \frac{1}{\sin x} - \frac{1}{3\sin^3 x}.$

453.35. $\displaystyle\int \frac{\cos^3 x\, dx}{\sin^5 x} = -\frac{1}{4}\operatorname{ctn}^4 x = \frac{1}{2\sin^2 x} - \frac{1}{4\sin^4 x}$
$+ \text{constant.}$

453.39. $\displaystyle\int \frac{\cos^3 x\, dx}{\sin^n x} = \frac{1}{(n-3)\sin^{n-3} x} - \frac{1}{(n-1)\sin^{n-1} x},$
$[n \neq 1 \text{ or } 3].$

453.41. $\int \dfrac{\cos^4 x \, dx}{\sin x} = \dfrac{\cos^3 x}{3} + \cos x + \log \left| \tan \dfrac{x}{2} \right|.$

453.7. $\int \dfrac{\cos^{n-2} x \, dx}{\sin^n x} = - \dfrac{\text{ctn}^{n-1} x}{n - 1}, \qquad\qquad [n \neq 1].$

453.8. $\int \dfrac{\cos^n x \, dx}{\sin^n x} = \int \text{ctn}^n x \, dx$

$$= - \dfrac{\text{ctn}^{n-1} x}{n - 1} - \int \text{ctn}^{n-2} x \, dx,$$

$$[n \neq 1. \quad \text{See } \mathbf{490.9}].$$

453.9. $\int \dfrac{\cos^n x \, dx}{\sin^m x}$

$$= - \dfrac{\cos^{n+1} x}{(m - 1) \sin^{m-1} x} - \dfrac{n - m + 2}{m - 1} \int \dfrac{\cos^n x \, dx}{\sin^{m-2} x},$$

$$[m \neq 1],$$

$$= \dfrac{\cos^{n-1} x}{(n - m) \sin^{m-1} x} + \dfrac{n - 1}{n - m} \int \dfrac{\cos^{n-2} x \, dx}{\sin^m x},$$

$$[m \neq n],$$

$$= - \dfrac{\cos^{n-1} x}{(m - 1) \sin^{m-1} x} - \dfrac{n - 1}{m - 1} \int \dfrac{\cos^{n-2} x \, dx}{\sin^{m-2} x},$$

$$[m \neq 1].$$

454.01. $\int \dfrac{\sin x \, dx}{1 + \cos x} = - \log (1 + \cos x).$

454.02. $\int \dfrac{\sin x \, dx}{1 - \cos x} = \log (1 - \cos x).$

454.03. $\int \dfrac{\cos x \, dx}{1 + \sin x} = \log (1 + \sin x).$

454.04. $\int \dfrac{\cos x \, dx}{1 - \sin x} = - \log (1 - \sin x).$

454.05. $\int \dfrac{dx}{\sin x (1 + \cos x)} = \dfrac{1}{2(1 + \cos x)} + \dfrac{1}{2} \log \left| \tan \dfrac{x}{2} \right|.$

454.06. $\int \dfrac{dx}{\sin x (1 - \cos x)} = - \dfrac{1}{2(1 - \cos x)} + \dfrac{1}{2} \log \left| \tan \dfrac{x}{2} \right|.$

454.07. $\displaystyle \int \frac{dx}{\cos x(1 + \sin x)} = -\frac{1}{2(1 + \sin x)}$

$$+ \frac{1}{2} \log \left| \tan \left(\frac{\pi}{4} + \frac{x}{2} \right) \right|.$$

454.08. $\displaystyle \int \frac{dx}{\cos x(1 - \sin x)} = \frac{1}{2(1 - \sin x)}$

$$+ \frac{1}{2} \log \left| \tan \left(\frac{\pi}{4} + \frac{x}{2} \right) \right|.$$

454.09. $\displaystyle \int \frac{\sin x \, dx}{\cos x(1 + \cos x)} = \log \left| \frac{1 + \cos x}{\cos x} \right|.$

454.10. $\displaystyle \int \frac{\sin x \, dx}{\cos x(1 - \cos x)} = \log \left| \frac{1 - \cos x}{\cos x} \right|.$

454.11. $\displaystyle \int \frac{\cos x \, dx}{\sin x(1 + \sin x)} = -\log \left| \frac{1 + \sin x}{\sin x} \right|.$

454.12. $\displaystyle \int \frac{\cos x \, dx}{\sin x(1 - \sin x)} = -\log \left| \frac{1 - \sin x}{\sin x} \right|.$

454.13. $\displaystyle \int \frac{\sin x \, dx}{\cos x(1 + \sin x)} = \frac{1}{2(1 + \sin x)}$

$$+ \frac{1}{2} \log \left| \tan \left(\frac{\pi}{4} + \frac{x}{2} \right) \right|.$$

454.14. $\displaystyle \int \frac{\sin x \, dx}{\cos x(1 - \sin x)} = \frac{1}{2(1 - \sin x)}$

$$- \frac{1}{2} \log \left| \tan \left(\frac{\pi}{4} + \frac{x}{2} \right) \right|.$$

454.15. $\displaystyle \int \frac{\cos x \, dx}{\sin x(1 + \cos x)} = -\frac{1}{2(1 + \cos x)} + \frac{1}{2} \log \left| \tan \frac{x}{2} \right|.$

454.16. $\displaystyle \int \frac{\cos x \, dx}{\sin x(1 - \cos x)} = -\frac{1}{2(1 - \cos x)} - \frac{1}{2} \log \left| \tan \frac{x}{2} \right|.$

455.01. $\displaystyle \int \frac{dx}{\sin x + \cos x} = \frac{1}{\sqrt{2}} \log \left| \tan \left(\frac{x}{2} + \frac{\pi}{8} \right) \right|.$

455.02. $\displaystyle \int \frac{dx}{\sin x - \cos x} = \frac{1}{\sqrt{2}} \log \left| \tan \left(\frac{x}{2} - \frac{\pi}{8} \right) \right|.$

455.03. $\displaystyle\int \frac{\sin x\, dx}{\sin x + \cos x} = \frac{x}{2} - \frac{1}{2}\log|\sin x + \cos x|.$

[See **482.2** and **492.1.**]

455.04. $\displaystyle\int \frac{\sin x\, dx}{\sin x - \cos x} = \frac{x}{2} + \frac{1}{2}\log|\sin x - \cos x|.$

[See **482.2** and **492.1.**]

455.05. $\displaystyle\int \frac{\cos x\, dx}{\sin x + \cos x} = \frac{x}{2} + \frac{1}{2}\log|\sin x + \cos x|.$

[See **482.1** and **492.2.**]

455.06. $\displaystyle\int \frac{\cos x\, dx}{\sin x - \cos x} = -\frac{x}{2} + \frac{1}{2}\log|\sin x - \cos x|.$

[See **482.1** and **492.2.**]

455.07. $\displaystyle\int \frac{dx}{(\sin x + \cos x)^2} = \frac{1}{2}\tan\left(x - \frac{\pi}{4}\right).$

455.08. $\displaystyle\int \frac{dx}{(\sin x - \cos x)^2} = \frac{1}{2}\tan\left(x + \frac{\pi}{4}\right).$

455.09. $\displaystyle\int \frac{dx}{1 + \cos x \pm \sin x} = \pm\log\left|1 \pm \tan\frac{x}{2}\right|.$

456.1. $\displaystyle\int \frac{dx}{b\cos x + c\sin x} = \frac{1}{r}\log\left|\tan\frac{x+\theta}{2}\right|$

where $r = \sqrt{(b^2 + c^2)}$, $\sin\theta = b/r$, $\cos\theta = c/r$.

[See **401.2** and **432.10.**]

456.2. $\displaystyle\int \frac{dx}{a + b\cos x + c\sin x} = \int \frac{d(x+\theta)}{a + r\sin(x+\theta)}$

where r and θ are given in **456.1.**   [See **436.00.**]

460.1. $\displaystyle\int \frac{dx}{a^2\cos^2 x + b^2\sin^2 x} = \frac{1}{ab}\tan^{-1}\left(\frac{b}{a}\tan x\right),$

$[a > 0,\quad b > 0].$ [See **436.5.**]

460.2. $\displaystyle\int \frac{dx}{a^2\cos^2 x - b^2\sin^2 x} = \frac{1}{2ab}\log\left|\frac{b\tan x + a}{b\tan x - a}\right|,$

$[a > 0,\quad b > 0].$ [See **436.7.**]

461. $\int \sin^m x \cos^n x \, dx$.   If either m or n is a positive odd integer, the other not necessarily positive nor an integer, put

$$\sin^2 x = 1 - \cos^2 x \quad \text{and} \quad \sin x \, dx = - \, d \cos x$$

or put

$$\cos^2 x = 1 - \sin^2 x \quad \text{and} \quad \cos x \, dx = d \sin x.$$

If both m and n are positive even integers, put

$$\sin^2 x = \tfrac{1}{2}(1 - \cos 2x), \quad \cos^2 x = \tfrac{1}{2}(1 + \cos 2x)$$

and

$$\sin x \cos x = \tfrac{1}{2} \sin 2x,$$

and similar expressions involving $2x$ instead of x, and so on. See also **450.9**.

465. $\displaystyle \int \sin mx \cos nx \, dx = - \frac{\cos (m-n)x}{2(m-n)} - \frac{\cos (m+n)x}{2(m+n)},$

$$[m^2 \neq n^2]. \qquad [\text{If } m^2 = n^2, \text{ see } \textbf{450.11}.]$$

470.1. $\displaystyle \int \frac{\cos x \, dx}{\sqrt{(1 + m^2 \sin^2 x)}} = \frac{1}{m} \log \{ m \sin x + \sqrt{(1 + m^2 \sin^2 x)} \}.$

470.2. $\displaystyle \int \frac{\cos x \, dx}{\sqrt{(1 - m^2 \sin^2 x)}} = \frac{1}{m} \sin^{-1} (m \sin x).$

470.3. $\displaystyle \int (\cos x) \sqrt{(1 + m^2 \sin^2 x)} \, dx$

$$= \frac{\sin x}{2} \sqrt{(1 + m^2 \sin^2 x)}$$

$$+ \frac{1}{2m} \log \{ m \sin x + \sqrt{(1 + m^2 \sin^2 x)} \}.$$

470.4. $\displaystyle \int (\cos x) \sqrt{(1 - m^2 \sin^2 x)} \, dx$

$$= \frac{\sin x}{2} \sqrt{(1 - m^2 \sin^2 x)} + \frac{1}{2m} \sin^{-1} (m \sin x).$$

475.1. $\displaystyle \int f(x, \sin x) \, dx = - \int f\left(\frac{\pi}{2} - y, \cos y \right) dy,$

where

$$y = \pi/2 - x.$$

475.2. $\displaystyle \int f(x, \cos x) \, dx = - \int f\left(\frac{\pi}{2} - y, \sin y \right) dy,$

where

$$y = \pi/2 - x.$$

Integrals Involving tan x

480.1. $\int \tan x \, dx = -\log |\cos x| = \log |\sec x|.$

[See **452.11** and **603.4.**]

480.2. $\int \tan^2 x \, dx = \tan x - x.$ [See **452.22.**]

480.3. $\int \tan^3 x \, dx = \frac{1}{2} \tan^2 x + \log |\cos x|.$ [See **452.33.**]

480.4. $\int \tan^4 x \, dx = \frac{1}{3} \tan^3 x - \tan x + x.$

480.9. $\int \tan^n x \, dx = \dfrac{\tan^{n-1} x}{n-1} - \int \tan^{n-2} x \, dx,$

[$n \neq 1.$ See **452.8**].

481.1. $\int x \tan x \, dx = \dfrac{x^3}{3} + \dfrac{x^5}{15} + \dfrac{2}{105} x^7 + \dfrac{17}{2835} x^9$

$+ \dfrac{62}{11 \times 2835} x^{11} + \cdots + \dfrac{2^{2n}(2^{2n} - 1)B_n}{(2n + 1)!} x^{2n+1} + \cdots,$

[$x^2 < \pi^2/4.$ See **415.03** and **45**].

481.2. $\int \dfrac{\tan x \, dx}{x} = x + \dfrac{x^3}{9} + \dfrac{2}{75} x^5 + \dfrac{17}{2205} x^7 + \dfrac{62}{9 \times 2835} x^9$

$+ \cdots + \dfrac{2^{2n}(2^{2n} - 1)B_n}{(2n - 1)(2n)!} x^{2n-1} + \cdots,$

[$x^2 < \pi^2/4.$ See **415.03** and **45**].

482.1. $\int \dfrac{dx}{\tan x \pm 1} = \pm \dfrac{x}{2} + \dfrac{1}{2} \log |\sin x \pm \cos x|.$

[See **455.05** and **.06.**]

482.2. $\int \dfrac{\tan x \, dx}{\tan x \pm 1} = \int \dfrac{dx}{1 \pm \operatorname{ctn} x} = \dfrac{x}{2} \mp \dfrac{1}{2} \log |\sin x \pm \cos x|.$

[See **455.03**, **455.04** and **492.1.**]

Integrals Involving ctn x

490.1. $\int ctn\, x\, dx = \log |\sin x|.$　　　[See **453.11** and **603.1.**]

490.2. $\int ctn^2\, x\, dx = -\, ctn\, x - x.$　　　[See **453.22.**]

490.3. $\int ctn^3\, x\, dx = -\, \frac{1}{2}\, ctn^2\, x - \log |\sin x|.$　　　[See **453.33.**]

490.4. $\int ctn^4\, x\, dx = -\, \frac{1}{3}\, ctn^3\, x + ctn\, x + x.$

490.9. $\int ctn^n\, x\, dx = -\, \dfrac{ctn^{n-1}\, x}{n-1} - \int ctn^{n-2}\, x\, dx,$

$$[n \neq 1. \quad \text{See } \mathbf{453.8}].$$

491.1. $\int x\, ctn\, x\, dx = x - \dfrac{x^3}{9} - \dfrac{x^5}{225} - \dfrac{2x^7}{6615} - \dfrac{x^9}{9 \times 4725}$

$$-\, \cdots - \frac{2^{2n}B_n}{(2n+1)!}\, x^{2n+1} - \cdots.$$

[See **415.04** and **45.**]

491.2. $\int \dfrac{ctn\, x\, dx}{x} = -\, \dfrac{1}{x} - \dfrac{x}{3} - \dfrac{x^3}{135} - \dfrac{2x^5}{4725} - \dfrac{x^7}{7 \times 4725}$

$$-\, \cdots - \frac{2^{2n}B_n}{(2n-1)(2n)!}\, x^{2n-1} - \cdots.$$

[See **415.04** and **45.**]

492.1. $\int \dfrac{dx}{1 \pm ctn\, x} = \int \dfrac{\tan x\, dx}{\tan x \pm 1}.$　　　[See **482.2.**]

492.2. $\int \dfrac{ctn\, x\, dx}{1 \pm ctn\, x} = \int \dfrac{dx}{\tan x \pm 1}.$　　　[See **482.1.**]

INVERSE TRIGONOMETRIC FUNCTIONS

500.

The following equations do not refer in general to the multiple values of the inverse trigonometric functions, but to the principal values. That is, $\sin^{-1} x$ and $\tan^{-1} x$ lie in the range from $-\pi/2$ to $\pi/2$ and $\cos^{-1} x$ and $\operatorname{ctn}^{-1} x$ in the range from 0 to π. Care should be taken in dealing with inverse functions and in integrating from one point to another. A process of curve plotting is frequently of assistance. Some of the graphs have more than one branch, and in general, integration should not be carried out from a point on one branch to a point on another branch.

501. $\quad \sin^{-1} x = x + \dfrac{x^3}{2 \cdot 3} + \dfrac{1 \cdot 3 x^5}{2 \cdot 4 \cdot 5} + \dfrac{1 \cdot 3 \cdot 5 x^7}{2 \cdot 4 \cdot 6 \cdot 7} + \cdots,$

$$[x^2 < 1. \quad -\pi/2 < \sin^{-1} x < \pi/2].$$

[Expand $1/\sqrt{(1 - x^2)}$ and then integrate it.]

502. $\quad \cos^{-1} x = \dfrac{\pi}{2} - \left(x + \dfrac{x^3}{2 \cdot 3} + \dfrac{1 \cdot 3 x^5}{2 \cdot 4 \cdot 5} + \dfrac{1 \cdot 3 \cdot 5 x^7}{2 \cdot 4 \cdot 6 \cdot 7} + \cdots \right),$

$$[x^2 < 1. \quad 0 < \cos^{-1} x < \pi].$$

503. $\quad \csc^{-1} x = \dfrac{1}{x} + \dfrac{1}{2 \cdot 3 x^3} + \dfrac{1 \cdot 3}{2 \cdot 4 \cdot 5 x^5} + \dfrac{1 \cdot 3 \cdot 5}{2 \cdot 4 \cdot 6 \cdot 7 x^7} + \cdots,$

$$[x^2 > 1. \quad -\pi/2 < \csc^{-1} x < \pi/2].$$

504. $\quad \sec^{-1} x = \dfrac{\pi}{2} - \left(\dfrac{1}{x} + \dfrac{1}{2 \cdot 3 x^3} + \dfrac{1 \cdot 3}{2 \cdot 4 \cdot 5 x^5} \right.$

$$\left. + \dfrac{1 \cdot 3 \cdot 5}{2 \cdot 4 \cdot 6 \cdot 7 x^7} + \cdots \right),$$

$$[x^2 > 1. \quad 0 < \sec^{-1} x < \pi].$$

505.1. $\quad \tan^{-1} x = x - \dfrac{x^3}{3} + \dfrac{x^5}{5} - \dfrac{x^7}{7} + \cdots, \qquad [x^2 < 1].$

[Expand $1/(1 + x^2)$ and then integrate it.]

505.2. $\quad \tan^{-1} x = \dfrac{\pi}{2} - \dfrac{1}{x} + \dfrac{1}{3 x^3} - \dfrac{1}{5 x^5} + \dfrac{1}{7 x^7} - \cdots, \qquad [x > 1].$

505.3. $\quad \tan^{-1} x = -\dfrac{\pi}{2} - \dfrac{1}{x} + \dfrac{1}{3 x^3} - \dfrac{1}{5 x^5} + \dfrac{1}{7 x^7} - \cdots,$

$$[x < -1].$$

505.4. $\tan^{-1} x = \dfrac{x}{1 + x^2}\left[1 + \dfrac{2}{3}\left(\dfrac{x^2}{1 + x^2}\right) + \dfrac{2 \cdot 4}{3 \cdot 5}\left(\dfrac{x^2}{1 + x^2}\right)^2\right.$

$$\left. + \dfrac{2 \cdot 4 \cdot 6}{3 \cdot 5 \cdot 7}\left(\dfrac{x^2}{1 + x^2}\right)^3 + \cdots\right], \qquad [\,x^2 < \infty\,].$$

[Ref. 31, p. 122.]

For these equations, $\tan^{-1} x$ is between $-\pi/2$ and $\pi/2$.

506.1. $\ctn^{-1} x = \dfrac{\pi}{2} - x + \dfrac{x^3}{3} - \dfrac{x^5}{5} + \dfrac{x^7}{7} - \cdots, \qquad [\,x^2 < 1\,].$

506.2. $\ctn^{-1} x = \dfrac{1}{x} - \dfrac{1}{3x^3} + \dfrac{1}{5x^5} - \dfrac{1}{7x^7} + \cdots, \qquad [\,x > 1\,].$

506.3. $\ctn^{-1} x = \pi + \dfrac{1}{x} - \dfrac{1}{3x^3} + \dfrac{1}{5x^5} - \dfrac{1}{7x^7} + \cdots,$

$$[\,x < -1\,].$$

507.10. $\sin^{-1}(x \pm iy) = n\pi + (-1)^n \sin^{-1}\dfrac{2x}{p + q}$

$$\pm i(-1)^n \cosh^{-1}\dfrac{p + q}{2}$$

taking the principal value of $\sin^{-1}$ (between $-\pi/2$ and $\pi/2$) and the positive values of $\cosh^{-1}$ and of p and q. The quantity $i = \sqrt{-1}$, and n is an integer or 0. The quantity x may be positive or negative but y is positive.

507.11. The quantity $p = \sqrt{(1 + x)^2 + y^2}$ (positive value),

and

507.12. $\qquad\qquad\qquad q = \sqrt{(1 - x)^2 + y^2}$ (positive value).

Note that if $y = 0$ and $x > 1$, $q = x - 1$ and $p + q = 2x$. If $y = 0$ and $x < 1$, $q = 1 - x$ and $p + q = 2$.

Alternative:

507.13a. $\sin^{-1} A = -i \log_\epsilon (\pm \sqrt{1 - A^2} + iA) + 2k\pi$

or

507.13b. $\qquad\qquad = i \log_\epsilon (\pm \sqrt{1 - A^2} - iA) + 2k\pi$

where A may be a complex quantity and k is an integer or 0.

For the square root of a complex quantity see **58** and for the logarithm see **604**. The two solutions a and b are identical. The one should be used, in any given case, which involves the numerical sum of two quantities instead of the difference, so as to obtain more convenient precise computation.

507.20. $\cos^{-1}(x + iy)$

$$= \pm \left(\cos^{-1}\frac{2x}{p+q} + 2k\pi - i\cosh^{-1}\frac{p+q}{2} \right).$$

507.21. $\cos^{-1}(x - iy)$

$$= \pm \left(\cos^{-1}\frac{2x}{p+q} + 2k\pi + i\cosh^{-1}\frac{p+q}{2} \right),$$

where y is positive, taking the principal value of $\cos^{-1}$ (between 0 and π) and the positive value of $\cosh^{-1}$. See **507.11** and **507.12**.

Alternative:

507.22a. $\cos^{-1}A = \mp\, i\log_\epsilon(A + \sqrt{A^2 - 1}) + 2k\pi$

or

507.22b. $= \pm\, i\log_\epsilon(A - \sqrt{A^2 - 1}) + 2k\pi$

where A may be a complex quantity. See note under **507.13**.

507.30. $\tan^{-1}(x + iy)$

$$= \frac{1}{2}\left\{ (2k+1)\pi - \tan^{-1}\frac{1+y}{x} - \tan^{-1}\frac{1-y}{x} \right\}$$

$$+ \frac{i}{4}\log_\epsilon\frac{(1+y)^2 + x^2}{(1-y)^2 + x^2},$$

where the principal values of $\tan^{-1}$ are taken (between $-\pi/2$ and $\pi/2$) and where x and y may be positive or negative.

Alternative:

507.31. $\tan^{-1}(x + iy) = \dfrac{i}{2}\log_\epsilon\dfrac{1 + y - ix}{1 - y + ix} + 2k\pi.$ [See **604**.]

[Ref. 46, Chap. XI.]

508. For small values of $\cos^{-1} x$,

$$\cos^{-1} x = \left[2 (1 - x) + \frac{1}{3} (1 - x)^2 + \frac{4}{45} (1 - x)^3 \right.$$
$$\left. + \frac{1}{35} (1 - x)^4 \ldots \right]^{1/2}$$

The last term used should be practically negligible. The numerical value of the square root may be taken from a large table of square roots, as in Refer. 65.

INVERSE TRIGONOMETRIC FUNCTIONS— DERIVATIVES

512.1. $\dfrac{d}{dx} \sin^{-1} \dfrac{x}{a} = \dfrac{1}{\sqrt{(a^2 - x^2)}}$, $\qquad \left[-\dfrac{\pi}{2} < \sin^{-1} \dfrac{x}{a} < \dfrac{\pi}{2} \right]$.

512.2. $\dfrac{d}{dx} \cos^{-1} \dfrac{x}{a} = \dfrac{-1}{\sqrt{(a^2 - x^2)}}$, $\qquad \left[0 < \cos^{-1} \dfrac{x}{a} < \pi \right]$.

512.3. $\dfrac{d}{dx} \tan^{-1} \dfrac{x}{a} = \dfrac{a}{a^2 + x^2}$.

512.4. $\dfrac{d}{dx} \operatorname{ctn}^{-1} \dfrac{x}{a} = \dfrac{-a}{a^2 + x^2}$.

512.5. $\dfrac{d}{dx} \sec^{-1} \dfrac{x}{a} = \dfrac{a}{x\sqrt{(x^2 - a^2)}}$, $\qquad \left[0 < \sec^{-1} \dfrac{x}{a} < \dfrac{\pi}{2} \right]$.

512.6. $\dfrac{d}{dx} \sec^{-1} \dfrac{x}{a} = \dfrac{-a}{x\sqrt{(x^2 - a^2)}}$, $\qquad \left[\dfrac{\pi}{2} < \sec^{-1} \dfrac{x}{a} < \pi \right]$.

512.7. $\dfrac{d}{dx} \csc^{-1} \dfrac{x}{a} = \dfrac{-a}{x\sqrt{(x^2 - a^2)}}$, $\qquad \left[0 < \csc^{-1} \dfrac{x}{a} < \dfrac{\pi}{2} \right]$.

512.8. $\dfrac{d}{dx} \csc^{-1} \dfrac{x}{a} = \dfrac{a}{x\sqrt{(x^2 - a^2)}}$, $\qquad \left[-\dfrac{\pi}{2} < \csc^{-1} \dfrac{x}{a} < 0 \right]$.

[Ref. 9, p. 3.]

[Except in 512.3 and 512.4, $a > 0$.]

INVERSE TRIGONOMETRIC FUNCTIONS— INTEGRALS $(a > 0)$

515. $\displaystyle\int \sin^{-1}\frac{x}{a}\,dx = x\sin^{-1}\frac{x}{a} + \sqrt{(a^2 - x^2)}.$

516. $\displaystyle\int \left(\sin^{-1}\frac{x}{a}\right)^2 dx = x\left(\sin^{-1}\frac{x}{a}\right)^2 - 2x + 2\sqrt{(a^2 - x^2)}\sin^{-1}\frac{x}{a}.$

517.1. $\displaystyle\int x\sin^{-1}\frac{x}{a}\,dx = \left(\frac{x^2}{2} - \frac{a^2}{4}\right)\sin^{-1}\frac{x}{a} + \frac{x}{4}\sqrt{(a^2 - x^2)}.$

517.2. $\displaystyle\int x^2\sin^{-1}\frac{x}{a}\,dx = \frac{x^3}{3}\sin^{-1}\frac{x}{a} + \frac{1}{9}(x^2 + 2a^2)\sqrt{(a^2 - x^2)}.$

517.3. $\displaystyle\int x^3\sin^{-1}\frac{x}{a}\,dx = \left(\frac{x^4}{4} - \frac{3a^4}{32}\right)\sin^{-1}\frac{x}{a}$
$$+ \frac{1}{32}(2x^3 + 3xa^2)\sqrt{(a^2 - x^2)}.$$

517.4. $\displaystyle\int x^4\sin^{-1}\frac{x}{a}\,dx = \frac{x^5}{5}\sin^{-1}\frac{x}{a}$
$$+ \frac{1}{75}(3x^4 + 4x^2a^2 + 8a^4)\sqrt{(a^2 - x^2)}.$$

517.5. $\displaystyle\int x^5\sin^{-1}\frac{x}{a}\,dx = \left(\frac{x^6}{6} - \frac{5a^6}{96}\right)\sin^{-1}\frac{x}{a}$
$$+ \frac{1}{288}(8x^5 + 10x^3a^2 + 15xa^4)\sqrt{(a^2 - x^2)}.$$

517.6. $\displaystyle\int x^6\sin^{-1}\frac{x}{a}\,dx = \frac{x^7}{7}\sin^{-1}\frac{x}{a}$
$$+ \frac{1}{245}(5x^6 + 6x^4a^2 + 8x^2a^4 + 16a^6)\sqrt{(a^2 - x^2)}.$$

517.9. $\displaystyle\int x^n\sin^{-1}\frac{x}{a}\,dx = \frac{x^{n+1}}{n+1}\sin^{-1}\frac{x}{a} - \frac{1}{n+1}\int \frac{x^{n+1}dx}{\sqrt{(a^2 - x^2)}},$
$$[n \neq -1]. \qquad [\text{See } 321\text{--}327.]$$

518.1. $\displaystyle\int \frac{1}{x}\sin^{-1}\frac{x}{a}\,dx = \frac{x}{a} + \frac{1}{2\cdot 3\cdot 3}\frac{x^3}{a^3} + \frac{1\cdot 3}{2\cdot 4\cdot 5\cdot 5}\frac{x^5}{a^5}$
$$+ \frac{1\cdot 3\cdot 5}{2\cdot 4\cdot 6\cdot 7\cdot 7}\frac{x^7}{a^7} + \cdots, \qquad [x^2 < a^2].$$

518.2. $\displaystyle\int \frac{1}{x^2} \sin^{-1}\frac{x}{a}\, dx = -\frac{1}{x}\sin^{-1}\frac{x}{a} - \frac{1}{a}\log\left|\frac{a + \sqrt{(a^2 - x^2)}}{x}\right|.$

518.3. $\displaystyle\int \frac{1}{x^3} \sin^{-1}\frac{x}{a}\, dx = -\frac{1}{2x^2}\sin^{-1}\frac{x}{a} - \frac{\sqrt{(a^2 - x^2)}}{2a^2 x}.$

518.4. $\displaystyle\int \frac{1}{x^4} \sin^{-1}\frac{x}{a}\, dx = -\frac{1}{3x^3}\sin^{-1}\frac{x}{a} - \frac{\sqrt{(a^2 - x^2)}}{6a^2 x^2}$

$$-\frac{1}{6a^3}\log\left|\frac{a + \sqrt{(a^2 - x^2)}}{x}\right|.$$

518.9. $\displaystyle\int \frac{1}{x^n} \sin^{-1}\frac{x}{a}\, dx = -\frac{1}{(n-1)x^{n-1}}\sin^{-1}\frac{x}{a}$

$$+\frac{1}{n-1}\int \frac{dx}{x^{n-1}\sqrt{(a^2 - x^2)}}, \qquad [n \neq 1].$$

$$[\text{See } \mathbf{341\text{--}346.}]$$

520. $\displaystyle\int \cos^{-1}\frac{x}{a}\, dx = x\cos^{-1}\frac{x}{a} - \sqrt{(a^2 - x^2)}.$

521. $\displaystyle\int \left(\cos^{-1}\frac{x}{a}\right)^2 dx = x\left(\cos^{-1}\frac{x}{a}\right)^2 - 2x - 2\sqrt{(a^2 - x^2)}\cos^{-1}\frac{x}{a}.$

522.1. $\displaystyle\int x\cos^{-1}\frac{x}{a}\, dx = \left(\frac{x^2}{2} - \frac{a^2}{4}\right)\cos^{-1}\frac{x}{a} - \frac{x}{4}\sqrt{(a^2 - x^2)}.$

522.2. $\displaystyle\int x^2\cos^{-1}\frac{x}{a}\, dx = \frac{x^3}{3}\cos^{-1}\frac{x}{a} - \frac{1}{9}(x^2 + 2a^2)\sqrt{(a^2 - x^2)}.$

522.3. $\displaystyle\int x^3\cos^{-1}\frac{x}{a}\, dx = \left(\frac{x^4}{4} - \frac{3a^4}{32}\right)\cos^{-1}\frac{x}{a}$

$$-\frac{1}{32}(2x^3 + 3xa^2)\sqrt{(a^2 - x^2)}.$$

522.4. $\displaystyle\int x^4\cos^{-1}\frac{x}{a}\, dx = \frac{x^5}{5}\cos^{-1}\frac{x}{a}$

$$-\frac{1}{75}(3x^4 + 4x^2 a^2 + 8a^4)\sqrt{(a^2 - x^2)}.$$

522.5. $\displaystyle\int x^5\cos^{-1}\frac{x}{a}\, dx = \left(\frac{x^6}{6} - \frac{5a^6}{96}\right)\cos^{-1}\frac{x}{a}$

$$-\frac{1}{288}(8x^5 + 10x^3 a^2 + 15xa^4)\sqrt{(a^2 - x^2)}.$$

522.6. $\displaystyle\int x^6 \cos^{-1}\frac{x}{a}\,dx = \frac{x^7}{7}\cos^{-1}\frac{x}{a}$

$$-\frac{1}{245}\,(5x^6 + 6x^4a^2 + 8x^2a^4 + 16a^6)\sqrt{(a^2 - x^2)}.$$

522.9. $\displaystyle\int x^n \cos^{-1}\frac{x}{a}\,dx = \frac{x^{n+1}}{n+1}\cos^{-1}\frac{x}{a} + \frac{1}{n+1}\int\frac{x^{n+1}dx}{\sqrt{(a^2-x^2)}},$

$$[n \neq -1]. \qquad [\text{See } 321\text{--}327.]$$

523.1. $\displaystyle\int \frac{1}{x}\cos^{-1}\frac{x}{a}\,dx = \frac{\pi}{2}\log|x| - \frac{x}{a} - \frac{1}{2\cdot 3\cdot 3}\frac{x^3}{a^3}$

$$-\frac{1\cdot 3}{2\cdot 4\cdot 5\cdot 5}\frac{x^5}{a^5} - \frac{1\cdot 3\cdot 5}{2\cdot 4\cdot 6\cdot 7\cdot 7}\frac{x^7}{a^7} - \cdots, \qquad [x^2 < a^2].$$

523.2. $\displaystyle\int \frac{1}{x^2}\cos^{-1}\frac{x}{a}\,dx = -\frac{1}{x}\cos^{-1}\frac{x}{a} + \frac{1}{a}\log\left|\frac{a + \sqrt{(a^2-x^2)}}{x}\right|.$

523.3. $\displaystyle\int \frac{1}{x^3}\cos^{-1}\frac{x}{a}\,dx = -\frac{1}{2x^2}\cos^{-1}\frac{x}{a} + \frac{\sqrt{(a^2-x^2)}}{2a^2x}.$

523.4. $\displaystyle\int \frac{1}{x^4}\cos^{-1}\frac{x}{a}\,dx = -\frac{1}{3x^3}\cos^{-1}\frac{x}{a} + \frac{\sqrt{(a^2-x^2)}}{6a^2x^2}$

$$+\frac{1}{6a^3}\log\left|\frac{a + \sqrt{(a^2-x^2)}}{x}\right|.$$

523.9. $\displaystyle\int \frac{1}{x^n}\cos^{-1}\frac{x}{a}\,dx = \frac{1}{(n-1)x^{n-1}}\cos^{-1}\frac{x}{a}$

$$-\frac{1}{n-1}\int\frac{dx}{x^{n-1}\sqrt{(a^2-x^2)}}, \qquad [n \neq 1].$$

$$[\text{See } 341\text{--}346.]$$

525. $\displaystyle\int \tan^{-1}\frac{x}{a}\,dx = x\tan^{-1}\frac{x}{a} - \frac{a}{2}\log(a^2 + x^2).$

525.1. $\displaystyle\int x\tan^{-1}\frac{x}{a}\,dx = \frac{1}{2}(x^2 + a^2)\tan^{-1}\frac{x}{a} - \frac{ax}{2}.$

525.2. $\displaystyle\int x^2\tan^{-1}\frac{x}{a}\,dx = \frac{x^3}{3}\tan^{-1}\frac{x}{a} - \frac{ax^2}{6} + \frac{a^3}{6}\log(a^2 + x^2).$

525.3. $\displaystyle\int x^3\tan^{-1}\frac{x}{a}\,dx = \frac{1}{4}(x^4 - a^4)\tan^{-1}\frac{x}{a} - \frac{ax^3}{12} + \frac{a^3x}{4}.$

525.4. $\int x^4 \tan^{-1} \frac{x}{a} \, dx = \frac{x^5}{5} \tan^{-1} \frac{x}{a} - \frac{ax^4}{20} + \frac{a^3 x^2}{10} - \frac{a^5}{10} \log (a^2 + x^2).$

525.5. $\int x^5 \tan^{-1} \frac{x}{a} \, dx = \frac{1}{6} (x^6 + a^6) \tan^{-1} \frac{x}{a} - \frac{ax^5}{30} + \frac{a^3 x^3}{18} - \frac{a^5 x}{6}.$

525.6. $\int x^6 \tan^{-1} \frac{x}{a} \, dx = \frac{x^7}{7} \tan^{-1} \frac{x}{a} - \frac{ax^6}{42} + \frac{a^3 x^4}{28} - \frac{a^5 x^2}{14}$

$$+ \frac{a^7}{14} \log (a^2 + x^2).$$

525.9. $\int x^n \tan^{-1} \frac{x}{a} \, dx = \frac{x^{n+1}}{n+1} \tan^{-1} \frac{x}{a} - \frac{a}{n+1} \int \frac{x^{n+1} dx}{a^2 + x^2},$

$$[n \neq -1]. \qquad [\text{See } 121\text{–}128.]$$

526.1. $\int \frac{1}{x} \tan^{-1} \frac{x}{a} \, dx = \frac{x}{a} - \frac{x^3}{3^2 a^3} + \frac{x^5}{5^2 a^5} - \frac{x^7}{7^2 a^7} + \cdots,$

$$[x^2 < a^2],$$

$$= \frac{\pi}{2} \log |x| + \frac{a}{x} - \frac{a^3}{3^2 x^3} + \frac{a^5}{5^2 x^5} - \frac{a^7}{7^2 x^7} + \cdots,$$

$$[x/a > 1],$$

$$= -\frac{\pi}{2} \log |x| + \frac{a}{x} - \frac{a^3}{3^2 x^3} + \frac{a^5}{5^2 x^5} - \frac{a^7}{7^2 x^7} + \cdots,$$

$$[x/a < -1],$$

For these equations, $\tan^{-1} (x/a)$ is between $-\pi/2$ and $\pi/2$.

526.2. $\int \frac{1}{x^2} \tan^{-1} \frac{x}{a} \, dx = -\frac{1}{x} \tan^{-1} \frac{x}{a} - \frac{1}{2a} \log \frac{a^2 + x^2}{x^2}.$

526.3. $\int \frac{1}{x^3} \tan^{-1} \frac{x}{a} \, dx = -\frac{1}{2} \left(\frac{1}{x^2} + \frac{1}{a^2} \right) \tan^{-1} \frac{x}{a} - \frac{1}{2ax}.$

526.4. $\int \frac{1}{x^4} \tan^{-1} \frac{x}{a} \, dx = -\frac{1}{3x^3} \tan^{-1} \frac{x}{a} - \frac{1}{6ax^2} + \frac{1}{6a^3} \log \frac{a^2 + x^2}{x^2}.$

526.5. $\int \frac{1}{x^5} \tan^{-1} \frac{x}{a} \, dx = \frac{1}{4} \left(\frac{1}{a^4} - \frac{1}{x^4} \right) \tan^{-1} \frac{x}{a} - \frac{1}{12ax^3} + \frac{1}{4a^3 x}.$

526.9. $\int \frac{1}{x^n} \tan^{-1} \frac{x}{a} \, dx = -\frac{1}{(n-1)x^{n-1}} \tan^{-1} \frac{x}{a}$

$$+ \frac{a}{n-1} \int \frac{dx}{x^{n-1}(a^2 + x^2)}, \qquad [n \neq 1].$$

$$[\text{See } 131\text{–}135.]$$

528. $\displaystyle\int \operatorname{ctn}^{-1}\frac{x}{a}\,dx = x\,\operatorname{ctn}^{-1}\frac{x}{a} + \frac{a}{2}\log(a^2+x^2).$

528.1. $\displaystyle\int x\,\operatorname{ctn}^{-1}\frac{x}{a}\,dx = \frac{1}{2}(x^2+a^2)\operatorname{ctn}^{-1}\frac{x}{a} + \frac{ax}{2}\cdot$

528.2. $\displaystyle\int x^2\,\operatorname{ctn}^{-1}\frac{x}{a}\,dx = \frac{x^3}{3}\operatorname{ctn}^{-1}\frac{x}{a} + \frac{ax^2}{6} - \frac{a^3}{6}\log(a^2+x^2).$

528.3. $\displaystyle\int x^3\,\operatorname{ctn}^{-1}\frac{x}{a}\,dx = \frac{1}{4}(x^4-a^4)\operatorname{ctn}^{-1}\frac{x}{a} + \frac{ax^3}{12} - \frac{a^3x}{4}\cdot$

528.4. $\displaystyle\int x^4\,\operatorname{ctn}^{-1}\frac{x}{a}\,dx = \frac{x^5}{5}\operatorname{ctn}^{-1}\frac{x}{a} + \frac{ax^4}{20} - \frac{a^3x^2}{10}$
$$+ \frac{a^5}{10}\log(a^2+x^2).$$

528.5. $\displaystyle\int x^5\,\operatorname{ctn}^{-1}\frac{x}{a}\,dx = \frac{1}{6}(x^6+a^6)\operatorname{ctn}^{-1}\frac{x}{a} + \frac{ax^5}{30} - \frac{a^3x^3}{18} + \frac{a^5x}{6}\cdot$

528.6. $\displaystyle\int x^6\,\operatorname{ctn}^{-1}\frac{x}{a}\,dx = \frac{x^7}{7}\operatorname{ctn}^{-1}\frac{x}{a} + \frac{ax^6}{42} - \frac{a^3x^4}{28} + \frac{a^5x^2}{14}$
$$- \frac{a^7}{14}\log(a^2+x^2).$$

528.9. $\displaystyle\int x^n\,\operatorname{ctn}^{-1}\frac{x}{a}\,dx = \frac{x^{n+1}}{n+1}\operatorname{ctn}^{-1}\frac{x}{a} + \frac{a}{n+1}\int \frac{x^{n+1}dx}{a^2+x^2},$
$$[n \neq -1].\qquad [\text{See } 121\text{–}128.]$$

529.1. $\displaystyle\int \frac{1}{x}\operatorname{ctn}^{-1}\frac{x}{a} = \frac{\pi}{2}\log|x| - \frac{x}{a} + \frac{x^3}{3^2a^3} - \frac{x^5}{5^2a^5} + \frac{x^7}{7^2a^7} - \cdots,$
$$[x^2 < a^2],$$

$$= -\frac{a}{x} + \frac{a^3}{3^2x^3} - \frac{a^5}{5^2x^5} + \frac{a^7}{7^2x^7} - \cdots,$$
$$[x/a > 1],$$

$$= \pi\log|x| - \frac{a}{x} + \frac{a^3}{3^2x^3} - \frac{a^5}{5^2x^5} + \frac{a^7}{7^2x^7} - \cdots,$$
$$[x/a < -1].$$

For these equations, $\operatorname{ctn}^{-1}(x/a)$ is between 0 and π.

529.2. $\displaystyle\int \frac{1}{x^2}\operatorname{ctn}^{-1}\frac{x}{a} = -\frac{1}{x}\operatorname{ctn}^{-1}\frac{x}{a} + \frac{1}{2a}\log\frac{a^2+x^2}{x^2}.$

529.3. $\int \frac{1}{x^3} \operatorname{ctn}^{-1} \frac{x}{a} = - \frac{1}{2x^2} \operatorname{ctn}^{-1} \frac{x}{a} + \frac{1}{2ax} + \frac{1}{2a^2} \tan^{-1} \frac{x}{a}$.

529.4. $\int \frac{1}{x^4} \operatorname{ctn}^{-1} \frac{x}{a} = - \frac{1}{3x^3} \operatorname{ctn}^{-1} \frac{x}{a} + \frac{1}{6ax^2} - \frac{1}{6a^3} \log \frac{a^2 + x^2}{x^2}$.

529.5. $\int \frac{1}{x^5} \operatorname{ctn}^{-1} \frac{x}{a} = - \frac{1}{4x^4} \operatorname{ctn}^{-1} \frac{x}{a} + \frac{1}{12ax^3} - \frac{1}{4a^3 x}$

$$- \frac{1}{4a^4} \tan^{-1} \frac{x}{a}.$$

529.9. $\int \frac{1}{x^n} \operatorname{ctn}^{-1} \frac{x}{a} = - \frac{1}{(n-1)x^{n-1}} \operatorname{ctn}^{-1} \frac{x}{a}$

$$- \frac{a}{n-1} \int \frac{dx}{x^{n-1}(a^2 + x^2)}, \qquad [n \neq 1]$$

$$[\text{See } 131\text{–}135.]$$

531. $\int \sec^{-1} \frac{x}{a} \, dx = x \sec^{-1} \frac{x}{a} - a \log |x + \sqrt{(x^2 - a^2)}|,$

$$[0 < \sec^{-1}(x/a) < \pi/2].$$

$$= x \sec^{-1} \frac{x}{a} + a \log |x + \sqrt{(x^2 - a^2)}|,$$

$$[\pi/2 < \sec^{-1}(x/a) < \pi].$$

531.1. $\int x \sec^{-1} \frac{x}{a} \, dx = \frac{x^2}{2} \sec^{-1} \frac{x}{a} - \frac{a}{2} \sqrt{(x^2 - a^2)},$

$$[0 < \sec^{-1}(x/a) < \pi/2].$$

$$= \frac{x^2}{2} \sec^{-1} \frac{x}{a} + \frac{a}{2} \sqrt{(x^2 - a^2)},$$

$$[\pi/2 < \sec^{-1}(x/a) < \pi].$$

531.2. $\int x^2 \sec^{-1} \frac{x}{a} \, dx$

$$= \frac{x^3}{3} \sec^{-1} \frac{x}{a} - \frac{ax}{6} \sqrt{(x^2 - a^2)} - \frac{a^3}{6} \log |x + \sqrt{(x^2 - a^2)}|,$$

$$[0 < \sec^{-1}(x/a) < \pi/2].$$

$$= \frac{x^3}{3} \sec^{-1} \frac{x}{a} + \frac{ax}{6} \sqrt{(x^2 - a^2)} + \frac{a^3}{6} \log |x + \sqrt{(x^2 - a^2)}|,$$

$$[\pi/2 < \sec^{-1}(x/a) < \pi].$$

531.9. $\displaystyle\int x^n \sec^{-1}\frac{x}{a}\,dx = \frac{x^{n+1}}{n+1}\sec^{-1}\frac{x}{a} - \frac{a}{n+1}\int \frac{x^n dx}{\sqrt{(x^2-a^2)}},$

$$[0 < \sec^{-1}(x/a) < \pi/2], \qquad [n \neq -1].$$

$$= \frac{x^{n+1}}{n+1}\sec^{-1}\frac{x}{a} + \frac{a}{n+1}\int \frac{x^n dx}{\sqrt{(x^2-a^2)}},$$

$$[\pi/2 < \sec^{-1}(x/a) < \pi], \qquad [n \neq -1].$$

532.1. $\displaystyle\int \frac{1}{x}\sec^{-1}\frac{x}{a}\,dx = \frac{\pi}{2}\log|x| + \frac{a}{x} + \frac{a^3}{2\cdot 3\cdot 3x^3} + \frac{1\cdot 3a^5}{2\cdot 4\cdot 5\cdot 5x^5}$

$$+ \frac{1\cdot 3\cdot 5a^7}{2\cdot 4\cdot 6\cdot 7\cdot 7x^7} + \cdots, \qquad [0 < \sec^{-1}(x/a) < \pi].$$

532.2. $\displaystyle\int \frac{1}{x^2}\sec^{-1}\frac{x}{a}\,dx = -\frac{1}{x}\sec^{-1}\frac{x}{a} + \frac{\sqrt{(x^2-a^2)}}{ax},$

$$[0 < \sec^{-1}(x/a) < \pi/2].$$

$$= -\frac{1}{x}\sec^{-1}\frac{x}{a} - \frac{\sqrt{(x^2-a^2)}}{ax},$$

$$[\pi/2 < \sec^{-1}(x/a) < \pi].$$

532.3. $\displaystyle\int \frac{1}{x^3}\sec^{-1}\frac{x}{a}\,dx$

$$= -\frac{1}{2x^2}\sec^{-1}\frac{x}{a} + \frac{\sqrt{(x^2-a^2)}}{4ax^2} + \frac{1}{4a^2}\cos^{-1}\left|\frac{a}{x}\right|,$$

$$[0 < \sec^{-1}(x/a) < \pi/2].$$

$$= -\frac{1}{2x^2}\sec^{-1}\frac{x}{a} - \frac{\sqrt{(x^2-a^2)}}{4ax^2} - \frac{1}{4a^2}\cos^{-1}\left|\frac{a}{x}\right|,$$

$$[\pi/2 < \sec^{-1}(x/a) < \pi].$$

532.4. $\displaystyle\int \frac{1}{x^4}\sec^{-1}\frac{x}{a}\,dx = -\frac{1}{3x^3}\sec^{-1}\frac{x}{a} + \frac{(2x^2+a^2)}{9a^3x^3}\sqrt{(x^2-a^2)},$

$$[0 < \sec^{-1}(x/a) < \pi/2].$$

$$= -\frac{1}{3x^3}\sec^{-1}\frac{x}{a} - \frac{(2x^2+a^2)}{9a^3x^3}\sqrt{(x^2-a^2)},$$

$$[\pi/2 < \sec^{-1}(x/a) < \pi].$$

532.9. $\displaystyle\int \frac{1}{x^n}\sec^{-1}\frac{x}{a}\,dx$

$$= -\frac{1}{(n-1)x^{n-1}}\sec^{-1}\frac{x}{a} + \frac{a}{n-1}\int \frac{dx}{x^n\sqrt{(x^2-a^2)}},$$

$$[0 < \sec^{-1}(x/a) < \pi/2], \qquad [n \neq 1].$$

$$= -\frac{1}{(n-1)x^{n-1}}\sec^{-1}\frac{x}{a} - \frac{a}{n-1}\int \frac{dx}{x^n\sqrt{(x^2-a^2)}},$$

$$[\pi/2 < \sec^{-1}(x/a) < \pi], \qquad [n \neq 1]$$

For 531–532.9, $x^2 > a^2$.

534. $\displaystyle\int \csc^{-1}\frac{x}{a}\,dx = x\,\csc^{-1}\frac{x}{a} + a\log|x + \sqrt{(x^2-a^2)}|,$

$$[0 < \csc^{-1}(x/a) < \pi/2].$$

$$= x\,\csc^{-1}\frac{x}{a} - a\log|x + \sqrt{(x^2-a^2)}|,$$

$$[-\pi/2 < \csc^{-1}(x/a) < 0].$$

534.1. $\displaystyle\int x\,\csc^{-1}\frac{x}{a}\,dx = \frac{x^2}{2}\csc^{-1}\frac{x}{a} + \frac{a}{2}\sqrt{(x^2-a^2)},$

$$[0 < \csc^{-1}(x/a) < \pi/2].$$

$$= \frac{x^2}{2}\csc^{-1}\frac{x}{a} - \frac{a}{2}\sqrt{(x^2-a^2)},$$

$$[-\pi/2 < \csc^{-1}(x/a) < 0].$$

534.2. $\displaystyle\int x^2\,\csc^{-1}\frac{x}{a}\,dx$

$$= \frac{x^3}{3}\csc^{-1}\frac{x}{a} + \frac{ax}{6}\sqrt{(x^2-a^2)} + \frac{a^3}{6}\log|x + \sqrt{(x^2-a^2)}|,$$

$$[0 < \csc^{-1}(x/a) < \pi/2].$$

$$= \frac{x^3}{3}\csc^{-1}\frac{x}{a} - \frac{ax}{6}\sqrt{(x^2-a^2)} - \frac{a^3}{6}\log|x + \sqrt{(x^2-a^2)}|,$$

$$[-\pi/2 < \csc^{-1}(x/a) < 0].$$

534.9. $\displaystyle\int x^n\,\csc^{-1}\frac{x}{a}\,dx = \frac{x^{n+1}}{n+1}\csc^{-1}\frac{x}{a} + \frac{a}{n+1}\int \frac{x^n\,dx}{\sqrt{(x^2-a^2)}},$

$$[0 < \csc^{-1}(x/a) < \pi/2],\qquad [n \neq -1].$$

$$= \frac{x^{n+1}}{n+1}\csc^{-1}\frac{x}{a} - \frac{a}{n+1}\int \frac{x^n\,dx}{\sqrt{(x^2-a^2)}},$$

$$[-\pi/2 < \csc^{-1}(x/a) < 0],\qquad [n \neq -1].$$

535.1. $\displaystyle\int \frac{1}{x}\csc^{-1}\frac{x}{a}\,dx = -\left(\frac{a}{x} + \frac{1}{2\cdot 3\cdot 3}\frac{a^3}{x^3} + \frac{1\cdot 3}{2\cdot 4\cdot 5\cdot 5}\frac{a^5}{x^5}\right.$

$$\left. + \frac{1\cdot 3\cdot 5}{2\cdot 4\cdot 6\cdot 7\cdot 7}\frac{a^7}{x^7} + \cdots\right),$$

$$[-\pi/2 < \csc^{-1}(x/a) < \pi/2].$$

535.2. $\displaystyle\int \frac{1}{x^2}\csc^{-1}\frac{x}{a}\,dx = -\frac{1}{x}\csc^{-1}\frac{x}{a} - \frac{\sqrt{(x^2-a^2)}}{ax},$

$$[0 < \csc^{-1}(x/a) < \pi/2].$$

$$= -\frac{1}{x}\csc^{-1}\frac{x}{a} + \frac{\sqrt{(x^2-a^2)}}{ax},$$

$$[-\pi/2 < \csc^{-1}(x/a) < 0].$$

535.3. $\int \frac{1}{x^3} \csc^{-1} \frac{x}{a} \, dx$

$$= -\frac{1}{2x^2} \csc^{-1} \frac{x}{a} - \frac{\sqrt{(x^2 - a^2)}}{4ax^2} - \frac{1}{4a^2} \cos^{-1} \left| \frac{a}{x} \right|,$$
$$[0 < \csc^{-1} (x/a) < \pi/2].$$

$$= -\frac{1}{2x^2} \csc^{-1} \frac{x}{a} + \frac{\sqrt{(x^2 - a^2)}}{4ax^2} + \frac{1}{4a^2} \cos^{-1} \left| \frac{a}{x} \right|,$$
$$[-\pi/2 < \csc^{-1} (x/a) < 0].$$

535.4. $\int \frac{1}{x^4} \csc^{-1} \frac{x}{a} \, dx = -\frac{1}{3x^3} \csc^{-1} \frac{x}{a} - \frac{(2x^2 + a^2)}{9a^3x^3} \sqrt{(x^2 - a^2)},$
$$[0 < \csc^{-1} (x/a) < \pi/2].$$

$$= -\frac{1}{3x^3} \csc^{-1} \frac{x}{a} + \frac{(2x^2 + a^2)}{9a^3x^3} \sqrt{(x^2 - a^2)},$$
$$[-\pi/2 < \csc^{-1} (x/a) < 0].$$

535.9. $\int \frac{1}{x^n} \csc^{-1} \frac{x}{a} \, dx$

$$= -\frac{1}{(n-1)x^{n-1}} \csc^{-1} \frac{x}{a} - \frac{a}{n-1} \int \frac{dx}{x^n \sqrt{(x^2 - a^2)}},$$
$$[0 < \csc^{-1} (x/a) < \pi/2], \qquad [n \neq 1].$$

$$= -\frac{1}{(n-1)x^{n-1}} \csc^{-1} \frac{x}{a} + \frac{a}{n-1} \int \frac{dx}{x^n \sqrt{(x^2 - a^2)}},$$
$$[-\pi/2 < \csc^{-1} (x/a) < 0], \qquad [n \neq 1].$$

For **534–535.9**, $x^2 > a^2$.

EXPONENTIAL FUNCTIONS

550. $\quad e^x = 1 + \dfrac{x}{1!} + \dfrac{x^2}{2!} + \dfrac{x^3}{3!} + \cdots + \dfrac{x^n}{n!} + \cdots, \qquad [x^2 < \infty]$.

550.1. $\quad a^x = e^{x \log a} = 1 + \dfrac{x \log a}{1!} + \dfrac{(x \log a)^2}{2!} + \cdots$

$$+ \dfrac{(x \log a)^n}{n!} + \cdots, \qquad [x^2 < \infty].$$

550.2. $\quad e^{-x} = 1 - \dfrac{x}{1!} + \dfrac{x^2}{2!} - \dfrac{x^3}{3!} + \dfrac{x^4}{4!} - \cdots, \qquad\qquad [x^2 < \infty]$.

551. $\quad \dfrac{x}{e^x - 1} = 1 - \dfrac{x}{2} + \dfrac{B_1 x^2}{2!} - \dfrac{B_2 x^4}{4!} + \dfrac{B_3 x^6}{6!} - \dfrac{B_4 x^8}{8!} + \cdots,$

$$[x^2 < 4\pi^2. \quad \text{See } \mathbf{45}]. \quad [\text{Ref. 34, p. 234.}]$$

552.1. $\quad e^{\sin u} = 1 + u + \dfrac{u^2}{2!} - \dfrac{3u^4}{4!} - \dfrac{8u^5}{5!} - \dfrac{3u^6}{6!} + \dfrac{56u^7}{7!} + \cdots,$

$$[u^2 < \infty].$$

552.2. $\quad e^{\cos u} = e \left[1 - \dfrac{u^2}{2!} + \dfrac{4u^4}{4!} - \dfrac{31u^6}{6!} + \cdots \right], \qquad [u^2 < \infty]$.

552.3. $\quad e^{\tan u} = 1 + u + \dfrac{u^2}{2!} + \dfrac{3u^3}{3!} + \dfrac{9u^4}{4!} + \dfrac{37u^5}{5!} + \cdots,$

$$[u^2 < \pi^2/4].$$

552.4. $\quad e^{\sin^{-1} u} = 1 + u + \dfrac{u^2}{2!} + \dfrac{2u^3}{3!} + \dfrac{5u^4}{4!} + \cdots, \qquad [u^2 < 1]$.

$$[\text{Ref. 5, p. 92–93.}]$$

552.5. $\quad e^{\tan^{-1} u} = 1 + u + \dfrac{u^2}{2!} - \dfrac{u^3}{3!} - \dfrac{7u^4}{4!} + \dfrac{5u^5}{5!} + \cdots,$

$$[u^2 < 1].$$

The term in u^n is $a_n u^n / n!$, where $a_{n+1} = a_n - n(n-1)a_{n-1}$.

$$[\text{Ref. 34, p. 164, No. 19.}]$$

552.6.　$e^{-x^2} + e^{-2^2 x^2} + e^{-3^2 x^2} + \cdots$

$$= -\frac{1}{2} + \frac{\sqrt{\pi}}{x}\left[\frac{1}{2} + e^{-\pi^2/x^2} + e^{-2^2\pi^2/x^2} + e^{-3^2\pi^2/x^2} + \cdots\right].$$

The second series may be more rapidly convergent than the first.

[Ref. 31, p. 129.]

553.　$\lim\limits_{x \to \infty} x^n e^{-x} = 0$, for all values of n.

[Ref. 8, p. 132.]

EXPONENTIAL FUNCTIONS—DERIVATIVES

563.　$\dfrac{de^x}{dx} = e^x$.　**563.1.**　$\dfrac{de^{ax}}{dx} = ae^{ax}$.　**563.2.**　$\dfrac{da^x}{dx} = a^x \log a$.

563.3.　$\dfrac{da^{cx}}{dx} = ca^{cx}\log a$.　　**563.4.**　$\dfrac{da^y}{dx} = a^y(\log a)\dfrac{dy}{dx}$,

where a is a constant.

563.5.　$\dfrac{du^y}{dx} = yu^{y-1}\dfrac{du}{dx} + u^y(\log u)\dfrac{dy}{dx}$.

563.6.　$\dfrac{dx^y}{dx} = yx^{y-1} + x^y(\log x)\dfrac{dy}{dx}$.

563.7.　$\dfrac{dx^x}{dx} = x^x(1 + \log x)$.

EXPONENTIAL FUNCTIONS—INTEGRALS

565.　$\displaystyle\int e^x dx = e^x$.　　**565.1.**　$\displaystyle\int e^{ax} dx = \frac{1}{a}e^{ax}$.

565.2.　$\displaystyle\int e^{-x} dx = -e^{-x}$.　**565.3.**　$\displaystyle\int a^x dx = a^x/\log a$.

566.　$\displaystyle\int f(e^{ax})dx = \frac{1}{a}\int \frac{f(z)dz}{z}$

where $z = e^{ax}$.　Note that

$$a^x = e^{x \log a}, \quad \text{and} \quad a^{cx} = e^{cx \log a}.$$

567.1.　$\displaystyle\int xe^{ax}dx = e^{ax}\left[\frac{x}{a} - \frac{1}{a^2}\right]$.

567.2. $\displaystyle\int x^2 e^{ax}dx = e^{ax}\left[\frac{x^2}{a} - \frac{2x}{a^2} + \frac{2}{a^3}\right].$

567.3. $\displaystyle\int x^3 e^{ax}dx = e^{ax}\left[\frac{x^3}{a} - \frac{3x^2}{a^2} + \frac{6x}{a^3} - \frac{6}{a^4}\right].$

567.8. $\displaystyle\int x^n e^{ax}dx = \frac{x^n e^{ax}}{a} - \frac{n}{a}\int x^{n-1}e^{ax}dx.$

567.9. $\displaystyle\int x^{\prime} e^{ax}dx = e^{ax}\left[\frac{x^n}{a} - \frac{nx^{n-1}}{a^2} + \frac{n(n-1)x^{n-2}}{a^3} - \cdots\right.$

$$\left. + (-1)^{n-1}\frac{n!\,x}{a^n} + (-1)^n\frac{n!}{a^{n+1}}\right], \qquad [n \geqq 0].$$

568.1. $\displaystyle\int \frac{e^{ax}dx}{x} = \log|x| + \frac{ax}{1!} + \frac{a^2 x^2}{2\cdot 2!} + \frac{a^3 x^3}{3\cdot 3!} + \cdots$

$$\cdots + \frac{a^n x^n}{n\cdot n!} + \cdots, \qquad [x^2 < \infty]$$

568.11. For $\displaystyle\int \frac{c^x dx}{x}$, note that $c^x = e^{x\log c}.$

568.2. $\displaystyle\int \frac{e^{ax}dx}{x^2} = -\frac{e^{ax}}{x} + a\int \frac{e^{ax}dx}{x}.$ [See 568.1.]

568.3. $\displaystyle\int \frac{e^{ax}dx}{x^3} = -\frac{e^{ax}}{2x^2} - \frac{ae^{ax}}{2x} + \frac{a^2}{2}\int \frac{e^{ax}dx}{x}.$ [See 568.1.]

568.8. $\displaystyle\int \frac{e^{ax}dx}{x^n} = -\frac{e^{ax}}{(n-1)x^{n-1}} + \frac{a}{n-1}\int \frac{e^{ax}dx}{x^{n-1}},$

$$[n > 1].$$

568.9. $\displaystyle\int \frac{e^{ax}dx}{x^n} = -\frac{e^{ax}}{(n-1)x^{n-1}} - \frac{ae^{ax}}{(n-1)(n-2)x^{n-2}} - \cdots$

$$-\frac{a^{n-2}e^{ax}}{(n-1)!\,x} + \frac{a^{n-1}}{(n-1)!}\int \frac{e^{ax}dx}{x},$$

$$[n > 1]. \qquad [\text{See } 568.1.]$$

569. $\displaystyle\int \frac{dx}{1+e^x} = x - \log(1+e^x) = \log\frac{e^x}{1+e^x}.$

569.1. $\displaystyle\int \frac{dx}{a+be^{px}} = \frac{x}{a} - \frac{1}{ap}\log|a+be^{px}|.$

570. $\displaystyle\int \frac{xe^x dx}{(1+x)^2} = \frac{e^x}{1+x}.$

570.1. $\displaystyle\int \frac{xe^{ax} dx}{(1+ax)^2} = \frac{e^{ax}}{a^2(1+ax)}.$

575.1. $\displaystyle\int e^{ax} \sin x \, dx = \frac{e^{ax}}{a^2+1} (a \sin x - \cos x).$

575.2. $\displaystyle\int e^{ax} \sin^2 x \, dx = \frac{e^{ax}}{a^2+4} \left(a \sin^2 x - 2 \sin x \cos x + \frac{2}{a} \right).$

575.3. $\displaystyle\int e^{ax} \sin^3 x \, dx = \frac{e^{ax}}{a^2+9} \left[a \sin^3 x - 3 \sin^2 x \cos x \right.$
$$\left. + \frac{6(a \sin x - \cos x)}{a^2+1} \right].$$

575.9. $\displaystyle\int e^{ax} \sin^n x \, dx = \frac{e^{ax} \sin^{n-1} x}{a^2+n^2} (a \sin x - n \cos x)$
$$+ \frac{n(n-1)}{a^2+n^2} \int e^{ax} \sin^{n-2} x \, dx.$$

576.1. $\displaystyle\int e^{ax} \cos x \, dx = \frac{e^{ax}}{a^2+1} (a \cos x + \sin x).$

576.2. $\displaystyle\int e^{ax} \cos^2 x \, dx = \frac{e^{ax}}{a^2+4} \left(a \cos^2 x + 2 \sin x \cos x + \frac{2}{a} \right).$

576.3. $\displaystyle\int e^{ax} \cos^3 x \, dx = \frac{e^{ax}}{a^2+9} \left[a \cos^3 x + 3 \sin x \cos^2 x \right.$
$$\left. + \frac{6(a \cos x + \sin x)}{a^2+1} \right].$$

576.9. $\displaystyle\int e^{ax} \cos^n x \, dx = \frac{e^{ax} \cos^{n-1} x}{a^2+n^2} (a \cos x + n \sin x)$
$$+ \frac{n(n-1)}{a^2+n^2} \int e^{ax} \cos^{n-2} x \, dx.$$
[Ref. 2, p. 141.]

577.1. $\displaystyle\int e^{ax} \sin nx \, dx = \frac{e^{ax}}{a^2+n^2} (a \sin nx - n \cos nx).$

577.2. $\displaystyle\int e^{ax} \cos nx \, dx = \frac{e^{ax}}{a^2+n^2} (a \cos nx + n \sin nx).$
[Ref. 7, p. 9.]

PROBABILITY INTEGRALS

585. Normal probability integral $= \dfrac{1}{\sqrt{(2\pi)}} \displaystyle\int_{-x}^{x} e^{-t^2/2}dt$

$$= \operatorname{erf} \frac{x}{\sqrt{2}} \qquad\qquad\qquad \text{[see 590]}$$

$$= x\left(\frac{2}{\pi}\right)^{1/2}\left[1 - \frac{x^2}{2\cdot 1!3} + \frac{x^4}{2^2\cdot 2!5} - \frac{x^6}{2^3\cdot 3!7} + \cdots\right]$$

$$[x^2 < \infty].$$

[See Table 1045.]

586. For large values of x, the following asymptotic series may be used:

$$\frac{1}{\sqrt{(2\pi)}}\int_{-x}^{x} e^{-t^2/2}dt$$

$$\approx 1 - \left(\frac{2}{\pi}\right)^{1/2}\frac{e^{-x^2/2}}{x}\left[1 - \frac{1}{x^2} + \frac{1\cdot 3}{x^4} - \frac{1\cdot 3\cdot 5}{x^6} + \frac{1\cdot 3\cdot 5\cdot 7}{x^8} - \cdots\right],$$

where $\approx$ denotes approximate equality. The error is less than the last term used.

590. Error function $= \operatorname{erf} x = \dfrac{2}{\sqrt{\pi}} \displaystyle\int_0^x e^{-t^2}dt$

$$= \frac{2x}{\sqrt{\pi}}\left[1 - \frac{x^2}{1!3} + \frac{x^4}{2!5} - \frac{x^6}{3!7} + \cdots\right] \quad [x^2 < \infty].$$

591. $\operatorname{Erf} x \approx 1 - \dfrac{e^{-x^2}}{x\sqrt{\pi}}\left[1 - \dfrac{1}{2x^2} + \dfrac{1\cdot 3}{2^2x^4} - \dfrac{1\cdot 3\cdot 5}{2^3x^6} + \cdots\right].$

592. Alternative form of the same series:

$$\operatorname{Erf} x \approx 1 - \frac{e^{-x^2}}{x\sqrt{\pi}}\left[1 - \frac{2!}{1!(2x)^2} + \frac{4!}{2!(2x)^4} - \frac{6!}{3!(2x)^6} + \cdots\right].$$

The error is less than the last term used.   [Ref. 9, p. 390.]

For tables of numerical values see Ref. 55e, Vols. I and II; Ref. 5. pp. 116–120; and Ref. 45, pp. 210–213.

LOGARITHMIC FUNCTIONS

In these algebraic expressions, log represents natural or Napierian logarithms. Other notations for natural logarithms are logn, ln and $\log_\epsilon$.

600. $\log_\epsilon a = 2.3026 \log_{10} a$.   **600.1.** $\log_{10} a = 0.43429 \log_\epsilon a$.

601. $\log (1 + x) = x - \dfrac{x^2}{2} + \dfrac{x^3}{3} - \dfrac{x^4}{4} + \dfrac{x^5}{5} - \cdots,$

$$[x^2 < 1 \text{ and } x = 1].$$

For $x = 1$, this gives a famous series:

601.01. $\log 2 = 1 - \dfrac{1}{2} + \dfrac{1}{3} - \dfrac{1}{4} + \dfrac{1}{5} - \cdots.$

601.1. $\log (1 - x) = - \left[x + \dfrac{x^2}{2} + \dfrac{x^3}{3} + \dfrac{x^4}{4} + \dfrac{x^5}{5} + \cdots \right],$

$$[x^2 < 1 \text{ and } x = -1].$$

601.2. $\log \left(\dfrac{1 + x}{1 - x} \right) = 2 \left[x + \dfrac{x^3}{3} + \dfrac{x^5}{5} + \dfrac{x^7}{7} + \cdots \right],$

$\qquad\qquad = 2 \tanh^{-1} x.$ $[x^2 < 1]$.   [See **708.**]

601.3. $\log \left(\dfrac{x + 1}{x - 1} \right) = 2 \left[\dfrac{1}{x} + \dfrac{1}{3x^3} + \dfrac{1}{5x^5} + \dfrac{1}{7x^7} + \cdots \right],$

$\qquad\qquad = 2 \operatorname{ctnh}^{-1} x.$ $[x^2 > 1]$.   [See **709.**]

601.4. $\log \left(\dfrac{x + 1}{x} \right) = 2 \left[\dfrac{1}{2x + 1} + \dfrac{1}{3(2x + 1)^3} \right.$

$$\left. + \dfrac{1}{5(2x + 1)^5} + \cdots \right],$$

$$[(2x + 1)^2 > 1]. \qquad [\text{Ref. 29, p. 6.}]$$

601.41. $\log (x + a) = \log x + 2 \left[\dfrac{a}{2x + a} + \dfrac{a^3}{3(2x + a)^3} \right.$

$$\left. + \dfrac{a^5}{5(2x + a)^5} + \cdots \right], \qquad [a^2 < (2x + a)^2].$$

601.5. $\log x = (x - 1) - \dfrac{(x - 1)^2}{2} + \dfrac{(x - 1)^3}{3}$

$$- \dfrac{(x - 1)^4}{4} + \cdots, \qquad [0 < x \leqq 2].$$

601.6. $\log x = \dfrac{x - 1}{x} + \dfrac{(x - 1)^2}{2x^2} + \dfrac{(x - 1)^3}{3x^3} + \cdots, \quad [x > \frac{1}{2}].$

601.7. $\log x = 2 \left[\dfrac{x-1}{x+1} + \dfrac{(x-1)^3}{3(x+1)^3} + \dfrac{(x-1)^5}{5(x+1)^5} + \cdots \right],$

$$[x > 0].$$

602.1. $\log \left[\dfrac{x}{a} + \sqrt{\left(\dfrac{x^2}{a^2} + 1 \right)} \right]$

$$= \dfrac{x}{a} - \dfrac{1}{2 \cdot 3} \dfrac{x^3}{a^3} + \dfrac{1 \cdot 3}{2 \cdot 4 \cdot 5} \dfrac{x^5}{a^5} - \dfrac{1 \cdot 3 \cdot 5}{2 \cdot 4 \cdot 6 \cdot 7} \dfrac{x^7}{a^7} + \cdots,$$

$$[x^2 < a^2].$$

$$= \log \dfrac{2x}{a} + \dfrac{1}{2 \cdot 2} \dfrac{a^2}{x^2} - \dfrac{1 \cdot 3}{2 \cdot 4 \cdot 4} \dfrac{a^4}{x^4} + \dfrac{1 \cdot 3 \cdot 5}{2 \cdot 4 \cdot 6 \cdot 6} \dfrac{a^6}{x^6} - \cdots,$$

$$[x/a > 1].$$

$$= - \log \left| \dfrac{2x}{a} \right| - \dfrac{1}{2 \cdot 2} \dfrac{a^2}{x^2} + \dfrac{1 \cdot 3}{2 \cdot 4 \cdot 4} \dfrac{a^4}{x^4} - \dfrac{1 \cdot 3 \cdot 5}{2 \cdot 4 \cdot 6 \cdot 6} \dfrac{a^6}{x^6} + \cdots,$$

$$[x/a < -1].$$

$$= \sinh^{-1} \dfrac{x}{a} = \operatorname{csch}^{-1} \dfrac{a}{x} \cdot \qquad \text{[See 706.]}$$

602.2. $\log \left[\sqrt{\left(\dfrac{x^2}{a^2} + 1 \right)} - \dfrac{x}{a} \right] = - \log \left[\dfrac{x}{a} + \sqrt{\left(\dfrac{x^2}{a^2} + 1 \right)} \right].$

Use the series in **602.1** and multiply by -1.

602.3. $\log \left[\dfrac{x}{a} + \sqrt{\left(\dfrac{x^2}{a^2} - 1 \right)} \right] - \log \dfrac{2x}{a} - \dfrac{1}{2 \cdot 2} \dfrac{a^2}{x^2} - \dfrac{1 \cdot 3}{2 \cdot 4 \cdot 4} \dfrac{a^4}{x^4}$

$$- \dfrac{1 \cdot 3 \cdot 5}{2 \cdot 4 \cdot 6 \cdot 6} \dfrac{a^6}{x^6} - \cdots, \qquad [x/a > 1].$$

$$\text{[See } \mathbf{260.01} \text{ and } \mathbf{707.}]$$

602.4. $\log \left[\dfrac{x}{a} - \sqrt{\left(\dfrac{x^2}{a^2} - 1 \right)} \right]$

$$= - \log \dfrac{2x}{a} + \dfrac{1}{2 \cdot 2} \dfrac{a^2}{x^2} + \dfrac{1 \cdot 3}{2 \cdot 4 \cdot 4} \dfrac{a^4}{x^4} + \dfrac{1 \cdot 3 \cdot 5}{2 \cdot 4 \cdot 6 \cdot 6} \dfrac{a^6}{x^6} + \cdots,$$

$$[x/a > 1].$$

$$= - \log \left[\dfrac{x}{a} + \sqrt{\left(\dfrac{x^2}{a^2} - 1 \right)} \right]. \qquad \text{[See } \mathbf{602.3} \text{ and } \mathbf{707.}]$$

602.5. $\log\left[\dfrac{a}{x} + \sqrt{\left(\dfrac{a^2}{x^2} + 1\right)}\right]$

$= \dfrac{a}{x} - \dfrac{1}{2\cdot 3}\dfrac{a^3}{x^3} + \dfrac{1\cdot 3}{2\cdot 4\cdot 5}\dfrac{a^5}{x^5} - \dfrac{1\cdot 3\cdot 5}{2\cdot 4\cdot 6\cdot 7}\dfrac{a^7}{x^7} + \cdots,$

$$[x^2 > a^2].$$

$= \log\dfrac{2a}{x} + \dfrac{1}{2\cdot 2}\dfrac{x^2}{a^2} - \dfrac{1\cdot 3}{2\cdot 4\cdot 4}\dfrac{x^4}{a^4} + \dfrac{1\cdot 3\cdot 5}{2\cdot 4\cdot 6\cdot 6}\dfrac{x^6}{a^6} - \cdots,$

$$[a/x > 1].$$

$= -\log\left|\dfrac{2a}{x}\right| - \dfrac{1}{2\cdot 2}\dfrac{x^2}{a^2} + \dfrac{1\cdot 3}{2\cdot 4\cdot 4}\dfrac{x^4}{a^4} - \dfrac{1\cdot 3\cdot 5}{2\cdot 4\cdot 6\cdot 6}\dfrac{x^6}{a^6} + \cdots,$

$$[a/x < -1].$$

$= \operatorname{csch}^{-1}\dfrac{x}{a} = \sinh^{-1}\dfrac{a}{x}.$ [See **602.1** and **711.**]

602.6. $\log\left[\sqrt{\left(\dfrac{a^2}{x^2} + 1\right)} - \dfrac{a}{x}\right] = -\log\left[\dfrac{a}{x} + \sqrt{\left(\dfrac{a^2}{x^2} + 1\right)}\right].$

Use the series in **602.5** and multiply by -1.

602.7. $\log\left[\dfrac{a}{x} + \sqrt{\left(\dfrac{a^2}{x^2} - 1\right)}\right] = \log\dfrac{2a}{x} - \dfrac{1}{2\cdot 2}\dfrac{x^2}{a^2} - \dfrac{1\cdot 3}{2\cdot 4\cdot 4}\dfrac{x^4}{a^4}$

$- \dfrac{1\cdot 3\cdot 5}{2\cdot 4\cdot 6\cdot 6}\dfrac{x^6}{a^6} - \cdots, \qquad [a/x > 1].$

602.8. $\log\left[\dfrac{a}{x} - \sqrt{\left(\dfrac{a^2}{x^2} - 1\right)}\right]$

$= -\log\dfrac{2a}{x} + \dfrac{1}{2\cdot 2}\dfrac{x^2}{a^2} + \dfrac{1\cdot 3}{2\cdot 4\cdot 4}\dfrac{x^4}{a^4} + \dfrac{1\cdot 3\cdot 5}{2\cdot 4\cdot 6\cdot 6}\dfrac{x^6}{a^6} + \cdots,$

$$[a/x > 1].$$

$= -\log\left[\dfrac{a}{x} + \sqrt{\left(\dfrac{a^2}{x^2} - 1\right)}\right].$ [See **710.**]

603.1. $\log|\sin x| = \log|x| - \dfrac{x^2}{6} - \dfrac{x^4}{180} - \dfrac{x^6}{2835} - \cdots$

$\cdots - \dfrac{2^{2n-1}B_n x^{2n}}{n(2n)!} - \cdots, \qquad [x^2 < \pi^2].$

[Integrate **415.04.** See **490.1** and **45.**]

603.2. $\log|\sin x| = -\log 2 - \cos 2x - \dfrac{\cos 4x}{2} - \dfrac{\cos 6x}{3} - \cdots,$

[Ref. 38, p. 275.] $[\sin x \neq 0].$

603.3. $\log \cos x = -\dfrac{x^2}{2} - \dfrac{x^4}{12} - \dfrac{x^6}{45} - \dfrac{17x^8}{2520} - \cdots$

$$\cdots - \frac{2^{2n-1}(2^{2n} - 1)B_n x^{2n}}{n(2n)!} - \cdots,$$

$[x^2 < \pi^2/4]$. [Integrate **415.03**. See **480.1** and **45**.]

603.4. $\log |\cos x| = -\log 2 + \cos 2x - \dfrac{\cos 4x}{2} + \dfrac{\cos 6x}{3} - \cdots,$

[Ref. 38, p. 275.] $[\cos x \neq 0]$.

603.5. $\log \cos x = -\dfrac{1}{2}\left[\sin^2 x + \dfrac{\sin^4 x}{2} + \dfrac{\sin^6 x}{3} \right.$

$$\left. + \frac{\sin^8 x}{4} + \cdots \right], \quad [x^2 < \pi^2/4].$$

603.6. $\log |\tan x| = \log |x| + \dfrac{x^2}{3} + \dfrac{7}{90}x^4 + \dfrac{62}{2835}x^6 + \cdots$

$$\cdots + \frac{2^{2n}(2^{2n-1} - 1)B_n x^{2n}}{n(2n)!} + \cdots,$$

$[x^2 < \pi^2/4]$. [See **415.06**, **432.10** and **45**.]

604. $\log (x + iy) = \log r + i(\theta + 2\pi k),$
where $r = \sqrt{(x^2 + y^2)}$, $\cos \theta = x/r$, $\sin \theta = y/r$, k is an integer
or 0, r is positive, $i = \sqrt{(-1)}$. [Ref. 5, p. 3.]

604.05. $x + iy = re^{i(\theta+2\pi k)}$. [$\theta$ in radians.] [See **604**.]

604.1. $\log (-1) - \log 1 + (2k + 1)\pi i$
$= (2k + 1)\pi i.$ [See **409.03**.]

605. $\lim\limits_{x \to 0} x \log x = 0.$ [See **72**.]

LOGARITHMIC FUNCTIONS—INTEGRALS

610. $\displaystyle\int \log x \, dx = x \log x - x.$

610.01. $\displaystyle\int \log (ax)dx = x \log (ax) - x.$

610.1. $\displaystyle\int x \log x \, dx = \dfrac{x^2}{2} \log x - \dfrac{x^2}{4} \cdot$

610.2. $\displaystyle\int x^2 \log x \, dx = \frac{x^3}{3} \log x - \frac{x^3}{9}.$

610.3. $\displaystyle\int x^3 \log x \, dx = \frac{x^4}{4} \log x - \frac{x^4}{16}.$

610.9. $\displaystyle\int x^p \log (ax) dx = \frac{x^{p+1}}{p+1} \log (ax) - \frac{x^{p+1}}{(p+1)^2},$
$$[p \neq -1].$$

611.1. $\displaystyle\int \frac{\log x}{x} \, dx = \frac{(\log x)^2}{2}.$

611.11. $\displaystyle\int \frac{\log (ax)}{x} \, dx = \frac{1}{2} \{\log (ax)\}^2.$

611.2. $\displaystyle\int \frac{\log x}{x^2} \, dx = -\frac{\log x}{x} - \frac{1}{x}.$

611.3. $\displaystyle\int \frac{\log x}{x^3} \, dx = -\frac{\log x}{2x^2} - \frac{1}{4x^2}.$

611.9. $\displaystyle\int \frac{\log (ax)}{x^p} \, dx = -\frac{\log (ax)}{(p-1)x^{p-1}} - \frac{1}{(p-1)^2 x^{p-1}},$
$$[p \neq 1]$$

612. $\displaystyle\int (\log x)^2 dx = x(\log x)^2 - 2x \log x + 2x.$

612.1. $\displaystyle\int x(\log x)^2 dx = \frac{x^2}{2} (\log x)^2 - \frac{x^2}{2} \log x + \frac{x^2}{4}.$

612.2. $\displaystyle\int x^2 (\log x)^2 dx = \frac{x^3}{3} (\log x)^2 - \frac{2x^3}{9} \log x + \frac{2x^3}{27}.$

612.9. $\displaystyle\int x^p (\log x)^2 dx = \frac{x^{p+1}}{p+1} (\log x)^2 - \frac{2x^{p+1}}{(p+1)^2} \log x$
$$+ \frac{2x^{p+1}}{(p+1)^3}, \qquad [p \neq -1]$$

613.1. $\displaystyle\int \frac{(\log x)^2 dx}{x} = \frac{(\log x)^3}{3}.$

613.2. $\displaystyle\int \frac{(\log x)^2 dx}{x^2} = -\frac{(\log x)^2}{x} - \frac{2 \log x}{x} - \frac{2}{x}.$

613.3. $\displaystyle\int \frac{(\log x)^2 dx}{x^3} = -\frac{(\log x)^2}{2x^2} - \frac{\log x}{2x^2} - \frac{1}{4x^2}.$

613.9. $\displaystyle\int \frac{(\log x)^2 dx}{x^p} = -\frac{(\log x)^2}{(p-1)x^{p-1}} - \frac{2\log x}{(p-1)^2 x^{p-1}}$

$$-\frac{2}{(p-1)^3 x^{p-1}}, \qquad [p \neq 1].$$

614. $\displaystyle\int (\log x)^3 dx = x(\log x)^3 - 3x(\log x)^2 + 6x\log x - 6x$

615. $\displaystyle\int (\log x)^q dx = x(\log x)^q - q\int (\log x)^{q-1} dx,$

$$[q \neq -1].$$

616.1. $\displaystyle\int \frac{(\log x)^q dx}{x} = \frac{(\log x)^{q+1}}{q+1}, \qquad [q \neq -1].$

616.2. $\displaystyle\int x^p (\log x)^q dx = \frac{x^{p+1}(\log x)^q}{p+1} - \frac{q}{p+1}\int x^p (\log x)^{q-1} dx,$

$$[p, q \neq -1].$$

616.3. $\displaystyle\int \frac{(\log x)^q dx}{x^p} = \frac{-(\log x)^q}{(p-1)x^{p-1}} + \frac{q}{p-1}\int \frac{(\log x)^{q-1} dx}{x^p},$

$$[p, -q \neq 1].$$

617. $\displaystyle\int \frac{dx}{\log x} = \log|\log x| + \log x + \frac{(\log x)^2}{2\cdot 2!}$

$$+\frac{(\log x)^3}{3\cdot 3!} + \cdots.$$

617.1. $\displaystyle\int \frac{x\,dx}{\log x} = \log|\log x| + 2\log x + \frac{(2\log x)^2}{2\cdot 2!}$

$$+\frac{(2\log x)^3}{3\cdot 3!} + \cdots.$$

617.2. $\displaystyle\int \frac{x^2 dx}{\log x} = \log|\log x| + 3\log x + \frac{(3\log x)^2}{2\cdot 2!}$

$$+\frac{(3\log x)^3}{3\cdot 3!} + \cdots.$$

617.9. $\displaystyle\int \frac{x^p dx}{\log x} = \log |\log x| + (p + 1) \log x$

$$+ \frac{(p + 1)^2 (\log x)^2}{2 \cdot 2!} + \frac{(p + 1)^3 (\log x)^3}{3 \cdot 3!} + \cdots,$$

$$\left[= \int \frac{e^y dy}{y} \text{ where } y = (p + 1) \log x. \quad \text{See } \mathbf{568.1} \right].$$

618.1. $\displaystyle\int \frac{dx}{x \log x} = \log |\log x|. \qquad [\text{Put } \log x = y, \, x = e^y.]$

618.2. $\displaystyle\int \frac{dx}{x^2 \log x} = \log |\log x| - \log x + \frac{(\log x)^2}{2 \cdot 2!}$

$$- \frac{(\log x)^3}{3 \cdot 3!} + \cdots.$$

618.3. $\displaystyle\int \frac{dx}{x^3 \log x} = \log |\log x| - 2 \log x + \frac{(2 \log x)^2}{2 \cdot 2!}$

$$- \frac{(2 \log x)^3}{3 \cdot 3!} + \cdots.$$

618.9. $\displaystyle\int \frac{dx}{x^p \log x} = \log |\log x| - (p - 1) \log x$

$$+ \frac{(p - 1)^2 (\log x)^2}{2 \cdot 2!} - \frac{(p - 1)^3 (\log x)^3}{3 \cdot 3!} + \cdots.$$

619.1. $\displaystyle\int \frac{dx}{x (\log x)^q} = \frac{-1}{(q - 1)(\log x)^{q-1}}, \qquad [q \neq 1].$

619.2. $\displaystyle\int \frac{x^p dx}{(\log x)^q} = \frac{-x^{p+1}}{(q - 1)(\log x)^{q-1}} + \frac{p + 1}{q - 1} \int \frac{x^p dx}{(\log x)^{q-1}},$

$$[q \neq 1].$$

619.3. $\displaystyle\int \frac{dx}{x^p (\log x)^q} = \frac{-1}{x^{p-1}(q - 1)(\log x)^{q-1}}$

$$- \frac{p - 1}{q - 1} \int \frac{dx}{x^p (\log x)^{q-1}}, \qquad [q \neq 1].$$

620. $\displaystyle\int \log (a + bx) dx = \frac{a + bx}{b} \log (a + bx) - x.$

620.1. $\displaystyle\int x \log (a + bx) dx = \frac{b^2 x^2 - a^2}{2b^2} \log (a + bx) + \frac{ax}{2b} - \frac{x^2}{4}.$

621.1. $\displaystyle\int \frac{\log{(a + bx)}dx}{x}$

$$= (\log a) \log x + \frac{bx}{a} - \frac{b^2x^2}{2^2a^2} + \frac{b^3x^3}{3^2a^3} - \frac{b^4x^4}{4^2a^4} + \cdots,$$

$$[b^2x^2 < a^2].$$

$$= \frac{(\log bx)^2}{2} - \frac{a}{bx} + \frac{a^2}{2^2b^2x^2} - \frac{a^3}{3^2b^3x^3} + \frac{a^4}{4^2b^4x^4} - \cdots,$$

$$[b^2x^2 > a^2].\qquad [\text{Ref. 5, No. 439.}]$$

621.2. $\displaystyle\int \frac{\log{(a + bx)}dx}{x^2} = \frac{b}{a}\log x - \left(\frac{1}{x} + \frac{b}{a}\right)\log{(a + bx)}.$

621.9. $\displaystyle\int \frac{\log{(a + bx)}dx}{x^p} = - \frac{\log{(a + bx)}}{(p - 1)x^{p-1}}$

$$+ \int \frac{b\,dx}{(p - 1)(a + bx)x^{p-1}},$$

$$[p \neq 1].\qquad [\text{See } 101\text{--}105.]$$

622. $\displaystyle\int \frac{\log x\,dx}{a + bx} = \frac{(\log x)\log{(a + bx)}}{b} - \int \frac{\log{(a + bx)}dx}{bx}.$

$$[\text{See } 621.1.]$$

623. $\displaystyle\int \log{(x^2 + a^2)}dx = x\log{(x^2 + a^2)} - 2x + 2a\tan^{-1}\frac{x}{a}.$

623.1. $\displaystyle\int x\log{(x^2 + a^2)}dx = \frac{1}{2}\left[(x^2 + a^2)\log{(x^2 + a^2)} - x^2\right].$

623.2. $\displaystyle\int x^2\log{(x^2 + a^2)}dx = \frac{1}{3}\left[x^3\log{(x^2 + a^2)} - \frac{2}{3}x^3\right.$

$$\left. + 2xa^2 - 2a^3\tan^{-1}\frac{x}{a}\right].$$

623.3. $\displaystyle\int x^3\log{(x^2 + a^2)}dx = \frac{1}{4}\left[(x^4 - a^4)\log{(x^2 + a^2)}\right.$

$$\left. - \frac{x^4}{2} + x^2a^2\right].$$

623.4. $\displaystyle\int x^4\log{(x^2 + a^2)}dx = \frac{1}{5}\left[x^5\log{(x^2 + a^2)} - \frac{2}{5}x^5\right.$

$$\left. + \frac{2}{3}x^3a^2 - 2xa^4 + 2a^5\tan^{-1}\frac{x}{a}\right].$$

623.5. $\displaystyle\int x^5 \log (x^2 + a^2) dx = \frac{1}{6}\Big[(x^6 + a^6) \log (x^2 + a^2)$

$$- \frac{x^6}{3} + \frac{x^4 a^2}{2} - x^2 a^4 \Big] \cdot$$

623.6. $\displaystyle\int x^6 \log (x^2 + a^2) dx = \frac{1}{7}\Big[x^7 \log (x^2 + a^2) - \frac{2}{7} x^7$

$$+ \frac{2}{5} x^5 a^2 - \frac{2}{3} x^3 a^4 + 2xa^6 - 2a^7 \tan^{-1} \frac{x}{a} \Big] \cdot$$

623.7. $\displaystyle\int x^7 \log (x^2 + a^2) dx = \frac{1}{8}\Big[(x^8 - a^8) \log (x^2 + a^2)$

$$- \frac{x^8}{4} + \frac{x^6 a^2}{3} - \frac{x^4 a^4}{2} + x^2 a^6 \Big] \cdot$$

624. $\displaystyle\int \log |x^2 - a^2| \, dx = x \log |x^2 - x^2| - 2x$

$$+ a \log \Big| \frac{x + a}{x - a} \Big| \cdot$$

524.1. $\displaystyle\int x \log |x^2 - a^2| \, dx = \frac{1}{2} \big[(x^2 - a^2) \log |x^2 - a^2| - x^2 \big].$

624.2. $\displaystyle\int x^2 \log |x^2 - a^2| \, dx = \frac{1}{3}\Big[x^3 \log |x^2 - a^2| - \frac{2}{3} x^3$

$$- 2xa^2 + a^3 \log \Big| \frac{x + a}{x - a} \Big| \Big] \cdot$$

624.3. $\displaystyle\int x^3 \log |x^2 - a^2| \, dx = \frac{1}{4}\Big[(x^4 - a^4) \log |x^2 - a^2|$

$$- \frac{x^4}{2} - x^2 a^2 \Big] \cdot$$

624.4. $\displaystyle\int x^4 \log |x^2 - a^2| \, dx = \frac{1}{5}\Big[x^5 \log |x^2 - a^2| - \frac{2}{5} x^5$

$$- \frac{2}{3} x^3 a^2 - 2xa^4 + a^5 \log \Big| \frac{x + a}{x - a} \Big| \Big] \cdot$$

624.5. $\displaystyle\int x^5 \log |x^2 - a^2| \, dx = \frac{1}{6}\Big[(x^6 - a^6) \log |x^2 - a^2|$

$$- \frac{x^6}{3} - \frac{x^4 a^2}{2} - x^2 a^4 \Big] \cdot$$

624.6. $\displaystyle\int x^6 \log |x^2 - a^2|\, dx = \frac{1}{7}\left[x^7 \log |x^2 - a^2| - \frac{2}{7}x^7 \right.$

$$\left. - \frac{2}{5}x^5 a^2 - \frac{2}{3}x^3 a^4 - 2xa^6 + a^7 \log\left|\frac{x+a}{x-a}\right| \right].$$

624.7. $\displaystyle\int x^7 \log |x^2 - a^2|\, dx = \frac{1}{8}\left[(x^8 - a^8) \log |x^2 - a^2| \right.$

$$\left. - \frac{x^8}{4} - \frac{x^6 a^2}{3} - \frac{x^4 a^4}{2} - x^2 a^6 \right].$$

When integrals of the type $\int x^p \log (a^2 - x^2)dx$ are required, these expressions can be used.

Integrals Involving $r = (x^2 + a^2)^{1/2}$

625. $\displaystyle\int \log (x + r)dx = x \log (x + r) - r.$ [See **730.**]

The positive value of r is to be taken.

625.1. $\displaystyle\int x \log (x + r)dx = \left(\frac{x^2}{2} + \frac{a^2}{4}\right) \log (x + r) - \frac{xr}{4}.$

[See **730.1.**]

625.2. $\displaystyle\int x^2 \log (x + r)dx = \frac{x^3}{3} \log (x + r) - \frac{r^3}{9} + \frac{a^2 r}{3}.$

[See **730.2.**]

625.3. $\displaystyle\int x^3 \log (x + r)dx = \left(\frac{x^4}{4} - \frac{3a^4}{32}\right) \log (x + r)$

$$- \frac{x^3 r}{16} + \frac{3}{32}a^2 xr.$$ [See **730.3.**]

625.4. $\displaystyle\int x^4 \log (x + r)dx = \frac{x^5}{5} \log (x + r) - \frac{r^5}{25}$

$$+ \frac{2}{15}a^2 r^3 - \frac{a^4 r}{5}.$$ [See **730.4.**]

625.9. $\displaystyle\int x^p \log (x + r)dx = \frac{x^{p+1}}{p + 1} \log (x + r)$

$$- \frac{1}{p + 1}\int \frac{x^{p+1}dx}{r},$$ $[p \neq -1]$.

[See **201.01–207.01** and **730.9.**]

626.1. $\int \frac{1}{x} \log \left[\frac{x}{a} + \sqrt{\left(\frac{x^2}{a^2} + 1 \right)} \right] dx$

$= \frac{x}{a} - \frac{1}{2 \cdot 3 \cdot 3} \frac{x^3}{a^3} + \frac{1 \cdot 3}{2 \cdot 4 \cdot 5 \cdot 5} \frac{x^5}{a^5} - \frac{1 \cdot 3 \cdot 5}{2 \cdot 4 \cdot 6 \cdot 7 \cdot 7} \frac{x^7}{a^7} + \cdots,$

$$[r^2 < a^2].$$

$= \frac{1}{2} \left(\log \frac{2x}{a} \right)^2 - \frac{1}{2^3} \frac{a^2}{x^2} + \frac{1 \cdot 3}{2 \cdot 4^3} \frac{a^4}{x^4} - \frac{1 \cdot 3 \cdot 5}{2 \cdot 4 \cdot 6^3} \frac{a^6}{x^6} + \cdots,$

$$[x/a > 1].$$

$= -\frac{1}{2} \left(\log \left| \frac{2x}{a} \right| \right)^2 + \frac{1}{2^3} \frac{a^2}{x^2} - \frac{1 \cdot 3}{2 \cdot 4^3} \frac{a^4}{x^4} + \frac{1 \cdot 3 \cdot 5}{2 \cdot 4 \cdot 6^3} \frac{a^6}{x^6} - \cdots,$

$$[x/a < -1]. \quad [\text{See } 731.1.]$$

626.2. $\int \frac{\log (x + r)}{x^2} = -\frac{\log (x + r)}{x} - \frac{1}{a} \log \left| \frac{a + r}{x} \right|,$

where $r = (x^2 + a^2)^{1/2}$. $\qquad\qquad$ [See **731.2.**]

626.3. $\int \frac{\log (x + r)}{x^3} = -\frac{\log (x + r)}{2x^2} - \frac{r}{2a^2 x}.$ $\quad$ [See **731.3.**]

626.9. $\int \frac{\log (x + r)}{x^p} = -\frac{\log (x + r)}{(p - 1)x^{p-1}} + \frac{1}{p - 1} \int \frac{dx}{x^{p-1}r},$

$$[p \neq 1]. \qquad [\text{See } 221.01\text{--}226.01 \text{ and } 731.9.]$$

Integrals Involving $s = (x^2 - a^2)^{1/2}$

627. $\int \log (x + s)dx = x \log (x + s) - s.$ $\qquad$ [See **732.**]

The positive value of s is to be taken.

627.1. $\int x \log (x + s)dx = \left(\frac{x^2}{2} - \frac{a^2}{4} \right) \log (x + s) - \frac{xs}{4}.$

$$[\text{See } 732.1.]$$

627.2. $\int x^2 \log (x + s)dx = \frac{x^3}{3} \log (x + s) - \frac{s^3}{9} - \frac{a^2 s}{3}.$

$$[\text{See } 732.2.]$$

627.3. $\int x^3 \log (x + s)dx = \left(\frac{x^4}{4} - \frac{3a^4}{32} \right) \log (x + s)$

$$- \frac{x^3 s}{16} - \frac{3}{32} a^2 xs. \quad [\text{See } 732.3.]$$

627.4. $\displaystyle\int x^4 \log (x + s)dx = \frac{x^5}{5} \log (x + s) - \frac{s^5}{25}$

$$- \frac{2}{15} a^2 s^3 - \frac{a^4 s}{5} . \qquad \text{[See 732.4.]}$$

627.9. $\displaystyle\int x^p \log (x + s)dx = \frac{x^{p+1}}{p + 1} \log (x + s)$

$$- \frac{1}{p + 1} \int \frac{x^{p+1}dx}{s}, \qquad [p \neq - 1].$$
$$\text{[See 261.01–267.01 and 732.9.]}$$

628.1. $\displaystyle\int \frac{1}{x} \log \left[\frac{x}{a} + \sqrt{\left(\frac{x^2}{a^2} - 1\right)} \right] dx$

$$= \frac{1}{2} \left(\log \frac{2x}{a} \right)^2 + \frac{1}{2^3} \frac{a^2}{x^2} + \frac{1\cdot 3}{2\cdot 4^3} \frac{a^4}{x^4} + \frac{1\cdot 3\cdot 5}{2\cdot 4\cdot 6^3} \frac{a^6}{x^6} + \cdots,$$
$$[x/a > 1]. \qquad \text{[See 733.1.]}$$

628.2. $\displaystyle\int \frac{\log (x + s)}{x^2} dx = - \frac{\log (x + s)}{x} + \frac{1}{a} \sec^{-1} \left| \frac{x}{a} \right|,$

$$[0 < \sec^{-1} |x/a| < \pi/2]. \qquad \text{[See 733.2.]}$$

628.3. $\displaystyle\int \frac{\log (x + s)}{x^3} dx = - \frac{\log (x + s)}{2x^2} + \frac{s}{2a^2 x} .$

$$\text{[See 733.3.]}$$

628.9. $\displaystyle\int \frac{\log (x + s)}{x^p} dx = - \frac{\log (x + s)}{(p - 1)x^{p-1}} + \frac{1}{p - 1} \int \frac{dx}{x^{p-1}s} ,$

$$[p \neq 1]. \qquad \text{[See 281.01–284.01 and 733.9.]}$$

630.1. $\displaystyle\int \log \sin x \, dx = x \log x - x - \frac{x^3}{18} - \frac{x^5}{900}$

$$- \frac{x^7}{19845} - \cdots - \frac{2^{2n-1}B_n x^{2n+1}}{n(2n + 1)!} - \cdots, \qquad [0 < x < \pi].$$
$$\text{[See 45.]} \qquad \text{[Integrate 603.1.]}$$

$$= - x \log 2 - \frac{\sin 2x}{2} - \frac{\sin 4x}{2\cdot 2^2} - \frac{\sin 6x}{2\cdot 3^2} - \cdots,$$
$$[0 < x < \pi]. \qquad \text{[Integrate 603.2.]}$$

630.2. $\displaystyle\int \log \cos x \, dx = -\frac{x^3}{6} - \frac{x^5}{60} - \frac{x^7}{315}$

$$-\frac{17x^9}{22680} - \cdots - \frac{2^{2n-1}(2^{2n}-1)B_n}{n(2n+1)!} x^{2n+1} - \cdots,$$

$[x^2 < \pi^2/4]$.   [See **45**.]   [Integrate **603.3**.]

$$= -x \log 2 + \frac{\sin 2x}{2} - \frac{\sin 4x}{2 \cdot 2^2} + \frac{\sin 6x}{2 \cdot 3^2} - \cdots,$$

$[x^2 < \pi^2/4]$.   [Integrate **603.4**.]

630.3. $\displaystyle\int \log \tan x \, dx = x \log x - x + \frac{x^3}{9} + \frac{7x^5}{450}$

$$+\frac{62x^7}{19845} + \cdots + \frac{2^{2n}(2^{2n-1}-1)B_n}{n(2n+1)!} x^{2n+1} + \cdots,$$

$[0 < x < \pi/2]$.   [See **45**.]   [Integrate **603.6**.]

631.1. $\displaystyle\int \sin \log x \, dx = \frac{1}{2} x \sin \log x - \frac{1}{2} x \cos \log x.$

631.2. $\displaystyle\int \cos \log x \, dx = \frac{1}{2} x \sin \log x + \frac{1}{2} x \cos \log x.$

632. $\displaystyle\int e^{ax} \log x \, dx = \frac{1}{a} e^{ax} \log x - \frac{1}{a} \int \frac{e^{ax}}{x} \, dx.$ [See **568.1**.]

[Ref. 20, p. 46, No. 106.]

Lambda Function and Gudermannian

640.   If $x = \log \tan \left(\dfrac{\pi}{4} + \dfrac{\theta}{2}\right) = \log (\sec \theta + \tan \theta)$

$\theta = \text{gd } x = $ the **gudermannian** of $x = 2 \tan^{-1} e^x - \dfrac{\pi}{2}$.

641. $x = \text{gd}^{-1} \theta = \lambda(\theta)$, the **lambda** function.

642.1. $\sinh x = \tan \theta.$ **642.2.** $\cosh x = \sec \theta.$

642.3. $\tanh x = \sin \theta.$ **642.4.** $\tanh (x/2) = \tan (\theta/2).$

642.5. $\dfrac{d \, \text{gd } x}{dx} = \text{sech } x.$ **642.6.** $\dfrac{d \, \text{gd}^{-1} x}{dx} = \sec x,$

$$[-\pi/2 < \theta < \pi/2].$$

If θ is tabulated for values of x, the hyperbolic functions may be obtained from a table of circular functions.

HYPERBOLIC FUNCTIONS

650.01. $\cosh^2 x - \sinh^2 x = 1.$

650.02. $\sinh x = \sqrt{(\cosh^2 x - 1)},$ $\qquad\qquad$ $[x > 0].$
$\qquad\quad = -\sqrt{(\cosh^2 x - 1)},$ $\qquad\qquad$ $[x < 0].$

650.03. $\cosh x = \sqrt{(1 + \sinh^2 x)}.$ $\quad$ **650.05.** $\operatorname{sech} x = 1/\cosh x.$

650.04. $\tanh x = \sinh x/\cosh x.$ $\qquad$ **650.06.** $\operatorname{csch} x = 1/\sinh x.$

650.07. $\tanh^2 x + \operatorname{sech}^2 x = 1.$

650.08. $\operatorname{ctnh}^2 x - \operatorname{csch}^2 x = 1.$

650.09. $\sinh (-x) = -\sinh x.$

650.10. $\cosh (-x) = \cosh x.$

650.11. $\tanh (-x) = -\tanh x.$

651.01. $\sinh (x \pm y) = \sinh x \cosh y \pm \cosh x \sinh y.$

651.02. $\cosh (x \pm y) = \cosh x \cosh y \pm \sinh x \sinh y.$

651.03. $2 \sinh x \cosh y = \sinh (x + y) + \sinh (x - y).$

651.04. $2 \cosh x \cosh y = \cosh (x + y) + \cosh (x - y).$

651.05. $2 \sinh x \sinh y = \cosh (x + y) - \cosh (x - y).$

651.06. $\sinh x + \sinh y = 2 \sinh \dfrac{x + y}{2} \cosh \dfrac{x - y}{2}.$

651.07. $\sinh x - \sinh y = 2 \sinh \dfrac{x - y}{2} \cosh \dfrac{x + y}{2}.$

651.08. $\cosh x + \cosh y = 2 \cosh \dfrac{x + y}{2} \cosh \dfrac{x - y}{2}.$

651.09. $\cosh x - \cosh y = 2 \sinh \dfrac{x + y}{2} \sinh \dfrac{x - y}{2}.$

651.10. $\sinh^2 x - \sinh^2 y = \sinh (x + y) \sinh (x - y)$
$\qquad\qquad\qquad\quad = \cosh^2 x - \cosh^2 y.$

651.11. $\sinh^2 x + \cosh^2 y = \cosh (x + y) \cosh (x - y)$
$\qquad\qquad\qquad\quad = \cosh^2 x + \sinh^2 y.$

651.12. $\operatorname{csch}^2 x - \operatorname{sech}^2 x = \operatorname{csch}^2 x \operatorname{sech}^2 x = \dfrac{1}{\sinh^2 x \cosh^2 x}.$

651.13. $(\sinh x + \cosh x)^n = \sinh nx + \cosh nx.$

651.14. $\dfrac{1}{\sinh x + \cosh x} = \cosh x - \sinh x.$

652.12. $\sinh 2x = 2 \sinh x \cosh x.$

652.13. $\sinh 3x = 3 \sinh x + 4 \sinh^3 x.$

652.22. $\cosh 2x = \cosh^2 x + \sinh^2 x$
$$= 2 \sinh^2 x + 1 = 2 \cosh^2 x - 1.$$

652.23. $\cosh 3x = 4 \cosh^3 x - 3 \cosh x.$

652.3. $\sinh^2 x = \frac{1}{2}(\cosh 2x - 1).$

652.4. $\cosh^2 x = \frac{1}{2}(\cosh 2x + 1).$

652.5. $\sinh \dfrac{x}{2} = \sqrt{\{\frac{1}{2}(\cosh x - 1)\}},$ $\qquad [x > 0].$

$\qquad\qquad = - \sqrt{\{\frac{1}{2}(\cosh x - 1)\}},$ $\qquad [x < 0].$

652.6. $\cosh \dfrac{x}{2} = \sqrt{\{\frac{1}{2}(\cosh x + 1)\}}.$

653.1. $\tanh (x \pm y) = \dfrac{\tanh x \pm \tanh y}{1 \pm \tanh x \tanh y}.$

653.2. $\tanh \left(\dfrac{x \pm y}{2}\right) = \dfrac{\sinh x \pm \sinh y}{\cosh x + \cosh y}.$

653.3. $\tanh 2x = \dfrac{2 \tanh x}{1 + \tanh^2 x}.$

653.4. $\tanh x \pm \tanh y = \dfrac{\sinh (x \pm y)}{\cosh x \cosh y}.$

653.5. $\tanh \dfrac{x}{2} = \dfrac{\cosh x - 1}{\sinh x} = \dfrac{\sinh x}{\cosh x + 1}.$

653.6. $\text{ctnh} (x \pm y) = \dfrac{\text{ctnh}\, x\, \text{ctnh}\, y \pm 1}{\text{ctnh}\, y \pm \text{ctnh}\, x}.$

653.7. $\text{ctnh}\, 2x = \dfrac{\text{ctnh}^2 x + 1}{2 \,\text{ctnh}\, x}.$

653.8. $\text{ctnh} \dfrac{x}{2} = \dfrac{\sinh x}{\cosh x - 1} = \dfrac{\cosh x + 1}{\sinh x}.$

654.1. $\quad \sinh x = \dfrac{1}{2}\left(e^x - e^{-x}\right)$

$$= \dfrac{1}{2}\left(\log_\epsilon^{-1} x - \dfrac{1}{\log_\epsilon^{-1} x}\right),$$

where $\log_\epsilon^{-1}$ denotes the natural anti-logarithm. This may be taken from a table of natural logarithms if series 550 is slowly convergent as with large values of x. By noting that $\log_\epsilon^{-1} x = \log_{10}^{-1}(.4343x)$, a table of common logarithms can be used.

654.2. $\quad \cosh x = \dfrac{1}{2}\left(e^x + e^{-x}\right)_{;}$

$$= \dfrac{1}{2}\left(\log_\epsilon^{-1} x + \dfrac{1}{\log_\epsilon^{-1} x}\right).$$

[See note under **654.1.**]

654.3. $\quad \tanh x = \dfrac{e^x - e^{-x}}{e^x + e^{-x}} = \dfrac{e^{2x} - 1}{e^{2x} + 1}.$

654.4. $\quad \cosh x + \sinh x = e^x.$ **654.5.** $\quad \cosh x - \sinh x = e^{-x}.$

654.6. $\quad \sinh(ix) = i\sin x.$ **654.7.** $\quad \cosh(ix) = \cos x.$

654.8. $\quad \tanh(ix) = i\tan x.$

655.1. $\quad \sinh(x \pm iy) = \sinh x \cos y \pm i \cosh x \sin y.$

655.2. $\quad \cosh(x \pm iy) = \cosh x \cos y \pm i \sinh x \sin y.$

655.3. $\quad \tanh(x \pm iy) = \dfrac{\sinh 2x \pm i \sin 2y}{\cosh 2x + \cos 2y}.$

655.4. $\quad \mathrm{ctnh}(x \pm iy) = \dfrac{\sinh 2x \mp i \sin 2y}{\cosh 2x - \cos 2y}.$

656.1. $\quad \sinh 0 = 0.$ **656.2.** $\quad \cosh 0 = 1.$ **656.3.** $\quad \tanh 0 = 0.$

657.1. $\quad \sinh x = x + \dfrac{x^3}{3!} + \dfrac{x^5}{5!} + \dfrac{x^7}{7!} + \cdots, \qquad [x^2 < \infty].$

657.2. $\quad \cosh x = 1 + \dfrac{x^2}{2!} + \dfrac{x^4}{4!} + \dfrac{x^6}{6!} + \cdots, \qquad [x^2 < \infty].$

657.3. $\quad \tanh x = x - \dfrac{x^3}{3} + \dfrac{2}{15}x^5 - \dfrac{17}{315}x^7 + \dfrac{62}{2835}x^9$

$$- \cdots + \dfrac{(-1)^{n-1}2^{2n}(2^{2n} - 1)}{(2n)!} B_n x^{2n-1} + \cdots,$$

$$[x^2 < \pi^2/4. \quad \text{See } \mathbf{45}].$$

657.4.   For large values of x,

$$\tanh x = 1 - \frac{2}{e^{2x}} + \frac{2}{e^{4x}} - \frac{2}{e^{6x}} + \cdots.$$

657.5. $\operatorname{ctnh} x = \dfrac{1}{x} + \dfrac{x}{3} - \dfrac{x^3}{45} + \dfrac{2x^5}{945} - \dfrac{x^7}{4725} + \cdots$

$$+ \frac{(-1)^{n-1}2^{2n}}{(2n)!} B_n x^{2n-1} + \cdots,$$

$$[x^2 < \pi^2. \quad \text{See } 45].$$

657.6.   For large values of x,

$$\operatorname{ctnh} x = 1 + \frac{2}{e^{2x}} + \frac{2}{e^{4x}} + \frac{2}{e^{6x}} + \cdots.$$

657.7. $\operatorname{sech} x = 1 - \dfrac{x^2}{2!} + \dfrac{5}{4!} x^4 - \dfrac{61}{6!} x^6 + \dfrac{1385}{8!} x^8 - \cdots$

$$+ \frac{(-1)^n}{(2n)!} E_n x^{2n} + \cdots,$$

$$[x^2 < \pi^2/4. \quad \text{See. } 45].$$

657.8. $\operatorname{csch} x = \dfrac{1}{x} - \dfrac{x}{6} + \dfrac{7x^3}{360} - \dfrac{31x^5}{15120} + \cdots$

$$+ \frac{2(-1)^n(2^{2n-1} - 1)}{(2n)!} B_n x^{2n-1} + \cdots,$$

$$[x^2 < \pi^2. \quad \text{See } 45].$$

HYPERBOLIC FUNCTIONS—DERIVATIVES

667.1. $\dfrac{d \sinh x}{dx} = \cosh x.$ **667.3.** $\dfrac{d \tanh x}{dx} = \operatorname{sech}^2 x.$

667.2. $\dfrac{d \cosh x}{dx} = \sinh x.$ **667.4.** $\dfrac{d \operatorname{ctnh} x}{dx} = - \operatorname{csch}^2 x.$

667.5. $\dfrac{d \operatorname{sech} x}{dx} = - \operatorname{sech} x \tanh x.$

667.6. $\dfrac{d \operatorname{csch} x}{dx} = - \operatorname{csch} x \operatorname{ctnh} x.$

HYPERBOLIC FUNCTIONS—INTEGRALS

670. An integral of a trigonometric function often can be changed into the corresponding integral of a hyperbolic function by changing x to ix and substituting

$$\sin (ix) = i \sinh x, \quad \cos (ix) = \cosh x, \quad \tan (ix) = i \tanh x, \text{ etc.}$$

[See 408.10–.15.]

This substitution is useful also with other classes of formulas.

Integrals Involving sinh x

671.10. $\displaystyle\int \sinh x \, dx = \cosh x.$

671.101. $\displaystyle\int \sinh \frac{x}{a} \, dx = a \cosh \frac{x}{a}.$

671.11. $\displaystyle\int x \sinh x \, dx = x \cosh x - \sinh x.$

671.12. $\displaystyle\int x^2 \sinh x \, dx = (x^2 + 2) \cosh x - 2x \sinh x.$

671.13. $\displaystyle\int x^3 \sinh x \, dx = (x^3 + 6x) \cosh x - (3x^2 + 6) \sinh x.$

671.19. $\displaystyle\int x^p \sinh x \, dx = x^p \cosh x - p \int x^{p-1} \cosh x \, dx.$

[See 677.1.]

671.20. $\displaystyle\int \sinh^2 x \, dx = \frac{\sinh 2x}{4} - \frac{x}{2}.$

671.21. $\displaystyle\int x \sinh^2 x \, dx = \frac{x \sinh 2x}{4} - \frac{\cosh 2x}{8} - \frac{x^2}{4}.$

671.30. $\displaystyle\int \sinh^3 x \, dx = \frac{\cosh^3 x}{3} - \cosh x.$

671.40. $\displaystyle\int \sinh^4 x \, dx = \frac{\sinh 4x}{32} - \frac{\sinh 2x}{4} + \frac{3x}{8}.$

671.90. $\displaystyle\int \sinh^p x \, dx = \frac{1}{p} \sinh^{p-1} x \cosh x - \frac{p-1}{p} \int \sinh^{p-2} x \, dx.$

672.11. $\int \dfrac{\sinh x}{x}\, dx = x + \dfrac{x^3}{3\cdot 3!} + \dfrac{x^5}{5\cdot 5!} + \dfrac{x^7}{7\cdot 7!} + \cdots.$

672.12. $\int \dfrac{\sinh x}{x^2}\, dx = -\dfrac{\sinh x}{x} + \int \dfrac{\cosh x}{x}\, dx.$ [See **678.11.**]

672.21. $\int \dfrac{\sinh^2 x}{x}\, dx = -\dfrac{1}{2}\log |x| + \dfrac{1}{2}\int \dfrac{\cosh 2x}{2x}\, d(2x).$

[See **678.11.**]

673.10. $\int \dfrac{dx}{\sinh x} = \int \operatorname{csch} x\, dx = \log \left| \tanh \dfrac{x}{2} \right|$

$\qquad\qquad = -\dfrac{1}{2}\log \dfrac{\cosh x + 1}{\cosh x - 1}.$

673.11. $\int \dfrac{x\, dx}{\sinh x} = x - \dfrac{x^3}{3\cdot 3!} + \dfrac{7x^5}{3\cdot 5\cdot 5!} - \dfrac{31x^7}{3\cdot 7\cdot 7!} + \dfrac{127x^9}{3\cdot 5\cdot 9!}$

$\qquad\qquad - \cdots + (-1)^n \dfrac{2(2^{2n-1} - 1)}{(2n + 1)!}\, B_n x^{2n+1} + \cdots,$

$\qquad\qquad\qquad\qquad\qquad [x^2 < \pi^2. \quad \text{See } \mathbf{45}].$

673.19. $\int \dfrac{x^p dx}{\sinh x}.$ Expand $\dfrac{1}{\sinh x}$ by **657.8**, multiply by x^p and

integrate, $\qquad\qquad\qquad\qquad\qquad\qquad [p \neq 0].$

673.20. $\int \dfrac{dx}{\sinh^2 x} = \int \operatorname{csch}^2 x\, dx = -\operatorname{ctnh} x.$

673.21. $\int \dfrac{x\, dx}{\sinh^2 x} = -x\operatorname{ctnh} x + \log |\sinh x|.$

673.30. $\int \dfrac{dx}{\sinh^3 x} = \int \operatorname{csch}^3 x\, dx$

$\qquad\qquad = -\dfrac{\cosh x}{2\sinh^2 x} - \dfrac{1}{2}\log \left| \tanh \dfrac{x}{2} \right|.$

673.40. $\int \dfrac{dx}{\sinh^4 x} = \operatorname{ctnh} x - \dfrac{\operatorname{ctnh}^3 x}{3}.$

673.90. $\int \dfrac{dx}{\sinh^p x} = -\dfrac{\cosh x}{(p - 1)\sinh^{p-1} x} - \dfrac{p - 2}{p - 1}\int \dfrac{dx}{\sinh^{p-2} x},$

$\qquad\qquad\qquad\qquad\qquad\qquad\qquad [p > 1].$

675. $\int \sinh mx \sinh nx\, dx = \dfrac{\sinh (m + n)x}{2(m + n)} - \dfrac{\sinh (m - n)x}{2(m - n)},$

$\qquad\qquad [m^2 \neq n^2. \quad \text{If } m^2 = n^2, \text{ see } \mathbf{671.20}].$

Integrals Involving cosh x

677.10. $\displaystyle\int \cosh x \, dx = \sinh x.$

677.101. $\displaystyle\int \cosh \frac{x}{a} \, dx = a \sinh \frac{x}{a}.$

677.11. $\displaystyle\int x \cosh x \, dx = x \sinh x - \cosh x.$

677.12. $\displaystyle\int x^2 \cosh x \, dx = (x^2 + 2) \sinh x - 2x \cosh x.$

677.13. $\displaystyle\int x^3 \cosh x \, dx = (x^3 + 6x) \sinh x - (3x^2 + 6) \cosh x.$

677.19. $\displaystyle\int x^p \cosh x \, dx = x^p \sinh x - p \int x^{p-1} \sinh x \, dx.$

[See **671.1.**]

677.20. $\displaystyle\int \cosh^2 x \, dx = \frac{\sinh 2x}{4} + \frac{x}{2}.$

677.21. $\displaystyle\int x \cosh^2 x \, dx = \frac{x \sinh 2x}{4} - \frac{\cosh 2x}{8} + \frac{x^2}{4}.$

677.30. $\displaystyle\int \cosh^3 x \, dx = \frac{\sinh^3 x}{3} + \sinh x.$

677.40. $\displaystyle\int \cosh^4 x \, dx = \frac{\sinh 4x}{32} + \frac{\sinh 2x}{4} + \frac{3x}{8}.$

677.90. $\displaystyle\int \cosh^p x \, dx = \frac{1}{p} \sinh x \cosh^{p-1} x$
$$+ \frac{p-1}{p} \int \cosh^{p-2} x \, dx.$$

678.11. $\displaystyle\int \frac{\cosh x}{x} \, dx = \log |x| + \frac{x^2}{2 \cdot 2!} + \frac{x^4}{4 \cdot 4!} + \frac{x^6}{6 \cdot 6!} + \cdots.$

678.12. $\displaystyle\int \frac{\cosh x}{x^2} \, dx = -\frac{\cosh x}{x} + \int \frac{\sinh x}{x} \, dx.$

[See **672.11.**]

678.21. $\displaystyle\int \frac{\cosh^2 x \, dx}{x} = \frac{1}{2} \log |x| + \frac{1}{2} \int \frac{\cosh 2x}{2x} \, d(2x).$

[See **678.11.**]

679.10. $\displaystyle\int \frac{dx}{\cosh x} = \int \operatorname{sech} x\, dx = \tan^{-1}(\sinh x)$

$$= 2\tan^{-1} e^x + \text{constant}.$$

679.11. $\displaystyle\int \frac{x\,dx}{\cosh x} = \frac{x^2}{2} - \frac{x^4}{4\cdot 2!} + \frac{5x^6}{6\cdot 4!} - \frac{61x^8}{8\cdot 6!} + \frac{1385x^{10}}{10\cdot 8!}$$

$$- \cdots + \frac{(-1)^n E_n}{(2n+2)(2n)!}\, x^{2n+2} + \cdots,$$

$$[x^2 < \pi^2/4. \quad \text{See } \mathbf{45}].$$

679.19. $\displaystyle\int \frac{x^p dx}{\cosh x}$.   Expand $\dfrac{1}{\cosh x}$ by **657.7**, multiply by x^p and integrate, $[p \neq 0]$.

679.20. $\displaystyle\int \frac{dx}{\cosh^2 x} = \int \operatorname{sech}^2 x\, dx = \tanh x.$

679.21. $\displaystyle\int \frac{x\,dx}{\cosh^2 x} = x\tanh x - \log\cosh x.$

679.30. $\displaystyle\int \frac{dx}{\cosh^3 x} = \frac{\sinh x}{2\cosh^2 x} + \frac{1}{2}\tan^{-1}(\sinh x).$

679.40. $\displaystyle\int \frac{dx}{\cosh^4 x} = \tanh x - \frac{\tanh^3 x}{3}.$

679.90. $\displaystyle\int \frac{dx}{\cosh^p x} = \frac{\sinh x}{(p-1)\cosh^{p-1} x} + \frac{p-2}{p-1}\int \frac{dx}{\cosh^{p-2} x},$

$$[p > 1].$$

681. $\displaystyle\int \cosh mx \cosh nx\, dx$

$$= \frac{\sinh(m+n)x}{2(m+n)} + \frac{\sinh(m-n)x}{2(m-n)},$$

$$[m^2 \neq n^2]. \quad [\text{If } m^2 = n^2, \text{ see } \mathbf{677.20}.]$$

682.01. $\displaystyle\int \frac{dx}{\cosh x + 1} = \tanh \frac{x}{2}.$

682.02. $\displaystyle\int \frac{dx}{\cosh x - 1} = -\operatorname{ctnh} \frac{x}{2}.$

682.03. $\displaystyle\int \frac{x\,dx}{\cosh x + 1} = x\tanh \frac{x}{2} - 2\log\cosh \frac{x}{2}.$

682.04. $\displaystyle\int \frac{x\,dx}{\cosh x - 1} = -\,x\,\operatorname{ctnh}\frac{x}{2} + 2\log\left|\sinh\frac{x}{2}\right|.$

682.05. $\displaystyle\int \frac{\cosh x\,dx}{\cosh x + 1} = x - \tanh\frac{x}{2}.$

682.06. $\displaystyle\int \frac{\cosh x\,dx}{\cosh x - 1} = x - \operatorname{ctnh}\frac{x}{2}.$

682.07. $\displaystyle\int \frac{dx}{\cosh x(\cosh x + 1)} = \tan^{-1}(\sinh x) - \tanh\frac{x}{2}.$

682.08. $\displaystyle\int \frac{dx}{\cosh x(\cosh x - 1)} = -\tan^{-1}(\sinh x) - \operatorname{ctnh}\frac{x}{2}.$

682.09. $\displaystyle\int \frac{dx}{(\cosh x + 1)^2} = \frac{1}{2}\tanh\frac{x}{2} - \frac{1}{6}\tanh^3\frac{x}{2}.$

682.10. $\displaystyle\int \frac{dx}{(\cosh x - 1)^2} = \frac{1}{2}\operatorname{ctnh}\frac{x}{2} - \frac{1}{6}\operatorname{ctnh}^3\frac{x}{2}.$

682.11. $\displaystyle\int \frac{dx}{\cosh^2 x + 1} = \frac{1}{2\sqrt{2}}\cosh^{-1}\left(\frac{3\cosh^2 x - 1}{\cosh^2 x + 1}\right).$

Use the positive value of the inverse cosh.

682.12. $\displaystyle\int \frac{dx}{\cosh^2 x - 1} = \int \frac{dx}{\sinh^2 x} = -\operatorname{ctnh} x.$ [See **673.20.**]

Integrals Involving sinh x and cosh x

685.11. $\displaystyle\int \sinh x \cosh x\,dx = \frac{\sinh^2 x}{2} = \frac{\cosh^2 x}{2} + \text{constant}$

$$= \frac{\cosh 2x}{4} + \text{constant}.$$

685.12. $\displaystyle\int \sinh x \cosh^2 x\,dx = \frac{\cosh^3 x}{3}.$

685.13. $\displaystyle\int \sinh x \cosh^3 x\,dx = \frac{\cosh^4 x}{4}.$

685.19. $\displaystyle\int \sinh x \cosh^p x\,dx = \frac{\cosh^{p+1} x}{p + 1},$ $[p \neq -1].$

685.21. $\displaystyle\int \sinh^2 x \cosh x\, dx = \frac{\sinh^3 x}{3}.$

685.22. $\displaystyle\int \sinh^2 x \cosh^2 x\, dx = \frac{\sinh 4x}{32} - \frac{x}{8}.$

685.31. $\displaystyle\int \sinh^3 x \cosh x\, dx = \frac{\sinh^4 x}{4}.$

685.91. $\displaystyle\int \sinh^p x \cosh x\, dx = \frac{\sinh^{p+1} x}{p+1},$ $\qquad [p \neq -1].$

686.11. $\displaystyle\int \frac{dx}{\sinh x \cosh x} = \log |\tanh x|.$

686.12. $\displaystyle\int \frac{dx}{\sinh x \cosh^2 x} = \frac{1}{\cosh x} + \log \left|\tanh \frac{x}{2}\right|.$

686.13. $\displaystyle\int \frac{dx}{\sinh x \cosh^3 x} = \frac{1}{2\cosh^2 x} + \log |\tanh x|.$

686.19. $\displaystyle\int \frac{dx}{\sinh x \cosh^p x} = \frac{1}{(p-1)\cosh^{p-1} x}$

$$+ \int \frac{dx}{\sinh x \cosh^{p-2} x}, \qquad [p \neq 1].$$

686.21. $\displaystyle\int \frac{dx}{\sinh^2 x \cosh x} = -\frac{1}{\sinh x} - \tan^{-1}(\sinh x).$

686.22. $\displaystyle\int \frac{dx}{\sinh^2 x \cosh^2 x} = -2\,\mathrm{ctnh}\, 2x.$

686.31. $\displaystyle\int \frac{dx}{\sinh^3 x \cosh x} = -\frac{1}{2\sinh^2 x} - \log |\tanh x|.$

686.91. $\displaystyle\int \frac{dx}{\sinh^p x \cosh x} = -\frac{1}{(p-1)\sinh^{p-1} x}$

$$- \int \frac{dx}{\sinh^{p-2} x \cosh x}, \qquad [p \neq 1].$$

687.11. $\displaystyle\int \frac{\sinh x\, dx}{\cosh x} = \int \tanh x\, dx = \log \cosh x.$

$$[\text{See } 691.01.]$$

687.12. $\displaystyle\int \frac{\sinh x\, dx}{\cosh^2 x} = -\frac{1}{\cosh x} = -\,\mathrm{sech}\, x.$

687.13. $\displaystyle \int \frac{\sinh x \, dx}{\cosh^3 x} = -\frac{1}{2 \cosh^2 x} = \frac{\tanh^2 x}{2} + \text{constant.}$

687.19. $\displaystyle \int \frac{\sinh x \, dx}{\cosh^p x} = -\frac{1}{(p-1) \cosh^{p-1} x},$ $[p \neq 1].$

687.21. $\displaystyle \int \frac{\sinh^2 x}{\cosh x} \, dx = \sinh x - \tan^{-1}(\sinh x).$

687.22. $\displaystyle \int \frac{\sinh^2 x}{\cosh^2 x} \, dx = \int \tanh^2 x = x - \tanh x.$ [See **691.02.**]

687.29. $\displaystyle \int \frac{\sinh^2 x}{\cosh^p x} \, dx = -\frac{\sinh x}{(p-1) \cosh^{p-1} x}$
$$+ \frac{1}{p-1} \int \frac{dx}{\cosh^{p-2} x}, \qquad [p \neq 1].$$

687.31. $\displaystyle \int \frac{\sinh^3 x}{\cosh x} \, dx = \frac{\sinh^2 x}{2} - \log \cosh x.$

687.32. $\displaystyle \int \frac{\sinh^3 x}{\cosh^2 x} \, dx = \cosh x + \operatorname{sech} x.$

687.33. $\displaystyle \int \frac{\sinh^3 x}{\cosh^3 x} \, dx = \int \tanh^3 x \, dx = -\frac{\tanh^2 x}{2} + \log \cosh x.$
$$\text{[See \textbf{691.03.}]}$$

687.34. $\displaystyle \int \frac{\sinh^3 x}{\cosh^4 x} \, dx = \frac{1}{3 \cosh^3 x} - \frac{1}{\cosh x}.$

687.39. $\displaystyle \int \frac{\sinh^3 x}{\cosh^p x} \, dx = \frac{1}{(p-1) \cosh^{p-1} x} - \frac{1}{(p-3) \cosh^{p-3} x},$
$$[p \neq 1 \text{ or } 3].$$

687.7. $\displaystyle \int \frac{\sinh^{p-2} x}{\cosh^p x} \, dx = \frac{\tanh^{p-1} x}{p-1},$ $[p \neq 1].$

688.11. $\displaystyle \int \frac{\cosh x}{\sinh x} \, dx = \int \operatorname{ctnh} x \, dx = \log |\sinh x|.$
$$\text{[See \textbf{692.01.}]}$$

688.12. $\displaystyle \int \frac{\cosh x}{\sinh^2 x} \, dx = -\frac{1}{\sinh x} = -\operatorname{csch} x.$

688.13. $\displaystyle \int \frac{\cosh x}{\sinh^3 x} \, dx = -\frac{1}{2 \sinh^2 x} = -\frac{\operatorname{ctnh}^2 x}{2} + \text{constant.}$

688.19. $\displaystyle\int \frac{\cosh x}{\sinh^p x}\,dx = -\frac{1}{(p-1)\sinh^{p-1} x},$ $[p \neq 1]$.

688.21. $\displaystyle\int \frac{\cosh^2 x}{\sinh x}\,dx = \cosh x + \log\left|\tanh \frac{x}{2}\right|.$

688.22. $\displaystyle\int \frac{\cosh^2 x}{\sinh^2 x}\,dx = \int \operatorname{ctnh}^2 x\,dx = x - \operatorname{ctnh} x.$

[See **692.02.**]

688.29. $\displaystyle\int \frac{\cosh^2 x}{\sinh^p x}\,dx = -\frac{\cosh x}{(p-1)\sinh^{p-1} x}$

$$+ \frac{1}{p-1}\int \frac{dx}{\sinh^{p-2} x},\qquad [p \neq 1].$$

688.31. $\displaystyle\int \frac{\cosh^3 x}{\sinh x}\,dx = \frac{\cosh^2 x}{2} + \log|\sinh x|.$

688.32. $\displaystyle\int \frac{\cosh^3 x}{\sinh^2 x}\,dx = \sinh x - \operatorname{csch} x.$

688.33. $\displaystyle\int \frac{\cosh^3 x}{\sinh^3 x}\,dx = \int \operatorname{ctnh}^3 x\,dx$

$$= -\frac{\operatorname{ctnh}^2 x}{2} + \log|\sinh x|.$$

[See **692.03.**]

688.34. $\displaystyle\int \frac{\cosh^3 x}{\sinh^4 x}\,dx = -\frac{1}{3\sinh^3 x} - \frac{1}{\sinh x}.$

688.39. $\displaystyle\int \frac{\cosh^3 x}{\sinh^p x}\,dx = -\frac{1}{(p-1)\sinh^{p-1} x}$

$$-\frac{1}{(p-3)\sinh^{p-3} x},\qquad [p \neq 1 \text{ or } 3].$$

688.7. $\displaystyle\int \frac{\cosh^{p-2} x}{\sinh^p x}\,dx = -\frac{\operatorname{ctnh}^{p-1} x}{p-1},\qquad [p \neq 1].$

689.01. $\displaystyle\int \frac{\sinh x\,dx}{\cosh x + 1} = \log(\cosh x + 1).$

689.02. $\displaystyle\int \frac{\sinh x\,dx}{\cosh x - 1} = \log(\cosh x - 1)$

689.03. $\displaystyle\int \frac{dx}{\sinh x(\cosh x + 1)} = -\frac{1}{2(\cosh x + 1)}$

$$+ \frac{1}{2} \log \left| \tanh \frac{x}{2} \right|.$$

689.04. $\displaystyle\int \frac{dx}{\sinh x(\cosh x - 1)} = \frac{1}{2(\cosh x - 1)}$

$$- \frac{1}{2} \log \left| \tanh \frac{x}{2} \right|.$$

689.05. $\displaystyle\int \frac{\sinh x\, dx}{\cosh x(\cosh x + 1)} = \log \left(\frac{\cosh x}{\cosh x + 1} \right).$

689.06. $\displaystyle\int \frac{\sinh x\, dx}{\cosh x(\cosh x - 1)} = \log \left(\frac{\cosh x - 1}{\cosh x} \right).$

689.07. $\displaystyle\int \sinh mx \cosh nx\, dx$

$$= \frac{\cosh (m + n)x}{2(m + n)} + \frac{\cosh (m - n)x}{2(m - n)},$$
$$[m^2 \neq n^2.\quad \text{If } m^2 = n^2, \text{ see } \mathbf{685.11}].$$

Integrals Involving tanh x and ctnh x

691.01. $\displaystyle\int \tanh x\, dx = \log \cosh x.$ [See **687.11**.]

691.02. $\displaystyle\int \tanh^2 x\, dx = x - \tanh x.$ [See **687.22**.]

691.03. $\displaystyle\int \tanh^3 x\, dx = -\frac{\tanh^2 x}{2} + \log \cosh x.$ [See **687.33**.]

691.09. $\displaystyle\int \tanh^p x\, dx = -\frac{\tanh^{p-1} x}{p - 1} + \int \tanh^{p-2} x\, dx,$

$$[p \neq 1].$$

693.01. $\displaystyle\int \ctnh x\, dx = \log |\sinh x|.$ [See **688.11**.]

693.02. $\displaystyle\int \ctnh^2 x\, dx = x - \ctnh x.$ [See **688.22**.]

693.03. $\displaystyle\int \ctnh^3 x\, dx = -\frac{\ctnh^2 x}{2} + \log |\sinh x|.$

$$[\text{See } \mathbf{688.33}.]$$

693.09 $\displaystyle\int \ctnh^p x\, dx = -\frac{\ctnh^{p-1} x}{p - 1} + \int \ctnh^{p-2} x\, dx,$

$$[p \neq 1].$$

INVERSE HYPERBOLIC FUNCTIONS

700. $\sinh^{-1} x = \cosh^{-1} \sqrt{(x^2 + 1)}$.

Use the positive value of $\cosh^{-1}$ when x is positive and the negative value when x is negative.

700.1. $\sinh^{-1} x = \tanh^{-1} \dfrac{x}{\sqrt{(x^2 + 1)}} = \operatorname{csch}^{-1} \dfrac{1}{x}$

$$= -\sinh^{-1}(-x) = \log\{x + \sqrt{(x^2 + 1)}\}.$$

[See **602.1** and **706.**]

701. $\cosh^{-1} x = \pm \sinh^{-1} \sqrt{(x^2 - 1)} = \pm \tanh^{-1} \dfrac{\sqrt{(x^2 - 1)}}{x}$

$$= \operatorname{sech}^{-1} \frac{1}{x} = \pm \log\{x + \sqrt{(x^2 - 1)}\},$$

$[x > 1]$.   [See **602.3** and **707.**]

702. $\tanh^{-1} x = \operatorname{ctnh}^{-1} \dfrac{1}{x} = \dfrac{1}{2} \log \dfrac{1 + x}{1 - x}$,

$[x^2 < 1]$.   [See **708.**]

703. $\operatorname{ctnh}^{-1} x = \tanh^{-1} \dfrac{1}{x} = \dfrac{1}{2} \log \dfrac{x + 1}{x - 1}$,

$[x^2 > 1]$.   [See **709.**]

704. $\operatorname{sech}^{-1} x = \pm \log\left\{\dfrac{1}{x} + \sqrt{\left(\dfrac{1}{x^2} - 1\right)}\right\}$,

$[0 < x < 1]$.   [See **710.**]

705. $\operatorname{csch}^{-1} x = \log\left\{\dfrac{1}{x} + \sqrt{\left(\dfrac{1}{x^2} + 1\right)}\right\}$.   [See **711.**]

706. $\sinh^{-1} x$

$$= x - \frac{1}{2\cdot 3} x^3 + \frac{1\cdot 3}{2\cdot 4\cdot 5} x^5 - \frac{1\cdot 3\cdot 5}{2\cdot 4\cdot 6\cdot 7} x^7 + \cdots,$$

$[x^2 < 1]$.

$$= \log(2x) + \frac{1}{2\cdot 2x^2} - \frac{1\cdot 3}{2\cdot 4\cdot 4x^4} + \frac{1\cdot 3\cdot 5}{2\cdot 4\cdot 6\cdot 6x^6} - \cdots,$$

$[x > 1]$.

$$= -\log|2x| - \frac{1}{2\cdot 2x^2} + \frac{1\cdot 3}{2\cdot 4\cdot 4x^4} - \frac{1\cdot 3\cdot 5}{2\cdot 4\cdot 6\cdot 6x^6} + \cdots,$$

$[x < -1]$.   [See **602.1.**]

707. $\cosh^{-1} x = \pm \left[\log (2x) - \dfrac{1}{2 \cdot 2x^2} - \dfrac{1 \cdot 3}{2 \cdot 4 \cdot 4x^4} \right.$

$$\left. - \dfrac{1 \cdot 3 \cdot 5}{2 \cdot 4 \cdot 6 \cdot 6x^6} - \cdots \right], \qquad [x > 1].$$

[See **602.3** and **602.4.**]

708. $\tanh^{-1} x = x + \dfrac{x^3}{3} + \dfrac{x^5}{5} + \dfrac{x^7}{7} + \cdots, \qquad [x^2 < 1].$

[See **601.2.**]

709. $\operatorname{ctnh}^{-1} x = \dfrac{1}{x} + \dfrac{1}{3x^3} + \dfrac{1}{5x^5} + \dfrac{1}{7x^7} + \cdots, \qquad [x^2 > 1].$

[See **601.3.**]

710. $\operatorname{sech}^{-1} x = \pm \left[\log \dfrac{2}{x} - \dfrac{1}{2 \cdot 2} x^2 - \dfrac{1 \cdot 3}{2 \cdot 4 \cdot 4} x^4 \right.$

$$\left. - \dfrac{1 \cdot 3 \cdot 5}{2 \cdot 4 \cdot 6 \cdot 6} x^6 - \cdots \right], \qquad [0 < x < 1].$$

[See **602.7** and **602.8.**]

711. $\operatorname{csch}^{-1} x$

$$= \dfrac{1}{x} - \dfrac{1}{2 \cdot 3x^3} + \dfrac{1 \cdot 3}{2 \cdot 4 \cdot 5x^5} - \dfrac{1 \cdot 3 \cdot 5}{2 \cdot 4 \cdot 6 \cdot 7x^7} + \cdots, \quad [x^2 > 1]$$

$$= \log \dfrac{2}{x} + \dfrac{1}{2 \cdot 2} x^2 - \dfrac{1 \cdot 3}{2 \cdot 4 \cdot 4} x^4 + \dfrac{1 \cdot 3 \cdot 5}{2 \cdot 4 \cdot 6 \cdot 6} x^6 - \cdots,$$

$$[0 < x < 1]$$

$$= - \log \left| \dfrac{2}{x} \right| - \dfrac{1}{2 \cdot 2} x^2 + \dfrac{1 \cdot 3}{2 \cdot 4 \cdot 4} x^4 - \dfrac{1 \cdot 3 \cdot 5}{2 \cdot 4 \cdot 6 \cdot 6} x^6 + \cdots,$$

$$[-1 < x < 0]. \qquad [\text{See } \mathbf{602.5.}]$$

720. $\sinh^{-1} (\pm x + iy) = \pm (-1)^n \cosh^{-1} \dfrac{s + t}{2}$

$$+ i(-1)^n \sin^{-1} \dfrac{2y}{s + t} + in\pi,$$

where the principal value of $\sin^{-1}$ (between $-\pi/2$ and $\pi/2$) and the positive value of $\cosh^{-1}$ are taken,

n is an integer or 0,

x is positive,

y is positive or negative,

and where

720.1. $s = \sqrt{(1 + y)^2 + x^2}$ (positive value),

720.2. $t = \sqrt{(1 - y)^2 + x^2}$ (positive value).

Note that if $x = 0$ and $y > 1$, $t = y - 1$ and $s + t = 2y$.

If $x = 0$ and $y < 1$, $t = 1 - y$ and $s + t = 2$.

Alternative:

720.3a. $\sinh^{-1} A = \log_\epsilon (\pm \sqrt{1 + A^2} + A) + i2k\pi$

or

720.3b. $= - \log_\epsilon (\pm \sqrt{1 + A^2} - A) + i2k\pi$

where A may be a complex quantity and k is an integer or 0.

For the square root of a complex quantity see **58** and for the logarithm see **604**. The two solutions a and b are identical. In any given case, the one should be used which involves the numerical sum of two quantities instead of the difference, so as to obtain more convenient precise computation.

721.1. $\cosh^{-1} (x + iy)$

$$= \pm \left(\cosh^{-1} \frac{p + q}{2} + i \cos^{-1} \frac{2x}{p + q} + i2k\pi \right).$$

721.2. $\cosh^{-1} (x - iy)$

$$= \pm \left(\cosh^{-1} \frac{p + q}{2} - i \cos^{-1} \frac{2x}{p + q} + i2k\pi \right),$$

where the positive value of $\cosh^{-1}$ and the principal value of $\cos^{-1}$ (between 0 and π) are taken,

x is positive or negative,

y is positive,

721.3. $p = \sqrt{(1 + x)^2 + y^2}$ (positive value),

721.4. $q = \sqrt{(1 - x)^2 + y^2}$ (positive value).

Alternative:

721.5a. $\cosh^{-1} A = \pm \log_\epsilon (A + \sqrt{A^2 - 1}) + i2k\pi$

or

721.5b. $= \mp \log_\epsilon (A - \sqrt{A^2 - 1}) + i2k\pi.$

See note following **720.3.**

722. $\tanh^{-1} (x + iy) = \dfrac{1}{4} \log_\epsilon \dfrac{(1 + x)^2 + y^2}{(1 - x)^2 + y^2}$

$$+ \frac{i}{2} \left\{ (2k + 1)\pi - \tan^{-1} \frac{1 + x}{y} - \tan^{-1} \frac{1 - x}{y} \right\},$$

where the principal values of $\tan^{-1}$ (between $- \pi/2$ and $\pi/2$) are taken and where x and y may be positive or negative.

[See formula for $\tanh^{-1} (x + iy)$ in Ref. 24, p. 115.]

Alternative:

722.1. $\tanh^{-1} (x + iy) = \dfrac{1}{2} \log_\epsilon \dfrac{1 + x + iy}{1 - x - iy}.$ [See **604.**]

[Ref. 46, Chap. XI.]

INVERSE HYPERBOLIC FUNCTIONS—DERIVATIVES

728.1. $\dfrac{d}{dx} \sinh^{-1} \dfrac{x}{a} = \dfrac{1}{\sqrt{(x^2 + a^2)}}$.

728.2. $\dfrac{d}{dx} \cosh^{-1} \dfrac{x}{a} = \dfrac{1}{\sqrt{(x^2 - a^2)}}$, $\left[\cosh^{-1} \dfrac{x}{a} > 0, \quad \dfrac{x}{a} > 1 \right]$.

728.3. $\dfrac{d}{dx} \cosh^{-1} \dfrac{x}{a} = \dfrac{-1}{\sqrt{(x^2 - a^2)}}$, $\left[\cosh^{-1} \dfrac{x}{a} < 0, \quad \dfrac{x}{a} > 1 \right]$.

728.4. $\dfrac{d}{dx} \tanh^{-1} \dfrac{x}{a} = \dfrac{a}{a^2 - x^2}$, $[x^2 < a^2]$.

728.5. $\dfrac{d}{dx} \operatorname{ctnh}^{-1} \dfrac{x}{a} = \dfrac{a}{a^2 - x^2}$, $[x^2 > a^2]$.

728.6. $\dfrac{d}{dx} \operatorname{sech}^{-1} \dfrac{x}{a} = \dfrac{-a}{x\sqrt{(a^2 - x^2)}}$,

$[\operatorname{sech}^{-1} (x/a) > 0, \quad 0 < x/a < 1]$.

728.7. $\dfrac{d}{dx} \operatorname{sech}^{-1} \dfrac{x}{a} = \dfrac{a}{x\sqrt{(a^2 - x^2)}}$,

$[\operatorname{sech}^{-1} (x/a) < 0, \quad 0 < x/a < 1]$.

728.8. $\dfrac{d}{dx} \operatorname{csch}^{-1} \dfrac{x}{a} = \dfrac{-a}{x\sqrt{(x^2 + a^2)}}$.

[Except in **728.4** and **728.5**, $a > 0$.]

INTEGRALS—($a > 0$)

730. $\displaystyle\int \sinh^{-1} \dfrac{x}{a} \, dx = x \sinh^{-1} \dfrac{x}{a} - \sqrt{(x^2 + a^2)}$.

730.1. $\displaystyle\int x \sinh^{-1} \dfrac{x}{a} \, dx = \left(\dfrac{x^2}{2} + \dfrac{a^2}{4} \right) \sinh^{-1} \dfrac{x}{a} - \dfrac{x}{4} \sqrt{(x^2 + a^2)}$.

730.2. $\displaystyle\int x^2 \sinh^{-1} \dfrac{x}{a} \, dx = \dfrac{x^3}{3} \sinh^{-1} \dfrac{x}{a} + \dfrac{2a^2 - x^2}{9} \sqrt{(x^2 + a^2)}$.

730.3. $\displaystyle\int x^3 \sinh^{-1} \dfrac{x}{a} \, dx = \left(\dfrac{x^4}{4} - \dfrac{3a^4}{32} \right) \sinh^{-1} \dfrac{x}{a}$

$+ \dfrac{3a^2 x - 2x^3}{32} \sqrt{(x^2 + a^2)}$.

730.4. $\displaystyle\int x^4 \sinh^{-1}\frac{x}{a}\,dx = \frac{x^5}{5}\sinh^{-1}\frac{x}{a}$

$$-\frac{8a^4 - 4a^2x^2 + 3x^4}{75}\sqrt{(x^2+a^2)}.$$

[See **625–625.4.**]

730.9. $\displaystyle\int x^p \sinh^{-1}\frac{x}{a}\,dx = \frac{x^{p+1}}{p+1}\sinh^{-1}\frac{x}{a}$

$$-\frac{1}{p+1}\int\frac{x^{p+1}dx}{\sqrt{(x^2+a^2)}}, \qquad [p \neq -1].$$

[See **201.01–207.01** and **625.9.**]

731.1. $\displaystyle\int \frac{1}{x}\sinh^{-1}\frac{x}{a}\,dx$

$$= \frac{x}{a} - \frac{1}{2\cdot 3\cdot 3}\frac{x^3}{a^3} + \frac{1\cdot 3}{2\cdot 4\cdot 5\cdot 5}\frac{x^5}{a^5} - \frac{1\cdot 3\cdot 5}{2\cdot 4\cdot 6\cdot 7\cdot 7}\frac{x^7}{a^7} + \cdots,$$

$$[x^2 < a^2].$$

$$= \frac{1}{2}\left(\log\frac{2x}{a}\right)^2 - \frac{1}{2^3}\frac{a^2}{x^2} + \frac{1\cdot 3}{2\cdot 4^3}\frac{a^4}{x^4} - \frac{1\cdot 3\cdot 5}{2\cdot 4\cdot 6^3}\frac{a^6}{x^6} + \cdots,$$

$$[x/a > 1].$$

$$= -\frac{1}{2}\left(\log\left|\frac{2x}{a}\right|\right)^2 + \frac{1}{2^3}\frac{a^2}{x^2} - \frac{1\cdot 3}{2\cdot 4^3}\frac{a^4}{x^4} + \frac{1\cdot 3\cdot 5}{2\cdot 4\cdot 6^3}\frac{a^6}{x^6} - \cdots,$$

$$[x/a < -1].$$

731.2. $\displaystyle\int \frac{1}{x^2}\sinh^{-1}\frac{x}{a}\,dx = -\frac{1}{x}\sinh^{-1}\frac{x}{a}$

$$-\frac{1}{a}\log\left|\frac{a+\sqrt{(x^2+a^2)}}{x}\right|.$$

731.3. $\displaystyle\int \frac{1}{x^3}\sinh^{-1}\frac{x}{a}\,dx = -\frac{1}{2x^2}\sinh^{-1}\frac{x}{a} - \frac{\sqrt{(x^2+a^2)}}{2a^2x}.$

[See **626.1 to .3.**]

731.9. $\displaystyle\int \frac{1}{x^p}\sinh^{-1}\frac{x}{a}\,dx = -\frac{1}{(p-1)x^{p-1}}\sinh^{-1}\frac{x}{a}$

$$+\frac{1}{p-1}\int\frac{dx}{x^{p-1}\sqrt{(x^2+a^2)}},$$

$$[p \neq 1]. \qquad [\text{See } \mathbf{221.01–226.01} \text{ and } \mathbf{626.9.}]$$

732.
$$\int \cosh^{-1}\frac{x}{a}\,dx = x\cosh^{-1}\frac{x}{a} - \sqrt{(x^2 - a^2)},$$
$$[\cosh^{-1}(x/a) > 0],$$
$$= x\cosh^{-1}\frac{x}{a} + \sqrt{(x^2 - a^2)},$$
$$[\cosh^{-1}(x/a) < 0].$$

732.1.
$$\int x\cosh^{-1}\frac{x}{a}\,dx = \left(\frac{x^2}{2} - \frac{a^2}{4}\right)\cosh^{-1}\frac{x}{a} - \frac{x}{4}\sqrt{(x^2 - a^2)},$$
$$[\cosh^{-1}(x/a) > 0],$$
$$= \left(\frac{x^2}{2} - \frac{a^2}{4}\right)\cosh^{-1}\frac{x}{a} + \frac{x}{4}\sqrt{(x^2 - a^2)},$$
$$[\cosh^{-1}(x/a) < 0].$$

732.2.
$$\int x^2\cosh^{-1}\frac{x}{a}\,dx = \frac{x^3}{3}\cosh^{-1}\frac{x}{a} - \frac{2a^2 + x^2}{9}\sqrt{(x^2 - a^2)},$$
$$[\cosh^{-1}(x/a) > 0].$$
$$= \frac{x^3}{3}\cosh^{-1}\frac{x}{a} + \frac{2a^2 + x^2}{9}\sqrt{(x^2 - a^2)},$$
$$[\cosh^{-1}(x/a) < 0].$$

732.3.
$$\int x^3\cosh^{-1}\frac{x}{a}\,dx$$
$$= \left(\frac{x^4}{4} - \frac{3a^4}{32}\right)\cosh^{-1}\frac{x}{a} - \frac{3a^2x + 2x^3}{32}\sqrt{(x^2 - a^2)},$$
$$[\cosh^{-1}(x/a) > 0],$$
$$= \left(\frac{x^4}{4} - \frac{3a^4}{32}\right)\cosh^{-1}\frac{x}{a} + \frac{3a^2x + 2x^3}{32}\sqrt{(x^2 - a^2)},$$
$$[\cosh^{-1}(x/a) < 0].$$

732.4.
$$\int x^4\cosh^{-1}\frac{x}{a}\,dx$$
$$= \frac{x^5}{5}\cosh^{-1}\frac{x}{a} - \frac{8a^4 + 4a^2x^2 + 3x^4}{75}\sqrt{(x^2 - a^2)},$$
$$[\cosh^{-1}(x/a) > 0],$$
$$= \frac{x^5}{5}\cosh^{-1}\frac{x}{a} + \frac{8a^4 + 4a^2x^2 + 3x^4}{75}\sqrt{(x^2 - a^2)},$$
$$[\cosh^{-1}(x/a) < 0].\quad [\text{See } \textbf{627–627.4.}]$$

732.9. $\displaystyle\int x^p \cosh^{-1}\frac{x}{a}\,dx$

$$= \frac{x^{p+1}}{p+1}\cosh^{-1}\frac{x}{a} - \frac{1}{p+1}\int \frac{x^{p+1}dx}{\sqrt{(x^2-a^2)}},$$
$$[\cosh^{-1}(x/a) > 0, \quad p \neq -1],$$
$$= \frac{x^{p+1}}{p+1}\cosh^{-1}\frac{x}{a} + \frac{1}{p+1}\int \frac{x^{p+1}dx}{\sqrt{(x^2-a^2)}},$$
$$[\cosh^{-1}(x/a) < 0, \quad p \neq -1]. \quad [\text{See } \mathbf{261.01\text{--}267.01} \text{ and } \mathbf{627.9.}]$$

733.1. $\displaystyle\int \frac{1}{x}\cosh^{-1}\frac{x}{a}\,dx$

$$= \frac{1}{2}\left(\log\frac{2x}{a}\right)^2 + \frac{1}{2^3}\frac{a^2}{x^2} + \frac{1\cdot 3}{2\cdot 4^3}\frac{a^4}{x^4} + \frac{1\cdot 3\cdot 5}{2\cdot 4\cdot 6^3}\frac{a^6}{x^6} + \cdots,$$
$$[\cosh^{-1}(x/a) > 0],$$
$$= -\left[\frac{1}{2}\left(\log\frac{2x}{a}\right)^2 + \frac{1}{2^3}\frac{a^2}{x^2} + \frac{1\cdot 3}{2\cdot 4^3}\frac{a^4}{x^4} + \frac{1\cdot 3\cdot 5}{2\cdot 4\cdot 6^5}\frac{a^6}{x^6} + \cdots\right],$$
$$[\cosh^{-1}(x/a) < 0].$$

733.2. $\displaystyle\int \frac{1}{x^2}\cosh^{-1}\frac{x}{a}\,dx = -\frac{1}{x}\cosh^{-1}\frac{x}{a} + \frac{1}{a}\sec^{-1}\left|\frac{x}{a}\right|,$

$$[\cosh^{-1}(x/a) > 0, \quad 0 < \sec^{-1}|x/a| < \pi/2],$$
$$= -\frac{1}{x}\cosh^{-1}\frac{x}{a} - \frac{1}{a}\sec^{-1}\left|\frac{x}{a}\right|,$$
$$[\cosh^{-1}(x/a) < 0, \quad 0 < \sec^{-1}|x/a| < \pi/2].$$

733.3. $\displaystyle\int \frac{1}{x^3}\cosh^{-1}\frac{x}{a}\,dx = -\frac{1}{2x^2}\cosh^{-1}\frac{x}{a} + \frac{\sqrt{(x^2-a^2)}}{2a^2x},$

$$[\cosh^{-1}(x/a) > 0],$$
$$= -\frac{1}{2x^2}\cosh^{-1}\frac{x}{a} - \frac{\sqrt{(x^2-a^2)}}{2a^2x},$$
$$[\cosh^{-1}(x/a) < 0]. \quad [\text{See } \mathbf{628.1\text{-}.3.}]$$

733.9. $\displaystyle\int \frac{1}{x^p}\cosh^{-1}\frac{x}{a}\,dx$

$$= -\frac{1}{(p-1)x^{p-1}}\cosh^{-1}\frac{x}{a} + \frac{1}{p-1}\int \frac{dx}{x^{p-1}\sqrt{(x^2-a^2)}},$$
$$[\cosh^{-1}(x/a) > 0, \quad p \neq 1],$$
$$= -\frac{1}{(p-1)x^{p-1}}\cosh^{-1}\frac{x}{a} - \frac{1}{p-1}\int \frac{dx}{x^{p-1}\sqrt{(x^2-a^2)}},$$
$$[\cosh^{-1}(x/a) < 0, \quad p \neq 1]. \quad [\text{See } \mathbf{281.01\text{--}284.01} \text{ and } \mathbf{628.9.}]$$

For **732** to **733.9**, $\dfrac{x}{a} > 1$.

734. $$\int \tanh^{-1}\frac{x}{a}\,dx = x\tanh^{-1}\frac{x}{a} + \frac{a}{2}\log(a^2 - x^2).$$

734.1. $$\int x\tanh^{-1}\frac{x}{a}\,dx = \frac{x^2 - a^2}{2}\tanh^{-1}\frac{x}{a} + \frac{ax}{2}.$$

734.2. $$\int x^2\tanh^{-1}\frac{x}{a}\,dx = \frac{x^3}{3}\tanh^{-1}\frac{x}{a} + \frac{ax^2}{6} + \frac{a^3}{6}\log(a^2 - x^2).$$

734.3. $$\int x^3\tanh^{-1}\frac{x}{a}\,dx = \frac{x^4 - a^4}{4}\tanh^{-1}\frac{x}{a} + \frac{ax^3}{12} + \frac{a^3 x}{4}.$$

734.9. $$\int x^p\tanh^{-1}\frac{x}{a}\,dx = \frac{x^{p+1}}{p + 1}\tanh^{-1}\frac{x}{a} - \frac{a}{p + 1}\int\frac{x^{p+1}dx}{a^2 - x^2},$$
$$[p \neq -1]. \qquad [\text{See } 141.1\text{--}148.1.]$$

735.1. $$\int \frac{1}{x}\tanh^{-1}\frac{x}{a}\,dx = \frac{x}{a} + \frac{x^3}{3^2 a^3} + \frac{x^5}{5^2 a^5} + \frac{x^7}{7^2 a^7} + \cdots.$$

735.2. $$\int \frac{1}{x^2}\tanh^{-1}\frac{x}{a}\,dx = -\frac{1}{x}\tanh^{-1}\frac{x}{a} - \frac{1}{2a}\log\left(\frac{a^2 - x^2}{x^2}\right).$$

735.3. $$\int \frac{1}{x^3}\tanh^{-1}\frac{x}{a}\,dx = \frac{1}{2}\left(\frac{1}{a^2} - \frac{1}{x^2}\right)\tanh^{-1}\frac{x}{a} - \frac{1}{2ax}.$$

735.4. $$\int \frac{1}{x^4}\tanh^{-1}\frac{x}{a}\,dx = -\frac{1}{3x^3}\tanh^{-1}\frac{x}{a} - \frac{1}{6ax^2}$$
$$- \frac{1}{6a^3}\log\left(\frac{a^2 - x^2}{x^2}\right).$$

735.5. $$\int \frac{1}{x^5}\tanh^{-1}\frac{x}{a}\,dx = \frac{1}{4}\left(\frac{1}{a^4} - \frac{1}{x^4}\right)\tanh^{-1}\frac{x}{a} - \frac{1}{12ax^3} - \frac{1}{4a^3 x}.$$

735.9. $$\int \frac{1}{x^p}\tanh^{-1}\frac{x}{a}\,dx = -\frac{1}{(p - 1)x^{p-1}}\tanh^{-1}\frac{x}{a}$$
$$+ \frac{a}{p - 1}\int\frac{dx}{x^{p-1}(a^2 - x^2)}, \qquad [p \neq 1].$$
$$[\text{See } 151.1\text{--}155.1.]$$

For **734–735.9**, $x^2 < a^2$.

736. $$\int \operatorname{ctnh}^{-1}\frac{x}{a}\,dx = x\operatorname{ctnh}^{-1}\frac{x}{a} + \frac{a}{2}\log(x^2 - a^2).$$

736.1. $$\int x\operatorname{ctnh}^{-1}\frac{x}{a}\,dx = \frac{x^2 - a^2}{2}\operatorname{ctnh}^{-1}\frac{x}{a} + \frac{ax}{2}.$$

736.2. $\displaystyle\int x^2 \operatorname{ctnh}^{-1} \frac{x}{a}\, dx = \frac{x^3}{3} \operatorname{ctnh}^{-1} \frac{x}{a} + \frac{ax^2}{6} + \frac{a^3}{6} \log (x^2 - a^2).$

736.3. $\displaystyle\int x^3 \operatorname{ctnh}^{-1} \frac{x}{a}\, dx = \frac{x^4 - a^4}{4} \operatorname{ctnh}^{-1} \frac{x}{a} + \frac{ax^3}{12} + \frac{a^3 x}{4}\, .$

736.9. $\displaystyle\int x^p \operatorname{ctnh}^{-1} \frac{x}{a}\, dx = \frac{x^{p+1}}{p+1} \operatorname{ctnh}^{-1} \frac{x}{a} - \frac{a}{p+1} \int \frac{x^{p+1} dx}{a^2 - x^2}\, ,$

$$[p \neq -1]. \qquad [\text{See } 141.1\text{--}148.1.]$$

737.1. $\displaystyle\int \frac{1}{x} \operatorname{ctnh}^{-1} \frac{x}{a}\, dx = -\frac{a}{x} - \frac{a^3}{3^2 x^3} - \frac{a^5}{5^2 x^5} - \frac{a^7}{7^2 x^7} - \cdots.$

737.2. $\displaystyle\int \frac{1}{x^2} \operatorname{ctnh}^{-1} \frac{x}{a}\, dx = -\frac{1}{x} \operatorname{ctnh}^{-1} \frac{x}{a} - \frac{1}{2a} \log \left(\frac{x^2 - a^2}{x^2} \right)\, .$

737.3. $\displaystyle\int \frac{1}{x^3} \operatorname{ctnh}^{-1} \frac{x}{a}\, dx = \frac{1}{2} \left(\frac{1}{a^2} - \frac{1}{x^2} \right) \operatorname{ctnh}^{-1} \frac{x}{a} - \frac{1}{2ax}\, .$

737.4. $\displaystyle\int \frac{1}{x^4} \operatorname{ctnh}^{-1} \frac{x}{a}\, dx = -\frac{1}{3x^3} \operatorname{ctnh}^{-1} \frac{x}{a}$

$$-\frac{1}{6ax^2} - \frac{1}{6a^3} \log \left(\frac{x^2 - a^2}{x^2} \right)\, .$$

737.5. $\displaystyle\int \frac{1}{x^5} \operatorname{ctnh}^{-1} \frac{x}{a}\, dx = \frac{1}{4} \left(\frac{1}{a^4} - \frac{1}{x^4} \right) \operatorname{ctnh}^{-1} \frac{x}{a} - \frac{1}{12ax^3} - \frac{1}{4a^3 x}\, .$

737.9. $\displaystyle\int \frac{1}{x^p} \operatorname{ctnh}^{-1} \frac{x}{a}\, dx = -\frac{1}{(p-1)x^{p-1}} \operatorname{ctnh}^{-1} \frac{x}{a}$

$$+\frac{a}{p-1} \int \frac{dx}{x^{p-1}(a^2 - x^2)}, \qquad [p \neq 1].$$

$$[\text{See } 151.1\text{--}155.1.]$$

For **736–737.9**, $x^2 > a^2$.

738. $\displaystyle\int \operatorname{sech}^{-1} \frac{x}{a}\, dx = x \operatorname{sech}^{-1} \frac{x}{a} + a \sin^{-1} \frac{x}{a},$

$$[\operatorname{sech}^{-1} (x/a) > 0],$$

$$= x \operatorname{sech}^{-1} \frac{x}{a} - a \sin^{-1} \frac{x}{a},$$

$$[\operatorname{sech}^{-1} (x/a) < 0].$$

738.1. $\displaystyle\int x \operatorname{sech}^{-1}\frac{x}{a}\,dx = \frac{x^2}{2}\operatorname{sech}^{-1}\frac{x}{a} - \frac{a}{2}\sqrt{(a^2 - x^2)},$

$$[\operatorname{sech}^{-1}(x/a) > 0],$$

$$= \frac{x^2}{2}\operatorname{sech}^{-1}\frac{x}{a} + \frac{a}{2}\sqrt{(a^2 - x^2)},$$

$$[\operatorname{sech}^{-1}(x/a) < 0].$$

738.2. $\displaystyle\int x^2 \operatorname{sech}^{-1}\frac{x}{a}\,dx = \frac{x^3}{3}\operatorname{sech}^{-1}\frac{x}{a} - \frac{ax}{6}\sqrt{(a^2 - x^2)}$

$$+ \frac{a^3}{6}\sin^{-1}\frac{x}{a}, \quad [\operatorname{sech}^{-1}(x/a) > 0],$$

$$= \frac{x^3}{3}\operatorname{sech}^{-1}\frac{x}{a} + \frac{ax}{6}\sqrt{(a^2 - x^2)}$$

$$- \frac{a^3}{6}\sin^{-1}\frac{x}{a}, \quad [\operatorname{sech}^{-1}(x/a) < 0].$$

738.9. $\displaystyle\int x^p \operatorname{sech}^{-1}\frac{x}{a}\,dx$

$$= \frac{x^{p+1}}{p+1}\operatorname{sech}^{-1}\frac{x}{a} + \frac{a}{p+1}\int \frac{x^p\,dx}{\sqrt{(a^2 - x^2)}},$$

$$[\operatorname{sech}^{-1}(x/a) > 0, \quad p \neq -1],$$

$$= \frac{x^{p+1}}{p+1}\operatorname{sech}^{-1}\frac{x}{a} - \frac{a}{p+1}\int \frac{x^p\,dx}{\sqrt{(a^2 - x^2)}},$$

$$[\operatorname{sech}^{-1}(x/a) < 0, \quad p \neq -1].$$

$$[\text{See } \mathbf{320.01\text{--}327.01.}]$$

739.1. $\displaystyle\int \frac{1}{x}\operatorname{sech}^{-1}\frac{x}{a}\,dx = -\frac{1}{2}\left(\log\frac{a}{x}\right)\log\frac{4a}{x} - \frac{1}{2^3}\frac{x^2}{a^2} - \frac{1\cdot3}{2\cdot4^3}\frac{x^4}{a^4}$

$$- \frac{1\cdot3\cdot5}{2\cdot4\cdot6^3}\frac{x^6}{a^6} - \cdots, \quad [\operatorname{sech}^{-1}(x/a) > 0],$$

$$= \frac{1}{2}\left(\log\frac{a}{x}\right)\log\frac{4a}{x} + \frac{1}{2^3}\frac{x^2}{a^2} + \frac{1\cdot3}{2\cdot4^3}\frac{x^4}{a^4}$$

$$+ \frac{1\cdot3\cdot5}{2\cdot4\cdot6^3}\frac{x^6}{a^6} + \cdots, \quad [\operatorname{sech}^{-1}(x/a) < 0].$$

739.2. $\displaystyle\int \frac{1}{x^2}\operatorname{sech}^{-1}\frac{x}{a}\,dx = -\frac{1}{x}\operatorname{sech}^{-1}\frac{x}{a} + \frac{\sqrt{(a^2 - x^2)}}{ax},$

$$[\operatorname{sech}^{-1}(x/a) > 0],$$

$$= -\frac{1}{x}\operatorname{sech}^{-1}\frac{x}{a} - \frac{\sqrt{(a^2 - x^2)}}{ax},$$

$$[\operatorname{sech}^{-1}(x/a) < 0].$$

739.9. $\displaystyle\int \frac{1}{x^p}\, \operatorname{sech}^{-1}\frac{x}{a}\, dx$

$$= -\frac{1}{(p-1)x^{p-1}}\, \operatorname{sech}^{-1}\frac{x}{a} - \frac{a}{p-1}\int \frac{dx}{x^p\sqrt{(a^2-x^2)}},$$
$$[\operatorname{sech}^{-1}(x/a) > 0, \quad p \neq 1],$$

$$= -\frac{1}{(p-1)x^{p-1}}\, \operatorname{sech}^{-1}\frac{x}{a} + \frac{a}{p-1}\int \frac{dx}{x^p\sqrt{(a^2-x^2)}},$$
$$[\operatorname{sech}^{-1}(x/a) < 0, \quad p \neq 1].$$
$$[\text{See } 342.01\text{–}346.01.]$$

For **738–739.9**, $0 < x/a < 1$.

740. $\displaystyle\int \operatorname{csch}^{-1}\frac{x}{a}\, dx = x\, \operatorname{csch}^{-1}\frac{x}{a} + a\, \sinh^{-1}\frac{x}{a}.$

740.1. $\displaystyle\int x\, \operatorname{csch}^{-1}\frac{x}{a}\, dx = \frac{x^2}{2}\, \operatorname{csch}^{-1}\frac{x}{a} + \frac{a}{2}\sqrt{(x^2+a^2)}.$

740.9. $\displaystyle\int x^p\, \operatorname{csch}^{-1}\frac{x}{a}\, dx = \frac{x^{p+1}}{p+1}\, \operatorname{csch}^{-1}\frac{x}{a}$

$$+ \frac{a}{p+1}\int \frac{x^p dx}{\sqrt{(x^2+a^2)}}, \quad [p \neq -1].$$
$$[\text{See } 200.01\text{–}207.01.]$$

741.1. $\displaystyle\int \frac{1}{x}\, \operatorname{csch}^{-1}\frac{x}{a}\, dx = -\frac{a}{x} + \frac{1}{2\cdot 3\cdot 3}\frac{a^3}{x^3} - \frac{1\cdot 3}{2\cdot 4\cdot 5\cdot 5}\frac{a^5}{x^5}$

$$+ \frac{1\cdot 3\cdot 5}{2\cdot 4\cdot 6\cdot 7\cdot 7}\frac{a^7}{x^7} - \cdots, \quad [x^2 > a^2],$$

$$= -\frac{1}{2}\left(\log\frac{a}{x}\right)\log\frac{4a}{x} + \frac{1}{2^3}\frac{x^2}{a^2} - \frac{1\cdot 3}{2\cdot 4^3}\frac{x^4}{a^4}$$

$$+ \frac{1\cdot 3\cdot 5}{2\cdot 4\cdot 6^3}\frac{x^6}{a^6} - \cdots, \quad [0 < x/a < 1],$$

$$= \frac{1}{2}\log\left|\frac{a}{x}\right|\log\left|\frac{4a}{x}\right| - \frac{1}{2^3}\frac{x^2}{a^2} + \frac{1\cdot 3}{2\cdot 4^3}\frac{x^4}{a^4}$$

$$- \frac{1\cdot 3\cdot 5}{2\cdot 4\cdot 6^3}\frac{x^6}{a^6} + \cdots, \quad [-1 < x/a < 0].$$

741.9. $\displaystyle\int \frac{1}{x^p}\, \operatorname{csch}^{-1}\frac{x}{a}\, dx$

$$= -\frac{1}{(p-1)x^{p-1}}\, \operatorname{csch}^{-1}\frac{x}{a} - \frac{a}{p-1}\int \frac{dx}{x^p\sqrt{(x^2+a^2)}},$$
$$[p \neq 1]. \qquad [\text{See } 222.01\text{–}226.01.]$$

ELLIPTIC FUNCTIONS

750. Let $u = \displaystyle\int_0^\varphi \frac{d\varphi}{\sqrt{(1 - k^2 \sin^2 \varphi)}}$, $\qquad [k^2 < 1]$,

$\qquad = \displaystyle\int_0^x \frac{dx}{\sqrt{(1 - x^2)}\sqrt{(1 - k^2 x^2)}}$, $\qquad [x = \sin \varphi]$,

$\qquad = F(\varphi, k) = $ **elliptic integral of the first kind.**

[See **770.**]

751.1. φ is the *amplitude*, and k the *modulus*.

751.2. $\varphi = \operatorname{am} u$.

751.3. $\sin \varphi = \operatorname{sn} u = x$.

751.4. $\cos \varphi = \operatorname{cn} u = \sqrt{(1 - x^2)}$.

751.5. $\Delta\varphi$ or $\Delta(\varphi, k) = \sqrt{(1 - k^2 \sin^2 \varphi)} = \operatorname{dn} u = \sqrt{(1 - k^2 x^2)}$.

751.6. $\tan \varphi = \operatorname{tn} u = \dfrac{x}{\sqrt{(1 - x^2)}}$.

751.7. The *complementary modulus* $= k' = \sqrt{(1 - k^2)}$.

752. $u = \operatorname{am}^{-1}(\varphi, k) = \operatorname{sn}^{-1}(x, k) = \operatorname{cn}^{-1}\{\sqrt{(1 - x^2)}, k)\}$

$\qquad = \operatorname{dn}^{-1}\{\sqrt{(1 - k^2 x^2)}, k\} = \operatorname{tn}^{-1}\left[\dfrac{x}{\sqrt{(1 - x^2)}}, k\right]$.

753.1. $\operatorname{am}(-u) = -\operatorname{am} u$.		**754.2.** $\operatorname{sn} 0 = 0$.	
753.2. $\operatorname{sn}(-u) = -\operatorname{sn} u$.		**754.3.** $\operatorname{cn} 0 = 1$.	
753.3. $\operatorname{cn}(-u) = \operatorname{cn} u$.		**754.4.** $\operatorname{dn} 0 = 1$.	
753.4. $\operatorname{dn}(-u) = \operatorname{dn} u$.		**755.1.** $\operatorname{sn}^2 u + \operatorname{cn}^2 u = 1$.	
753.5. $\operatorname{tn}(-u) = -\operatorname{tn} u$.		**755.2.** $\operatorname{dn}^2 u + k^2 \operatorname{sn}^2 u = 1$.	
754.1. $\operatorname{am} 0 = 0$.		**755.3.** $\operatorname{dn}^2 u - k^2 \operatorname{cn}^2 u = k'^2$.	

756.1. $\operatorname{sn}(u \pm v) = \dfrac{\operatorname{sn} u \operatorname{cn} v \operatorname{dn} v \pm \operatorname{cn} u \operatorname{sn} v \operatorname{dn} u}{1 - k^2 \operatorname{sn}^2 u \operatorname{sn}^2 v}$.

756.2. $\operatorname{cn}(u \pm v) = \dfrac{\operatorname{cn} u \operatorname{cn} v \mp \operatorname{sn} u \operatorname{sn} v \operatorname{dn} u \operatorname{dn} v}{1 - k^2 \operatorname{sn}^2 u \operatorname{sn}^2 v}$.

756.3. $\operatorname{dn}(u \pm v) = \dfrac{\operatorname{dn} u \operatorname{dn} v \mp k^2 \operatorname{sn} u \operatorname{sn} v \operatorname{cn} u \operatorname{cn} v}{1 - k^2 \operatorname{sn}^2 u \operatorname{sn}^2 v}$.

756.4. $\operatorname{tn}(u \pm v) = \dfrac{\operatorname{tn} u \operatorname{dn} v \pm \operatorname{tn} v \operatorname{dn} u}{1 \mp \operatorname{tn} u \operatorname{tn} v \operatorname{dn} u \operatorname{dn} v}$.

757.1. $\operatorname{sn} 2u = \dfrac{2 \operatorname{sn} u \operatorname{cn} u \operatorname{dn} u}{1 - k^2 \operatorname{sn}^4 u}.$

757.2. $\operatorname{cn} 2u = \dfrac{\operatorname{cn}^2 u - \operatorname{sn}^2 u \operatorname{dn}^2 u}{1 - k^2 \operatorname{sn}^4 u} = \dfrac{2 \operatorname{cn}^2 u}{1 - k^2 \operatorname{sn}^4 u} - 1.$

757.3. $\operatorname{dn} 2u = \dfrac{\operatorname{dn}^2 u - k^2 \operatorname{sn}^2 u \operatorname{cn}^2 u}{1 - k^2 \operatorname{sn}^4 u} = \dfrac{2 \operatorname{dn}^2 u}{1 - k^2 \operatorname{sn}^4 u} - 1.$

757.4. $\operatorname{tn} 2u = \dfrac{2 \operatorname{tn} u \operatorname{dn} u}{1 - \operatorname{tn}^2 u \operatorname{dn}^2 u}.$

758.1. $\operatorname{sn} \dfrac{u}{2} = \sqrt{\left(\dfrac{1 - \operatorname{cn} u}{1 + \operatorname{dn} u} \right)}.$

758.2. $\operatorname{cn} \dfrac{u}{2} = \sqrt{\left(\dfrac{\operatorname{cn} u + \operatorname{dn} u}{1 + \operatorname{dn} u} \right)}.$

758.3. $\operatorname{dn} \dfrac{u}{2} = \sqrt{\left(\dfrac{\operatorname{cn} u + \operatorname{dn} u}{1 + \operatorname{cn} u} \right)}.$

759.1. $\operatorname{sn} (iu, k) = i \operatorname{tn} (u, k').$

759.2. $\operatorname{cn} (iu, k) = \dfrac{1}{\operatorname{cn} (u, k')}.$

759.3. $\operatorname{dn} (iu, k) = \dfrac{\operatorname{dn} (u, k')}{\operatorname{cn} (u, k')}.$

760.1. $\operatorname{sn} u = u - (1 + k^2) \dfrac{u^3}{3!} + (1 + 14k^2 + k^4) \dfrac{u^5}{5!}$

$$- (1 + 135k^2 + 135k^4 + k^6) \dfrac{u^7}{7!} + \cdots.$$

760.2. $\operatorname{cn} u = 1 - \dfrac{u^2}{2!} + (1 + 4k^2) \dfrac{u^4}{4!} - (1 + 44k^2 + 16k^4) \dfrac{u^6}{6!}$

$$+ (1 + 408k^2 + 912k^4 + 64k^6) \dfrac{u^8}{8!} - \cdots.$$

760.3. $\operatorname{dn} u = 1 - k^2 \dfrac{u^2}{2!} + (4 + k^2)k^2 \dfrac{u^4}{4!} - (16 + 44k^2 + k^4)k^2 \dfrac{u^6}{6!}$

$$+ (64 + 912k^2 + 408k^4 + k^6)k^2 \dfrac{u^8}{8!} - \cdots.$$

760.4. $\operatorname{am} u = u - k^2 \dfrac{u^3}{3!} + (4 + k^2)k^2 \dfrac{u^5}{5!} - (16 + 44k^2 + k^4)k^2 \dfrac{u^7}{7!}$

$$+ (64 + 912k^2 + 408k^4 + k^6)k^2 \dfrac{u^9}{9!} - \cdots.$$

[Ref. 21, p. 156.]

ELLIPTIC FUNCTIONS—DERIVATIVES

768.1. $\dfrac{d}{du} \operatorname{sn} u = \operatorname{cn} u \operatorname{dn} u.$

768.2. $\dfrac{d}{du} \operatorname{cn} u = - \operatorname{sn} u \operatorname{dn} u.$

768.3. $\dfrac{d}{du} \operatorname{dn} u = - k^2 \operatorname{sn} u \operatorname{cn} u.$ [Ref. 36, p. 25.]

ELLIPTIC FUNCTIONS—INTEGRALS

770. Elliptic Integral of the First Kind.

$$F(\varphi, k) = \int_0^\varphi \frac{d\varphi}{\sqrt{(1 - k^2 \sin^2 \varphi)}}, \qquad [k^2 < 1],$$

$$= \int_0^x \frac{dx}{\sqrt{(1 - x^2)}\sqrt{(1 - k^2 x^2)}}, \quad [x = \sin \varphi].$$

[See 750.]

771. Elliptic Integral of the Second Kind.

$$E(\varphi, k) = \int_0^\varphi \sqrt{(1 - k^2 \sin^2 \varphi)}\,d\varphi$$

$$= \int_0^x \frac{\sqrt{(1 - k^2 x^2)}}{\sqrt{(1 - x^2)}}\, dx, \qquad [x = \sin \varphi].$$

772. Elliptic Integral of the Third Kind.

$$\Pi(\varphi, n, k) = \int_0^\varphi \frac{d\varphi}{(1 + n \sin^2 \varphi)\sqrt{(1 - k^2 \sin^2 \varphi)}}$$

$$= \int_0^x \frac{dx}{(1 + nx^2)\sqrt{(1 - x^2)}\sqrt{(1 - k^2 x^2)}},$$

[x = \sin \varphi].

The letter n is called the parameter.

Complete Elliptic Integrals (See Tables 1040–1041)

773.1. $K = \displaystyle\int_0^{\pi/2} \dfrac{d\varphi}{\sqrt{(1 - k^2 \sin^2 \varphi)}}$

$\qquad = \dfrac{\pi}{2}\left(1 + \dfrac{1^2}{2^2} k^2 + \dfrac{1^2 \cdot 3^2}{2^2 \cdot 4^2} k^4 + \dfrac{1^2 \cdot 3^2 \cdot 5^2}{2^2 \cdot 4^2 \cdot 6^2} k^6 + \cdots\right),$

$\hfill [k^2 < 1].$

773.2. $K = \dfrac{\pi}{2}(1 + m)\left[1 + \dfrac{1^2}{2^2} m^2 + \dfrac{1^2 \cdot 3^2}{2^2 \cdot 4^2} m^4\right.$

$\hfill \left. + \dfrac{1^2 \cdot 3^2 \cdot 5^2}{2^2 \cdot 4^2 \cdot 6^2} m^6 + \cdots\right],$

where $m = (1 - k')/(1 + k')$. $\hfill$ [Ref. 31, p. 135.]

This series is more rapidly convergent than **773.1** since $m^2 < k^2$.

773.3. $K = \log \dfrac{4}{k'} + \dfrac{1^2}{2^2}\left(\log \dfrac{4}{k'} - \dfrac{2}{1 \cdot 2}\right) k'^2$

$\qquad + \dfrac{1^2 \cdot 3^2}{2^2 \cdot 4^2}\left(\log \dfrac{4}{k'} - \dfrac{2}{1 \cdot 2} - \dfrac{2}{3 \cdot 4}\right) k'^4$

$\qquad + \dfrac{1^2 \cdot 3^2 \cdot 5^2}{2^2 \cdot 4^2 \cdot 6^2}\left(\log \dfrac{4}{k'} - \dfrac{2}{1 \cdot 2} - \dfrac{2}{3 \cdot 4} - \dfrac{2}{5 \cdot 6}\right) k'^6 + \cdots,$

where $k' = \sqrt{(1 - k^2)}$, and log denotes natural logarithm.

$\hfill$ [Ref. 33, pp. 46 and 54.]

774.1. $E = \displaystyle\int_0^{\pi/2} \sqrt{(1 - k^2 \sin^2 \varphi)}\, d\varphi$

$\qquad = \dfrac{\pi}{2}\left(1 - \dfrac{1}{2^2} k^2 - \dfrac{1^2 \cdot 3}{2^2 \cdot 4^2} k^4 - \dfrac{1^2 \cdot 3^2 \cdot 5}{2^2 \cdot 4^2 \cdot 6^2} k^6 - \cdots\right),$

$\hfill [k^2 < 1].$

774.2. $E = \dfrac{\pi}{2(1 + m)}\left[1 + \dfrac{m^2}{2^2} + \dfrac{1^2}{2^2 \cdot 4^2} m^4 + \dfrac{1^2 \cdot 3^2}{2^2 \cdot 4^2 \cdot 6^2} m^6 + \cdots\right]$

where $m = (1 - k')/(1 + k')$. $\hfill$ [Ref. 31, p. 136.]

This series is more rapidly convergent than **774.1** since $m^2 < k^2$.

774.3. $E = 1 + \dfrac{1}{2}\left(\log \dfrac{4}{k'} - \dfrac{1}{1 \cdot 2}\right) k'^2$

$\qquad + \dfrac{1^2 \cdot 3}{2^2 \cdot 4}\left(\log \dfrac{4}{k'} - \dfrac{2}{1 \cdot 2} - \dfrac{1}{3 \cdot 4}\right) k'^4$

$\qquad + \dfrac{1^2 \cdot 3^2 \cdot 5}{2^2 \cdot 4^2 \cdot 6}\left(\log \dfrac{4}{k'} - \dfrac{2}{1 \cdot 2} - \dfrac{2}{3 \cdot 4} - \dfrac{1}{5 \cdot 6}\right) k'^6 + \cdots.$

$\hfill$ [Ref. 33, pp. 46 and 54.]

775. $\displaystyle F(\varphi, k) = \int_0^\varphi \frac{d\varphi}{\sqrt{(1 - k^2 \sin^2 \varphi)}}$

$$= \frac{2\varphi}{\pi} K - \sin \varphi \cos \varphi \left(\frac{1}{2} A_2 k^2 + \frac{1 \cdot 3}{2 \cdot 4} A_4 k^4 \right.$$

$$\left. + \frac{1 \cdot 3 \cdot 5}{2 \cdot 4 \cdot 6} A_6 k^6 + \cdots \right),$$

where

$$A_2 = \frac{1}{2}, \qquad A_4 = \frac{3}{2 \cdot 4} + \frac{1}{4} \sin^2 \varphi,$$

$$A_6 = \frac{3 \cdot 5}{2 \cdot 4 \cdot 6} + \frac{5}{4 \cdot 6} \sin^2 \varphi + \frac{1}{6} \sin^4 \varphi,$$

$$A_8 = \frac{3 \cdot 5 \cdot 7}{2 \cdot 4 \cdot 6 \cdot 8} + \frac{5 \cdot 7}{4 \cdot 6 \cdot 8} \sin^2 \varphi + \frac{7}{6 \cdot 8} \sin^4 \varphi + \frac{1}{8} \sin^6 \varphi,$$

and K is found by **773** or from tables.   [Ref. 5, No. 526.]

776. $\displaystyle F(\varphi, k) = \varphi + \frac{1}{2} v_2 k^2 + \frac{1 \cdot 3}{2 \cdot 4} v_4 k^4 + \frac{1 \cdot 3 \cdot 5}{2 \cdot 4 \cdot 6} v_6 k^6 + \cdots,$
where

$$v_{2n} = \int \sin^{2n} \varphi \, d\varphi. \qquad \text{[See } \mathbf{430}.] \quad \text{[Ref. 36, p. 26.]}$$

777. $\displaystyle E(\varphi, k) = \int_0^\varphi \sqrt{(1 - k^2 \sin^2 \varphi)} d\varphi$

$$= \frac{2\varphi}{\pi} E + \sin \varphi \cos \varphi \left(\frac{1}{2} A_2 k^2 + \frac{1}{2 \cdot 4} A_4 k^4 \right.$$

$$\left. + \frac{1 \cdot 3}{2 \cdot 4 \cdot 6} A_6 k^6 + \cdots \right)$$

where $A_2, A_4, \cdots$ are given in **775**, and where E may be obtained by **774** or from tables.   [Ref. 5, No. 527.]

780.1. $\displaystyle \int_0^x \frac{dx}{\sqrt{(1 + x^2)}\sqrt{(1 + k'^2 x^2)}} = \operatorname{tn}^{-1}(x, k) = F(\tan^{-1} x, k),$

$$[0 < x < 1]. \quad \text{[Ref. 36, p. 42, eq. (4).]}$$

780.2. $\displaystyle \int_0^x \frac{dx}{\sqrt{(a^2 - x^2)}\sqrt{(b^2 - x^2)}} = \frac{1}{a} \operatorname{sn}^{-1}\left(\frac{x}{b}, \frac{b}{a}\right),$

$$[a > b > x > 0]. \quad \text{[Ref. 5, No. 536.]}$$

780.3. $\displaystyle \int_0^x \frac{dx}{\sqrt{(a^2 + x^2)}\sqrt{(b^2 + x^2)}} = \frac{1}{a} \operatorname{tn}^{-1}\left[\frac{x}{b}, \sqrt{\left(\frac{a^2 - b^2}{a^2}\right)}\right],$

$$[a^2 > b^2, \quad x > 0]. \quad \text{[Ref. 5, No. 541.]}$$

785.1. $\int \operatorname{sn} u \, du = -\frac{1}{k} \cosh^{-1}\left(\frac{\operatorname{dn} u}{k'}\right).$ 　　[Ref. 36, p. 58.]

785.2. $\int \operatorname{cn} u \, du = \frac{1}{k} \cos^{-1}(\operatorname{dn} u).$

785.3. $\int \operatorname{dn} u \, du = \sin^{-1}(\operatorname{sn} u) = \operatorname{am} u.$

786.1. $\int \frac{du}{\operatorname{sn} u} = \log\left(\frac{\operatorname{sn} u}{\operatorname{cn} u + \operatorname{dn} u}\right).$

786.2. $\int \frac{du}{\operatorname{cn} u} = \frac{1}{k'} \log\left(\frac{k' \operatorname{sn} u + \operatorname{dn} u}{\operatorname{cn} u}\right).$

786.3. $\int \frac{du}{\operatorname{dn} u} = \frac{1}{k'} \tan^{-1}\left(\frac{k' \operatorname{sn} u - \operatorname{cn} u}{k' \operatorname{sn} u + \operatorname{cn} u}\right).$ 　[Ref. 5, No. 563.]

787.1. $\int_0^u \operatorname{sn}^2 u \, du = \frac{1}{k^2}\{u - E(\operatorname{am} u, k)\}.$

787.2. $\int_0^u \operatorname{cn}^2 u \, du = \frac{1}{k^2}\{E(\operatorname{am} u, k) - k'^2 u\}.$

787.3. $\int_0^u \operatorname{dn}^2 u \, du = E(\operatorname{am} u, k).$

787.4. $\int_0^u \operatorname{tn}^2 u \, du = \frac{1}{k'^2}\{\operatorname{dn} u \operatorname{tn} u - E(\operatorname{am} u, k)\}.$

788.1. $\int \operatorname{sn}^{-1} x \, dx = x \operatorname{sn}^{-1} x + \frac{1}{k} \cosh\left[\frac{\sqrt{(1 - k^2 x^2)}}{k'}\right].$

788.2. $\int \operatorname{cn}^{-1} x \, dx = x \operatorname{cn}^{-1} x - \frac{1}{k} \cos^{-1} \sqrt{(k'^2 + k^2 x^2)}.$

788.3. $\int \operatorname{dn}^{-1} x \, dx = x \operatorname{dn}^{-1} x - \sin^{-1}\left[\frac{\sqrt{(1 - x^2)}}{k}\right].$
　　　　　　　　　　　　　　　　　[Ref. 36, Chap. III.]

789.1. $\frac{\partial E}{\partial k} = \frac{1}{k}(E - K).$

789.2. $\frac{\partial K}{\partial k} = \frac{1}{k}\left(\frac{E}{k'^2} - K\right).$

BESSEL FUNCTIONS

800.   Bessel's differential equation is

$$\frac{d^2u}{dx^2} + \frac{1}{x}\frac{du}{dx} + \left(1 - \frac{n^2}{x^2}\right)u = 0.$$

[Ref. 12, p. 7, eq. (7).]

Bessel Function of the First Kind, $J_n(x)$

Denote $\dfrac{d}{dx}J_n(x)$ by J_n', etc.

801.1. $xJ_n' = nJ_n - xJ_{n+1}.$ **801.3.** $2nJ_n = xJ_{n-1} + xJ_{n+1}.$

801.2. $xJ_n' = -nJ_n + xJ_{n-1}.$ **801.4.** $2J_n' = J_{n-1} - J_{n+1}.$

801.5. $4J_n'' = J_{n-2} - 2J_n + J_{n+2}.$

801.6. $\dfrac{d}{dx}(x^nJ_n) = x^nJ_{n-1}.$ **801.7.** $\dfrac{d}{dx}(x^{-n}J_n) = -x^{-n}J_{n+1}.$

801.82. $J_2 = \dfrac{2J_1}{x} - J_0.$

801.83. $J_3 = \left(\dfrac{8}{x^2} - 1\right)J_1 - \dfrac{4J_0}{x}.$

801.84. $J_4 = \left(1 - \dfrac{24}{x^2}\right)J_0 + \dfrac{8}{x}\left(\dfrac{6}{x^2} - 1\right)J_1.$

801.85. $J_5 = \dfrac{12}{x}\left(1 - \dfrac{16}{x^2}\right)J_0 + \left(\dfrac{384}{x^4} - \dfrac{72}{x^2} + 1\right)J_1.$

801.90. $J_0' = -J_1$

801.91. $J_1' = J_0 - \dfrac{J_1}{x}.$

801.92. $J_2' = \dfrac{2J_0}{x} + \left(1 - \dfrac{4}{x^2}\right)J_1.$

801.93. $J_3' = \left(\dfrac{12}{x^2} - 1\right)J_0 + \left(5 - \dfrac{24}{x^2}\right)\dfrac{J_1}{x}.$

801.94. $J_4' = \dfrac{8}{x}\left(\dfrac{12}{x^2} - 1\right) J_0 - \left(\dfrac{192}{x^4} - \dfrac{40}{x^2} + 1\right) J_1.$

801.95. $J_5' = \left(\dfrac{960}{x^4} - \dfrac{84}{x^2} + 1\right) J_0 - \left(\dfrac{1920}{x^4} - \dfrac{408}{x^2} + 13\right) \dfrac{J_1}{x}.$

For tables of $J_0(x)$ and $J_1(x)$ see Ref. 50; Ref. 12, p. 267, Ref. 13, p. 666, and Ref. 17.

Bessel Function of the Second Kind, $N_n(x)$

$N_n(x)$ as in Ref. 17 and Ref. 62, pp. 357–358, and same as $Y_n(x)$ in Ref. 13 (not boldface Y_n) and Ref. 50.

802.1. $xN_n' = nN_n - xN_{n+1}.$

802.2. $xN_n' = -nN_n + xN_{n-1}.$

802.3. $2nN_n = xN_{n-1} + xN_{n+1}.$

802.4. $2N_n' = N_{n-1} - N_{n+1}.$

802.5. $4N_n'' = N_{n-2} - 2N_n + N_{n+2}.$

802.6. $\dfrac{d}{dx}(x^n N_n) = x^n N_{n-1}.$

802.7. $\dfrac{d}{dx}(x^{-n} N_n) = -x^{-n} N_{n+1}.$

802.82. $N_2 = \dfrac{2N_1}{x} - N_0.$

802.83. $N_3 = \left(\dfrac{8}{x^2} - 1\right) N_1 - \dfrac{4N_0}{x}.$

802.84. $N_4 = \left(1 - \dfrac{24}{x^2}\right) N_0 + \dfrac{8}{x}\left(\dfrac{6}{x^2} - 1\right) N_1.$

802.85. $N_5 = \dfrac{12}{x}\left(1 - \dfrac{16}{x^2}\right) N_0 + \left(\dfrac{384}{x^4} - \dfrac{72}{x^2} + 1\right) N_1.$

802.90. $N_0' = -N_1.$

802.91. $N_1' = N_0 - \dfrac{N_1}{x}.$

802.92. $N_2' = \dfrac{2N_0}{x} + \left(1 - \dfrac{4}{x^2}\right) N_1.$

802.93. $N_3' = \left(\dfrac{12}{x^2} - 1\right) N_0 + \left(5 - \dfrac{24}{x^2}\right) \dfrac{N_1}{x}.$

802.94. $N_4' = \dfrac{8}{x}\left(\dfrac{12}{x^2} - 1\right) N_0 - \left(\dfrac{192}{x^4} - \dfrac{40}{x^2} + 1\right) N_1.$

802.95. $N_5' = \left(\dfrac{960}{x^4} - \dfrac{84}{x^2} + 1\right) N_0 - \left(\dfrac{1920}{x^4} - \dfrac{408}{x^2} + 13\right) \dfrac{N_1}{x}.$

For tables of $N_0(x)$ and $N_1(x)$ see Ref. 50, Ref. 13, p. 666, and Ref. 17.

Modified Bessel Function of the First Kind, $I_n(x)$

803.1. $xI_n' = nI_n + xI_{n+1}.$ **803.3.** $2nI_n = xI_{n-1} - xI_{n+1}.$

803.2. $xI_n' = -nI_n + xI_{n-1}.$ **803.4.** $2I_n' = I_{n-1} + I_{n+1}.$

803.5. $4I_n'' = I_{n-2} + 2I_n + I_{n+2}.$

803.6. $\dfrac{d}{dx}(x^n I_n) = x^n I_{n-1}.$ **803.7.** $\dfrac{d}{dx}(x^{-n} I_n) = x^{-n} I_{n+1}.$

803.82. $I_2 = I_0 - \dfrac{2I_1}{x}.$

803.83. $I_3 = \left(\dfrac{8}{x^2} + 1\right) I_1 - \dfrac{4I_0}{x}.$

803.84. $I_4 = \left(\dfrac{24}{x^2} + 1\right) I_0 - \dfrac{8}{x}\left(\dfrac{6}{x^2} + 1\right) I_1.$

803.85. $I_5 = \left(\dfrac{384}{x^4} + \dfrac{72}{x^2} + 1\right) I_1 - \dfrac{12}{x}\left(\dfrac{16}{x^2} + 1\right) I_0.$

803.90. $I_0' = I_1.$

803.91. $I_1' = I_0 - \dfrac{I_1}{x}.$

803.92. $I_2' = I_1\left(\dfrac{4}{x^2} + 1\right) - \dfrac{2I_0}{x}.$

803.93. $I_3' = \left(\dfrac{12}{x^2} + 1\right) I_0 - \left(\dfrac{24}{x^2} + 5\right) \dfrac{I_1}{x}.$

803.94. $I_4' = \left(\dfrac{192}{x^4} + \dfrac{40}{x^2} + 1\right) I_1 - \dfrac{8}{x}\left(\dfrac{12}{x^2} + 1\right) I_0.$

803.95. $I_5' = \left(\dfrac{960}{x^4} + \dfrac{84}{x^2} + 1\right) I_0 - \left(\dfrac{1920}{x^4} + \dfrac{408}{x^2} + 13\right) \dfrac{I_1}{x}.$

For tables of $I_0(x)$ and $I_1(x)$ see Ref. 50, p. 214, Ref. 12, p. 303, and Ref. 17.   Tables of $e^{-x}I_0(x)$ and $e^{-x}I_1(x)$, Ref. 13.

Modified Bessel Function of the Second Kind, $K_n(x)$

804.1. $xK_n' = nK_n - xK_{n+1}.$

804.2. $xK_n' = -nK_n - xK_{n-1}.$

804.3. $2nK_n = xK_{n+1} - xK_{n-1}.$

804.4. $2K_n' = -K_{n-1} - K_{n+1}.$

804.5. $4K_n'' = K_{n-2} + 2K_n + K_{n+2}.$

804.6. $\dfrac{d}{dx}(x^n K_n) = -x^n K_{n-1}.$

804.7. $\dfrac{d}{dx}(x^{-n} K_n) = -x^{-n} K_{n+1}.$

804.82. $K_2 = K_0 + \dfrac{2K_1}{x}.$

804.83. $K_3 = \dfrac{4K_0}{x} + \left(\dfrac{8}{x^2} + 1\right)K_1.$

804.84. $K_4 = \left(\dfrac{24}{x^2} + 1\right)K_0 + \dfrac{8}{x}\left(\dfrac{6}{x^2} + 1\right)K_1.$

804.85. $K_5 = \dfrac{12}{x}\left(\dfrac{16}{x^2} + 1\right)K_0 + \left(\dfrac{384}{x^4} + \dfrac{72}{x^2} + 1\right)K_1.$

804.90. $K_0' = -K_1$

804.91. $K_1' = -K_0 - \dfrac{K_1}{x}.$

804.92. $K_2' = -\dfrac{2K_0}{x} - \left(\dfrac{4}{x^2} + 1\right)K_1.$

804.93. $K_3' = -\left(\dfrac{12}{x^2} + 1\right)K_0 - \left(\dfrac{24}{x^2} + 5\right)\dfrac{K_1}{x}.$

804.94. $K_4' = -\dfrac{8}{x}\left(\dfrac{12}{x^2} + 1\right)K_0 - \left(\dfrac{192}{x^4} + \dfrac{40}{x^2} + 1\right)K_1.$

804.95. $K_5' = -\left(\dfrac{960}{x^4} + \dfrac{84}{x^2} + 1\right)K_0 - \left(\dfrac{1920}{x^4} + \dfrac{408}{x^2} + 13\right)\dfrac{K_1}{x}.$

For tables of $K_0(x)$ and $K_1(x)$ see Ref. 50, p. 266, and Ref. 12, p. 313. Tables of $e^x K_0(x)$ and $e^x K_1(x)$, Ref. 13. Tables of $(2/\pi)K_0(x)$ and $(2/\pi)K_1(x)$, Ref. 17.

807.1. $J_0(x) = 1 - (\tfrac{1}{2}x)^2 + \dfrac{(\tfrac{1}{2}x)^4}{1^2 \cdot 2^2} - \dfrac{(\tfrac{1}{2}x)^6}{1^2 \cdot 2^2 \cdot 3^2} + \cdots.$

807.21. $J_1(x) = -J_0'(x) = \tfrac{1}{2}x - \dfrac{(\tfrac{1}{2}x)^3}{1^2 \cdot 2} + \dfrac{(\tfrac{1}{2}x)^5}{1^2 \cdot 2^2 \cdot 3} - \cdots.$

807.22. $J_2(x) = \dfrac{x^2}{2^2 2!} - \dfrac{x^4}{2^4 1! 3!} + \dfrac{x^6}{2^6 2! 4!} - \dfrac{x^8}{2^8 3! 5!} + \cdots.$

807.3.   When n is a positive integer,

$$J_n(x) = \frac{(\tfrac{1}{2}x)^n}{n!}\left[1 - \frac{(\tfrac{1}{2}x)^2}{1(n+1)} + \frac{(\tfrac{1}{2}x)^4}{1 \cdot 2(n+1)(n+2)} - \cdots \right].$$

807.4.   When n is an integer,

$$J_{-n}(x) = (-1)^n J_n(x).$$

807.5.   When n is not a positive integer, replace $n!$ in **807.3** by
$\mathrm{II}(n)$.   [See **853.1.**]   [Ref. 12, p. 14, eq. (16).]

807.61. $J_1'(x) = \dfrac{1}{2} - \dfrac{3x^2}{2^3 1! 2!} + \dfrac{5x^4}{2^5 2! 3!} - \dfrac{7x^6}{2^7 3! 4!} + \cdots.$

807.62. $J_2'(x) = \dfrac{x}{4} - \dfrac{4x^3}{2^4 1! 3!} + \dfrac{6x^5}{2^6 2! 4!} - \dfrac{8x^7}{2^8 3! 5!} + \cdots.$

807.69. $J_n'(x) = \dfrac{x^{n-1}}{2^n (n-1)!} - \dfrac{(n+2)x^{n+1}}{2^{n+2} 1!(n+1)!}$

$$+ \frac{(n+4)x^{n+3}}{2^{n+4} 2!(n+2)!} - \frac{(n+6)x^{n+5}}{2^{n+6} 3!(n+3)!} + \cdots,$$
$$[n \text{ an integer} > 0].$$

Asymptotic Series for Large Values of x

808.1. $J_0(x) = \left(\dfrac{2}{\pi x}\right)^{1/2}\left[P_0(x) \cos\left(x - \dfrac{\pi}{4}\right)\right.$

$$\left. - Q_0(x) \sin\left(x - \frac{\pi}{4}\right)\right],$$

where

808.11. $P_0(x) \approx 1 - \dfrac{1^2 \cdot 3^2}{2!(8x)^2} + \dfrac{1^2 \cdot 3^2 \cdot 5^2 \cdot 7^2}{4!(8x)^4}$

$$- \frac{1^2 \cdot 3^2 \cdot 5^2 \cdot 7^2 \cdot 9^2 \cdot 11^2}{6!(8x)^6} + \cdots.$$

808.12. $Q_0(x) \approx -\dfrac{1^2}{1! 8x} + \dfrac{1^2 \cdot 3^2 \cdot 5^2}{3!(8x)^3} - \dfrac{1^2 \cdot 3^2 \cdot 5^2 \cdot 7^2 \cdot 9^2}{5!(8x)^5} + \cdots.$

808.2. $J_1(x) = \left(\dfrac{2}{\pi x}\right)^{1/2}\left[P_1(x) \cos\left(x - \dfrac{3\pi}{4}\right)\right.$

$$\left. - Q_1(x) \sin\left(x - \frac{3\pi}{4}\right)\right],$$

where

808.21. $P_1(x) \approx 1 + \dfrac{1^2 \cdot 3 \cdot 5}{2!(8x)^2} - \dfrac{1^2 \cdot 3^2 \cdot 5^2 \cdot 7 \cdot 9}{4!(8x)^4}$

$$+ \dfrac{1^2 \cdot 3^2 \cdot 5^2 \cdot 7^2 \cdot 9^2 \cdot 11 \cdot 13}{6!(8x)^6} - \cdots;$$

The signs are alternately $+$ and $-$ after the first term.

808.22. $Q_1(x) \approx \dfrac{1 \cdot 3}{1!8x} - \dfrac{1^2 \cdot 3^2 \cdot 5 \cdot 7}{3!(8x)^3} + \dfrac{1^2 \cdot 3^2 \cdot 5^2 \cdot 7^2 \cdot 9 \cdot 11}{5!(8x)^5} - \cdots.$

808.3. $J_n(x) = \left(\dfrac{2}{\pi x}\right)^{1/2} \left[P_n(x) \cos\left(x - \dfrac{n\pi}{2} - \dfrac{\pi}{4}\right)\right.$

$$\left. - Q_n(x) \sin\left(x - \dfrac{n\pi}{2} - \dfrac{\pi}{4}\right) \right],$$

where

808.31. $P_n(x) \approx 1 - \dfrac{(4n^2 - 1^2)(4n^2 - 3^2)}{2!(8x)^2}$

$$+ \dfrac{(4n^2 - 1^2)(4n^2 - 3^2)(4n^2 - 5^2)(4n^2 - 7^2)}{4!(8x)^4} - \cdots.$$

808.32. $Q_n(x) \approx \dfrac{4n^2 - 1^2}{1!8x}$

$$- \dfrac{(4n^2 - 1^2)(4n^2 - 3^2)(4n^2 - 5^2)}{3!(8x)^3} + \cdots.$$

808.4. $J_n'(x) = -\left(\dfrac{2}{\pi x}\right)^{1/2} \left[P_n^{(1)}(x) \sin\left(x - \dfrac{n\pi}{2} - \dfrac{\pi}{4}\right)\right.$

$$\left. + Q_n^{(1)}(x) \cos\left(x - \dfrac{n\pi}{2} - \dfrac{\pi}{4}\right) \right],$$

where, from **801.4,**

808.41. $P_n^{(1)}(x) \approx 1 - \dfrac{(4n^2 - 1^2)(4n^2 + 3 \times 5)}{2!(8x)^2}$

$$+ \dfrac{(4n^2 - 1^2)(4n^2 - 3^2)(4n^2 - 5^2)(4n^2 + 7 \times 9)}{4!(8x)^4} - \cdots.$$

808.42. $Q_n^{(1)}(x) \approx \dfrac{4n^2 + 1 \times 3}{1!8x}$

$$- \dfrac{(4n^2 - 1^2)(4n^2 - 3^2)(4n^2 + 5 \times 7)}{3!(8x)^3} + \cdots.$$

Extension of these series can be made by inspection. The sign $\approx$ denotes approximate equality. Note that the various series for large values of x are asymptotic expansions and there is a limit to the amount of precision which they will give.

809.01. $J_{\frac{1}{2}}(x) = \left(\dfrac{2}{\pi x}\right)^{1/2} \sin x.$

809.03. $J_{\frac{3}{2}}(x) = \left(\dfrac{2}{\pi x}\right)^{1/2} \left(\dfrac{\sin x}{x} - \cos x\right).$

809.05. $J_{\frac{5}{2}}(x) = \left(\dfrac{2}{\pi x}\right)^{1/2} \left\{\left(\dfrac{3}{x^2} - 1\right) \sin x - \dfrac{3}{x} \cos x\right\}.$

809.21. $J_{-\frac{1}{2}}(x) = \left(\dfrac{2}{\pi x}\right)^{1/2} \cos x.$

809.23. $J_{-\frac{3}{2}}(x) = \left(\dfrac{2}{\pi x}\right)^{1/2} \left(-\sin x - \dfrac{\cos x}{x}\right).$

809.25. $J_{-\frac{5}{2}}(x) = \left(\dfrac{2}{\pi x}\right)^{1/2} \left\{\dfrac{3}{x} \sin x + \left(\dfrac{3}{x^2} - 1\right) \cos x\right\}.$

[For higher orders see Ref. 12, p. 17.]

811.1. $N_0(x) = \dfrac{2}{\pi}\left(\gamma + \log_\epsilon \dfrac{x}{2}\right) J_0(x) + \dfrac{2}{\pi} \dfrac{(\frac{1}{2}x)^2}{(1!)^2}$

$$- \dfrac{2}{\pi} \dfrac{(\frac{1}{2}x)^4}{(2!)^2}\left(1 + \tfrac{1}{2}\right) + \dfrac{2}{\pi} \dfrac{(\frac{1}{2}x)^6}{(3!)^2}\left(1 + \tfrac{1}{2} + \tfrac{1}{3}\right) - \cdots,$$

where γ is Euler's constant 0.577 2157.   [See **851.1.**]

[See note preceding **802.1.**]

811.2. $N_1(x) = \dfrac{2}{\pi}\left(\gamma + \log_\epsilon \dfrac{x}{2}\right) J_1(x) - \dfrac{2}{\pi x}$

$$- \dfrac{1}{\pi} \sum_{p=0}^{\infty} \dfrac{(-1)^p}{p!(p+1)!}\left(\dfrac{x}{2}\right)^{2p+1}\left\{2\left(1 + \tfrac{1}{2} + \cdots + \dfrac{1}{p}\right) + \dfrac{1}{p+1}\right\}.$$

811.3. $N_n(x) = \dfrac{2}{\pi}\left(\gamma + \log_\epsilon \dfrac{x}{2}\right) J_n(x)$

$$- \dfrac{1}{\pi} \sum_{p=0}^{n-1} \dfrac{(n-p-1)!}{p!}\left(\dfrac{x}{2}\right)^{2p-n} - \dfrac{1}{\pi} \sum_{p=0}^{\infty} \dfrac{(-1)^p}{p!(n+p)!}\left(\dfrac{x}{2}\right)^{2p+n}$$

$$\times \left(1 + \dfrac{1}{2} + \dfrac{1}{3} + \cdots + \dfrac{1}{p} + 1 + \dfrac{1}{2} + \cdots + \dfrac{1}{n+p}\right),$$

where n is a positive integer. The last quantity in parentheses is $\left(1 + \frac{1}{2} + \cdots + \frac{1}{n}\right)$ when $p = 0$.

[Ref. 49, p. 161, eq. (61)
and Ref. 50, p. 174.]

Asymptotic Series for Large Values of x

812.1. $\quad N_0(x) = \left(\dfrac{2}{\pi x}\right)^{1/2} \left[P_0(x) \sin\left(x - \dfrac{\pi}{4}\right) \right.$
$$\left. + Q_0(x) \cos\left(x - \dfrac{\pi}{4}\right)\right].$$

812.2. $\quad N_1(x) = \left(\dfrac{2}{\pi x}\right)^{1/2} \left[P_1(x) \sin\left(x - \dfrac{3\pi}{4}\right) \right.$
$$\left. + Q_1(x) \cos\left(x - \dfrac{3\pi}{4}\right)\right].$$

812.3. $\quad N_n(x) = \left(\dfrac{2}{\pi x}\right)^{1/2} \left[P_n(x) \sin\left(x - \dfrac{n\pi}{2} - \dfrac{\pi}{4}\right) \right.$
$$\left. + Q_n(x) \cos\left(x - \dfrac{n\pi}{2} - \dfrac{\pi}{4}\right)\right].$$

[For the P and Q series see **808.**]

812.4. $\quad N_n'(x) = \left(\dfrac{2}{\pi x}\right)^{1/2} \left[P_n^{(1)}(x) \cos\left(x - \dfrac{n\pi}{2} - \dfrac{\pi}{4}\right) \right.$
$$\left. - Q_n^{(1)}(x) \sin\left(x - \dfrac{n\pi}{2} - \dfrac{\pi}{4}\right)\right].$$

[For $P_n^{(1)}(x)$ and $Q_n^{(1)}(x)$ see **808.41** and **808.42.**]

813.1. $\quad I_0(x) = J_0(ix) = 1 + (\tfrac{1}{2}x)^2 + \dfrac{(\tfrac{1}{2}x)^4}{1^2 \cdot 2^2} + \dfrac{(\tfrac{1}{2}x)^6}{1^2 \cdot 2^2 \cdot 3^2} + \cdots$
where $i = \sqrt{(-1)}$.

813.2. $\quad I_1(x) = i^{-1}J_1(ix) = I_0'(x) = \tfrac{1}{2}x + \dfrac{(\tfrac{1}{2}x)^3}{1^2 \cdot 2} + \dfrac{(\tfrac{1}{2}x)^5}{1^2 \cdot 2^2 \cdot 3} + \cdots.$

813.3. $\quad$ When n is a positive integer,

$$I_n(x) = i^{-n}J_n(ix)$$
$$= \dfrac{(\tfrac{1}{2}x)^n}{n!} \left[1 + \dfrac{(\tfrac{1}{2}x)^2}{1(n+1)} + \dfrac{(\tfrac{1}{2}x)^4}{1 \cdot 2(n+1)(n+2)} + \cdots \right]$$
$$= \sum_{p=0}^{\infty} \dfrac{(\tfrac{1}{2}x)^{n+2p}}{p!(n+p)!}.$$

813.4.   When n is an integer,

$$I_{-n}(x) = I_n(x).$$

813.5.   When n is not a positive integer, replace $n!$ in **813.3** by
$\Pi(n)$.   [See **853.1.**]   [Ref. 12, p. 20.]

Asymptotic Series for Large Values of x

814.1. $I_0(x) \approx \dfrac{e^x}{\sqrt{(2\pi x)}} \left[1 + \dfrac{1^2}{1!8x} + \dfrac{1^2 \cdot 3^2}{2!(8x)^2} + \cdots \right].$

814.2. $I_n(x) \approx \dfrac{e^x}{\sqrt{(2\pi x)}} \left[1 - \dfrac{4n^2 - 1^2}{1!8x} \right.$

$$\left. + \dfrac{(4n^2 - 1^2)(4n^2 - 3^2)}{2!(8x)^2} - \cdots \right].$$

814.3. $I_n'(x) \approx \dfrac{e^x}{\sqrt{(2\pi x)}}$

$$\times \left[1 - \dfrac{4n^2 + 1 \times 3}{1!8x} + \dfrac{(4n^2 - 1^2)(4n^2 + 3 \times 5)}{2!(8x)^2} \right.$$

$$\left. - \dfrac{(4n^2 - 1^2)(4n^2 - 3^2)(4n^2 + 5 \times 7)}{3!(8x)^3} + \cdots \right].$$

The terms of the series in **814.3** are similar to those in **808.41** and **808.42**.

815.1. $K_0(x) = - \left(\gamma + \log_\epsilon \dfrac{x}{2} \right) I_0(x)$

$$+ \dfrac{(\frac{1}{2}x)^2}{(1!)^2} + \dfrac{(\frac{1}{2}x)^4}{(2!)^2} \left(1 + \tfrac{1}{2}\right) + \dfrac{(\frac{1}{2}x)^6}{(3!)^2} \left(1 + \tfrac{1}{2} + \tfrac{1}{3}\right) + \cdots,$$

where γ is Euler's constant, 0.577 2157.

815.2. $K_n(x) = (-1)^{n+1} \left(\gamma + \log_\epsilon \dfrac{x}{2} \right) I_n(x)$

$$+ \dfrac{1}{2} \sum_{p=0}^{n-1} \dfrac{(-1)^p (n - p - 1)!}{p!} \left(\dfrac{x}{2}\right)^{2p-n}$$

$$+ \dfrac{(-1)^n}{2} \sum_{p=0}^{\infty} \dfrac{1}{p!(n+p)!} \left(\dfrac{x}{2}\right)^{2p+n}$$

$$\times \left(1 + \dfrac{1}{2} + \dfrac{1}{3} + \cdots + \dfrac{1}{p} + 1 + \dfrac{1}{2} + \cdots + \dfrac{1}{n+p}\right),$$

where n is a positive integer. The last quantity in parentheses is $\left(1 + \dfrac{1}{2} + \cdots + \dfrac{1}{n}\right)$ when $p = 0$.

[Ref. 13, p. 80, and Ref. 50, p. 264.]

Note that the letter K is sometimes, particularly in earlier writings, used to denote other expressions in connection with Bessel functions.

815.3. When n is an integer,

$$K_{-n}(x) = K_n(x).$$

815.4. When n is not an integer,

$$K_n(x) = \frac{\pi}{2 \sin n\pi} \{I_{-n}(x) - I_n(x)\}.$$

Asymptotic Series for Large Values of x
816.1.

$$K_0(x) \approx \left(\frac{\pi}{2x}\right)^{1/2} e^{-x} \left[1 - \frac{1^2}{1!\,8x} + \frac{1^2 \cdot 3^2}{2!\,(8x)^2} - \frac{1^2 \cdot 3^2 \cdot 5^2}{3!\,(8x)^3} + \cdots\right]$$

where $\approx$ denotes approximate equality.

816.2. $K_n(x) \approx \left(\dfrac{\pi}{2x}\right)^{1/2} e^{-x} \left[1 + \dfrac{4n^2 - 1^2}{1!\,8x}\right.$

$$\left. + \frac{(4n^2 - 1^2)(4n^2 - 3^2)}{2!\,(8x)^2} + \cdots\right].$$

[Ref. 12, p. 55, eq. (50).]

816.3. $K_n{}'(x) \approx -\left(\dfrac{\pi}{2x}\right)^{1/2} e^{-x}$

$$\times \left[1 + \frac{4n^2 + 1 \times 3}{1!\,8x} + \frac{(4n^2 - 1^2)(4n^2 + 3 \times 5)}{2!\,(8x)^2}\right.$$

$$\left. + \frac{(4n^2 - 1^2)(4n^2 - 3^2)(4n^2 + 5 \times 7)}{3!\,(8x)^3} + \cdots\right]$$

[from **804.4**.]

The series can be extended by inspection.

817.1. $H_0{}^{(1)}(z) = J_0(z) + iN_0(z).$

817.2. $K_0(z) = \dfrac{\pi i}{2} H_0{}^{(1)}(iz).$

817.3. $H_n^{(1)}(z) = J_n(z) + iN_n(z)$.

817.4. $H_n^{(2)}(z) = J_n(z) - iN_n(z)$.   [Ref. 13, p. 73.]

817.5. $K_n(z) = \dfrac{\pi i}{2} e^{in\pi/2} H_n^{(1)}(iz)$.   [Ref. 13, p. 78.]

For all values of x and φ,

818.1. $\cos(x \sin \varphi) = J_0(x) + 2J_2(x) \cos 2\varphi$
$$+ 2J_4(x) \cos 4\varphi + \cdots.$$

818.2. $\sin(x \sin \varphi) = 2J_1(x) \sin \varphi + 2J_3(x) \sin 3\varphi$
$$+ 2J_5(x) \sin 5\varphi + \cdots.$$

818.3. $\cos(x \cos \varphi) = J_0(x) - 2J_2(x) \cos 2\varphi$
$$+ 2J_4(x) \cos 4\varphi - \cdots.$$

818.4. $\sin(x \cos \varphi) = 2J_1(x) \cos \varphi - 2J_3(x) \cos 3\varphi$
$$+ 2J_5(x) \cos 5\varphi - \cdots.$$
$$[\text{Ref. 12, p. 32.}]$$

Bessel Functions of Argument $xi\sqrt{i}$, of the First Kind
(For numerical values see Table 1050.)

820.1. $\operatorname{ber} x + i \operatorname{bei} x = J_0(xi\sqrt{i}) = I_0(x\sqrt{i})$
$$= \operatorname{ber}_0 x + i \operatorname{bei}_0 x.$$

820.2. $\operatorname{ber}' x = \dfrac{d}{dx} \operatorname{ber} x$, etc.

820.3. $\operatorname{ber} x = 1 - \dfrac{(\frac{1}{2}x)^4}{(2!)^2} + \dfrac{(\frac{1}{2}x)^8}{(4!)^2} - \cdots$.

820.4. $\operatorname{bei} x = \dfrac{(\frac{1}{2}x)^2}{(1!)^2} - \dfrac{(\frac{1}{2}x)^6}{(3!)^2} + \dfrac{(\frac{1}{2}x)^{10}}{(5!)^2} - \cdots$.

820.5. $\operatorname{ber}' x = - \dfrac{(\frac{1}{2}x)^3}{1!\,2!} + \dfrac{(\frac{1}{2}x)^7}{3!\,4!} - \dfrac{(\frac{1}{2}x)^{11}}{5!\,6!} + \cdots$.

820.6. $\operatorname{bei}' x = \tfrac{1}{2}x - \dfrac{(\frac{1}{2}x)^5}{2!\,3!} + \dfrac{(\frac{1}{2}x)^9}{4!\,5!} - \cdots$.

821.1.   For large values of x,

$$\operatorname{ber} x \approx \dfrac{e^{x/\sqrt{2}}}{\sqrt{(2\pi x)}} \left[L_0(x) \cos\left(\dfrac{x}{\sqrt{2}} - \dfrac{\pi}{8}\right) - M_0(x) \sin\left(\dfrac{x}{\sqrt{2}} - \dfrac{\pi}{8}\right) \right].$$

821.2. $\text{bei } x \approx \dfrac{e^{x/\sqrt{2}}}{\sqrt{(2\pi x)}} \left[M_0(x) \cos \left(\dfrac{x}{\sqrt{2}} - \dfrac{\pi}{8} \right) \right.$

$$\left. + L_0(x) \sin \left(\dfrac{x}{\sqrt{2}} - \dfrac{\pi}{8} \right) \right],$$

where

821.3. $L_0(x) = 1 + \dfrac{1^2}{1!\, 8x} \cos \dfrac{\pi}{4} + \dfrac{1^2 \cdot 3^2}{2!\, (8x)^2} \cos \dfrac{2\pi}{4}$

$$+ \dfrac{1^2 \cdot 3^2 \cdot 5^2}{3!\, (8x)^3} \cos \dfrac{3\pi}{4} + \cdots,$$

821.4. $M_0(x) = - \dfrac{1^2}{1!\, 8x} \sin \dfrac{\pi}{4} - \dfrac{1^2 \cdot 3^2}{2!\, (8x)^2} \sin \dfrac{2\pi}{4}$

$$- \dfrac{1^2 \cdot 3^2 \cdot 5^2}{3!\, (8x)^3} \sin \dfrac{3\pi}{4} - \cdots.$$

821.5. $\text{ber}' x \approx \dfrac{e^{x/\sqrt{2}}}{\sqrt{(2\pi x)}} \left[S_0(x) \cos \left(\dfrac{x}{\sqrt{2}} + \dfrac{\pi}{8} \right) \right.$

$$\left. - T_0(x) \sin \left(\dfrac{x}{\sqrt{2}} + \dfrac{\pi}{8} \right) \right],$$

821.6. $\text{bei}' x \approx \dfrac{e^{x/\sqrt{2}}}{\sqrt{(2\pi x)}} \left[T_0(x) \cos \left(\dfrac{x}{\sqrt{2}} + \dfrac{\pi}{8} \right) \right.$

$$\left. + S_0(x) \sin \left(\dfrac{x}{\sqrt{2}} + \dfrac{\pi}{8} \right) \right]$$

where

821.7. $S_0(x) = 1 - \dfrac{1 \cdot 3}{1!\, 8x} \cos \dfrac{\pi}{4} - \dfrac{1^2 \cdot 3 \cdot 5}{2!\, (8x)^2} \cos \dfrac{2\pi}{4}$

$$- \dfrac{1^2 \cdot 3^2 \cdot 5 \cdot 7}{3!\, (8x)^3} \cos \dfrac{3\pi}{4} - \dfrac{1^2 \cdot 3^2 \cdot 5^2 \cdot 7 \cdot 9}{4!\, (8x)^4} \cos \dfrac{4\pi}{4} - \cdots,$$

821.8. $T_0(x) = \dfrac{1 \cdot 3}{1!\, 8x} \sin \dfrac{\pi}{4} + \dfrac{1^2 \cdot 3 \cdot 5}{2!\, (8x)^2} \sin \dfrac{2\pi}{4}$

$$+ \dfrac{1^2 \cdot 3^2 \cdot 5 \cdot 7}{3!\, (8x)^3} \sin \dfrac{3\pi}{4} + \dfrac{1^2 \cdot 3^2 \cdot 5^2 \cdot 7 \cdot 9}{4!\, (8x)^4} \sin \dfrac{4\pi}{4} + \cdots.$$

[Ref. 14.]

822.1.   When n is a positive integer,

$$\text{ber}_n x + i \, \text{bei}_n x = J_n(xi\sqrt{i}) = i^n I_n(x\sqrt{i}).$$

822.2. $\quad \text{ber}_n x = \sum\limits_{p=0}^{\infty} \dfrac{(-1)^{n+p}(\frac{1}{2}x)^{n+2p}}{p!\,(n+p)!} \cos \dfrac{(n+2p)\pi}{4}$

where

$$p = 0, 1, 2, 3, \cdots.$$

822.3. $\quad \text{bei}_n x = \sum\limits_{p=0}^{\infty} \dfrac{(-1)^{n+p+1}(\frac{1}{2}x)^{n+2p}}{p!\,(n+p)!} \sin \dfrac{(n+2p)\pi}{4}.$

822.4. $\text{ber}_n' x$

$$= \sum\limits_{p=0}^{\infty} \dfrac{(-1)^{n+p}\left(\dfrac{n}{2}+p\right)\left(\dfrac{1}{2}x\right)^{n+2p-1}}{p!\,(n+p)!} \cos \dfrac{(n+2p)\pi}{4}.$$

822.5. $\text{bei}_n' x$

$$= \sum\limits_{p=0}^{\infty} \dfrac{(-1)^{n+p+1}\left(\dfrac{n}{2}+p\right)\left(\dfrac{1}{2}x\right)^{n+2p-1}}{p!\,(n+p)!} \sin \dfrac{(n+2p)\pi}{4}.$$

823.1. For large values of x, when n is a positive integer,

$$\text{ber}_n x \approx \dfrac{e^{x/\sqrt{2}}}{\sqrt{(2\pi x)}} \left[L_n(x) \cos \left(\dfrac{x}{\sqrt{2}} - \dfrac{\pi}{8} + \dfrac{n\pi}{2} \right) \right.$$
$$\left. - M_n(x) \sin \left(\dfrac{x}{\sqrt{2}} - \dfrac{\pi}{8} + \dfrac{n\pi}{2} \right) \right],$$

823.2. $\quad \text{bei}_n x \approx \dfrac{e^{x/\sqrt{2}}}{\sqrt{(2\pi x)}} \left[M_n(x) \cos \left(\dfrac{x}{\sqrt{2}} - \dfrac{\pi}{8} + \dfrac{n\pi}{2} \right) \right.$
$$\left. + L_n(x) \sin \left(\dfrac{x}{\sqrt{2}} - \dfrac{\pi}{8} + \dfrac{n\pi}{2} \right) \right],$$

where

823.3. $\quad L_n(x) = 1 - \dfrac{4n^2 - 1^2}{1!\,8x} \cos \dfrac{\pi}{4}$

$$+ \dfrac{(4n^2 - 1^2)(4n^2 - 3^2)}{2!\,(8x)^2} \cos \dfrac{2\pi}{4} - \cdots,$$

823.4. $\quad M_n(x) = \dfrac{4n^2 - 1^2}{1!\,8x} \sin \dfrac{\pi}{4}$

$$- \dfrac{(4n^2 - 1^2)(4n^2 - 3^2)}{2!\,(8x)^2} \sin \dfrac{2\pi}{4} + \cdots.$$

823.5. $\quad \text{ber}_n' x \approx \dfrac{e^{x/\sqrt{2}}}{\sqrt{(2\pi x)}} \left[S_n(x) \cos \left(\dfrac{x}{\sqrt{2}} + \dfrac{\pi}{8} + \dfrac{n\pi}{2} \right) \right.$
$$\left. - T_n(x) \sin \left(\dfrac{x}{\sqrt{2}} + \dfrac{\pi}{8} + \dfrac{n\pi}{2} \right) \right],$$

823.6. $\mathrm{bei}_n' x \approx \dfrac{e^{x/\sqrt{2}}}{\sqrt{(2\pi x)}} \left[T_n(x) \cos\left(\dfrac{x}{\sqrt{2}} + \dfrac{\pi}{8} + \dfrac{n\pi}{2}\right) \right.$

$$\left. + S_n(x) \sin\left(\dfrac{x}{\sqrt{2}} + \dfrac{\pi}{8} + \dfrac{n\pi}{2}\right) \right].$$

where

823.7. $S_n(x) = 1 - \dfrac{4n^2 + 1 \times 3}{1!\,8x} \cos\dfrac{\pi}{4}$

$$+ \dfrac{(4n^2 - 1^2)(4n^2 + 3 \times 5)}{2!\,(8x)^2} \cos\dfrac{2\pi}{4}$$

$$- \dfrac{(4n^2 - 1^2)(4n^2 - 3^2)(4n^2 + 5 \times 7)}{3!\,(8x)^3} \cos\dfrac{3\pi}{4} + \cdots,$$

823.8. $T_n(x) = \dfrac{4n^2 + 1 \times 3}{1!\,8x} \sin\dfrac{\pi}{4}$

$$- \dfrac{(4n^2 - 1^2)(4n^2 + 3 \times 5)}{2!\,(8x)^2} \sin\dfrac{2\pi}{4}$$

$$+ \dfrac{(4n^2 - 1^2)(4n^2 - 3^2)(4n^2 + 5 \times 7)}{3!\,(8x)^3} \sin\dfrac{3\pi}{4} - \cdots.$$

Bessel Functions of Argument $xi\sqrt{i}$, of the Second Kind
(For numerical values see Table 1050.)

824.1. $\mathrm{ker}\, x + i\,\mathrm{kei}\, x = K_0(x\sqrt{i}).$

824.2. $\mathrm{ker}'\, x = \dfrac{d}{dx}\,\mathrm{ker}\, x$, etc.

824.3. $\mathrm{ker}\, x = \left(\log\dfrac{2}{x} - \gamma\right)\mathrm{ber}\, x + \dfrac{\pi}{4}\,\mathrm{bei}\, x$

$$- (1 + \tfrac{1}{2})\dfrac{(\tfrac{1}{2}x)^4}{(2!)^2} + (1 + \tfrac{1}{2} + \tfrac{1}{3} + \tfrac{1}{4})\dfrac{(\tfrac{1}{2}x)^8}{(4!)^2}$$

$$- (1 + \tfrac{1}{2} + \tfrac{1}{3} + \tfrac{1}{4} + \tfrac{1}{5} + \tfrac{1}{6})\dfrac{(\tfrac{1}{2}x)^{12}}{(6!)^2} + \cdots,$$

where
$$\gamma = 0.577\,2157.$$

824.4. $\mathrm{kei}\, x = \left(\log\dfrac{2}{x} - \gamma\right)\mathrm{bei}\, x - \dfrac{\pi}{4}\,\mathrm{ber}\, x$

$$+ \dfrac{(\tfrac{1}{2}x)^2}{(1!)^2} - (1 + \tfrac{1}{2} + \tfrac{1}{3})\dfrac{(\tfrac{1}{2}x)^6}{(3!)^2}$$

$$+ (1 + \tfrac{1}{2} + \tfrac{1}{3} + \tfrac{1}{4} + \tfrac{1}{5})\dfrac{(\tfrac{1}{2}x)^{10}}{(5!)^2} - \cdots.$$

824.5. $\ker' x = \left(\log \dfrac{2}{x} - \gamma\right) \operatorname{ber}' x - \dfrac{1}{x} \operatorname{ber} x + \dfrac{\pi}{4} \operatorname{bei}' x$

$$- \left(1 + \tfrac{1}{2}\right) \frac{(\tfrac{1}{2}x)^3}{1!\,2!} + \left(1 + \tfrac{1}{2} + \tfrac{1}{3} + \tfrac{1}{4}\right) \frac{(\tfrac{1}{2}x)^7}{3!\,4!} - \cdots.$$

824.6. $\operatorname{kei}' x = \left(\log \dfrac{2}{x} - \gamma\right) \operatorname{bei}' x - \dfrac{1}{x} \operatorname{bei} x - \dfrac{\pi}{4} \operatorname{ber}' x$

$$+ \tfrac{1}{2}x - \left(1 + \tfrac{1}{2} + \tfrac{1}{3}\right) \frac{(\tfrac{1}{2}x^5)}{2!\,3!}$$

$$+ \left(1 + \tfrac{1}{2} + \tfrac{1}{3} + \tfrac{1}{4} + \tfrac{1}{5}\right) \frac{(\tfrac{1}{2}x)^9}{4!\,5!} - \cdots.$$

825.1. For large values of x,

$$\ker x \approx \left(\frac{\pi}{2x}\right)^{1/2} e^{-x/\sqrt{2}} \left[L_0(-x) \cos\left(\frac{x}{\sqrt{2}} + \frac{\pi}{8}\right) \right.$$

$$\left. + M_0(-x) \sin\left(\frac{x}{\sqrt{2}} + \frac{\pi}{8}\right) \right].$$

825.2. $\operatorname{kei} x \approx \left(\dfrac{\pi}{2x}\right)^{1/2} e^{-x/\sqrt{2}} \left[M_0(-x) \cos\left(\dfrac{x}{\sqrt{2}} + \dfrac{\pi}{8}\right) \right.$

$$\left. - L_0(-x) \sin\left(\frac{x}{\sqrt{2}} + \frac{\pi}{8}\right) \right].$$

See **821.3** and **821.4**, changing x to $-x$.

825.3. $\ker' x \approx - \left(\dfrac{\pi}{2x}\right)^{1/2} e^{-x/\sqrt{2}} \left[S_0(-x) \cos\left(\dfrac{x}{\sqrt{2}} - \dfrac{\pi}{8}\right) \right.$

$$\left. + T_0(-x) \sin\left(\frac{x}{\sqrt{2}} - \frac{\pi}{8}\right) \right].$$

825.4. $\operatorname{kei}' x \approx - \left(\dfrac{\pi}{2x}\right)^{1/2} e^{-x/\sqrt{2}} \left[T_0(-x) \cos\left(\dfrac{x}{\sqrt{2}} - \dfrac{\pi}{8}\right) \right.$

$$\left. - S_0(-x) \sin\left(\frac{x}{\sqrt{2}} - \frac{\pi}{8}\right) \right].$$

See **821.7** and **821.8**, changing x to $-x$.

826.1. When n is a positive integer,

$$\ker_n x + i \operatorname{kei}_n x = i^{-n} K_n(x\sqrt{i}).$$

826.2. $\ker_n x = \left(\log\dfrac{2}{x} - \gamma\right)\mathrm{ber}_n x + \dfrac{\pi}{4}\,\mathrm{bei}_n x$

$$+ \frac{1}{2}\sum_{p=0}^{n-1}\frac{(-1)^{n+p}(n-p-1)!}{p!}\left(\frac{x}{2}\right)^{2p-n}\cos\frac{(n+2p)\pi}{4}$$

$$+ \frac{1}{2}\sum_{p=0}^{\infty}\left(1 + \frac{1}{2} + \frac{1}{3} + \cdots + \frac{1}{p} + 1 + \frac{1}{2} + \frac{1}{3} + \cdots + \frac{1}{n+p}\right)$$

$$\frac{(-1)^{n+p}(\tfrac{1}{2}x)^{n+2p}}{p!\,(n+p)!}\cos\frac{(n+2p)\pi}{4}.$$

826.3. $\mathrm{kei}_n x = \left(\log\dfrac{2}{x} - \gamma\right)\mathrm{bei}_n x - \dfrac{\pi}{4}\,\mathrm{ber}_n x$

$$+ \frac{1}{2}\sum_{p=0}^{n-1}\frac{(-1)^{n+p}(n-p-1)!}{p!}\left(\frac{x}{2}\right)^{2p-n}\sin\frac{(n+2p)\pi}{4}$$

$$- \frac{1}{2}\sum_{p=0}^{\infty}\left(1 + \frac{1}{2} + \frac{1}{3} + \cdots + \frac{1}{p} + 1 + \frac{1}{2} + \frac{1}{3} + \cdots + \frac{1}{n+p}\right)$$

$$\frac{(-1)^{n+p}(\tfrac{1}{2}x)^{n+2p}}{p!\,(n+p)!}\sin\frac{(n+2p)\pi}{4}.$$

826.4. $\ker_n' x = \left(\log\dfrac{2}{x} - \gamma\right)\mathrm{ber}_n' x - \dfrac{\mathrm{ber}_n x}{x} + \dfrac{\pi}{4}\,\mathrm{bei}_n' x$

$$+ \frac{1}{4}\sum_{p=0}^{n-1}\frac{(-1)^{n+p}(2p-n)(n-p-1)!}{p!}\left(\frac{x}{2}\right)^{2p-n-1}\cos\frac{(n+2p)\pi}{4}$$

$$+ \frac{1}{4}\sum_{p=0}^{\infty}\left(1 + \frac{1}{2} + \frac{1}{3} + \cdots + \frac{1}{p} + 1 + \frac{1}{2} + \frac{1}{3} + \cdots + \frac{1}{n+p}\right)$$

$$\frac{(-1)^{n+p}(n+2p)(\tfrac{1}{2}x)^{n+2p-1}}{p!\,(n+p)!}\cos\frac{(n+2p)\pi}{4}.$$

826.5. $\mathrm{kei}_n' x = \left(\log\dfrac{2}{x} - \gamma\right)\mathrm{bei}_n' x - \dfrac{\mathrm{bei}_n x}{x} - \dfrac{\pi}{4}\,\mathrm{ber}_n' x$

$$+ \frac{1}{4}\sum_{p=0}^{n-1}\frac{(-1)^{n+p}(2p-n)(n-p-1)!}{p!}\left(\frac{x}{2}\right)^{2p-n-1}\sin\frac{(n+2p)\pi}{4}$$

$$- \frac{1}{4}\sum_{p=0}^{\infty}\left(1 + \frac{1}{2} + \frac{1}{3} + \cdots + \frac{1}{p} + 1 + \frac{1}{2} + \frac{1}{3} + \cdots + \frac{1}{n+p}\right)$$

$$\frac{(-1)^{n+p}(n+2p)(\tfrac{1}{2}x)^{n+2p-1}}{p!\,(n+p)!}\sin\frac{(n+2p)\pi}{4}.$$

827.1. For large values of x, when n is a positive integer,

$$\ker_n x \approx \left(\frac{\pi}{2x}\right)^{1/2} e^{-x/\sqrt{2}}\left[L_n(-x)\cos\left(\frac{x}{\sqrt{2}} + \frac{\pi}{8} + \frac{n\pi}{2}\right)\right.$$

$$\left. + M_n(-x)\sin\left(\frac{x}{\sqrt{2}} + \frac{\pi}{8} + \frac{n\pi}{2}\right)\right].$$

827.2. $\quad \text{kei}_n x \approx \left(\dfrac{\pi}{2x}\right)^{1/2} e^{-x/\sqrt{2}} \left[M_n(-x) \cos\left(\dfrac{x}{\sqrt{2}} + \dfrac{\pi}{8} + \dfrac{n\pi}{2}\right) \right.$

$$\left. - L_n(-x) \sin\left(\dfrac{x}{\sqrt{2}} + \dfrac{\pi}{8} + \dfrac{n\pi}{2}\right) \right].$$

[See **823.3** and **823.4.**]

827.3. $\quad \text{ker}_n' x \approx - \left(\dfrac{\pi}{2x}\right)^{1/2} e^{-x/\sqrt{2}} \left[S_n(-x) \cos\left(\dfrac{x}{\sqrt{2}} - \dfrac{\pi}{8} + \dfrac{n\pi}{2}\right) \right.$

$$\left. + T_n(-x) \sin\left(\dfrac{x}{\sqrt{2}} - \dfrac{\pi}{8} + \dfrac{n\pi}{2}\right) \right].$$

827.4. $\quad \text{kei}_n' x \approx - \left(\dfrac{\pi}{2x}\right)^{1/2} e^{-x/\sqrt{2}} \left[T_n(-x) \cos\left(\dfrac{x}{\sqrt{2}} - \dfrac{\pi}{8} + \dfrac{n\pi}{2}\right) \right.$

$$\left. - S_n(-x) \sin\left(\dfrac{x}{\sqrt 2} - \dfrac{\pi}{8} + \dfrac{n\pi}{2}\right) \right]. \qquad [\text{See } \textbf{823.7} \text{ and } \textbf{823.8.}]$$

Note that the series for large values of x are asymptotic expansions and there is a limit to the amount of precision which they will give.

Recurrence Formulas

828.1. $\quad \text{ber}_1 x = \dfrac{1}{\sqrt{2}} (\text{ber}' x - \text{bei}' x).$

828.2. $\quad \text{bei}_1 x = \dfrac{1}{\sqrt{2}} (\text{ber}' x + \text{bei}' x).$

828.3. $\quad \text{ber}_2 x = \dfrac{2 \, \text{bei}' x}{x} - \text{ber} \, x.$

828.4. $\quad \text{bei}_2 x = - \dfrac{2 \, \text{ber}' x}{x} - \text{bei} \, x.$

828.5. $\quad \text{ber}_2' x = - \text{ber}' x - \dfrac{2 \, \text{ber}_2 x}{x}.$

828.6. $\quad \text{bei}_2' x = - \text{bei}' x - \dfrac{2 \, \text{bei}_2 x}{x}.$

829.1. $\quad \text{ber}_{n+1} x = - \dfrac{n\sqrt{2}}{x} (\text{ber}_n x - \text{bei}_n x) - \text{ber}_{n-1} x.$

829.2. $\quad \text{bei}_{n+1} x = - \dfrac{n\sqrt{2}}{x} (\text{ber}_n x + \text{bei}_n x) - \text{bei}_{n-1} x.$

829.3. $\quad \text{ber}_n' x = - \dfrac{1}{\sqrt{2}} (\text{ber}_{n-1} x + \text{bei}_{n-1} x) - \dfrac{n \, \text{ber}_n x}{x}.$

829.4. $\text{bei}_n' x = \dfrac{1}{\sqrt{2}} (\text{ber}_{n-1} x - \text{bei}_{n-1} x) - \dfrac{n \, \text{bei}_n x}{x}.$

330.   The formulas of **828–829** are applicable to Bessel functions of the second kind by changing ber to ker and bei to kei.

[Ref. 14, eq. (1)–(60).]

BESSEL FUNCTIONS—INTEGRALS

835.1. $\displaystyle\int x^n J_{n-1}(x)dx = x^n J_n(x).$

835.2. $\displaystyle\int x^{-n} J_{n+1}(x)dx = - x^{-n} J_n(x).$

835.3. $\displaystyle\int x^n I_{n-1}(x)dx = x^n I_n(x).$

835.4. $\displaystyle\int x^{-n} I_{n+1}(x)dx = x^{-n} I_n(x).$

835.5. $\displaystyle\int x^n K_{n-1}(x)dx = - x^n K_n(x).$

335.6. $\displaystyle\int x^{-n} K_{n+1}(x)dx = - x^{-n} K_n(x).$

836.1. $\displaystyle\int_0^x x \, \text{ber} \, x \, dx = x \, \text{bei}' \, x.$

836.2. $\displaystyle\int_0^x x \, \text{bei} \, x \, dx = - x \, \text{ber}' \, x.$

836.3. $\displaystyle\int_0^x x \, \text{ker} \, x \, dx = x \, \text{kei}' \, x.$

836.4. $\displaystyle\int_0^x x \, \text{kei} \, x \, dx = - x \, \text{ker}' \, x.$ [Ref. 12, p. 27.]

837.1. $\displaystyle\int x(\text{ber}_n^2 x + \text{bei}_n^2 x)dx = x(\text{ber}_n x \, \text{bei}_n' x - \text{bei}_n x \, \text{ber}_n' x).$

837.2. $\displaystyle\int x(\text{ber}_n'^2 x + \text{bei}_n'^2 x)dx = x \, (\text{ber}_n x \, \text{ber}_n' x + \text{bei}_n x \, \text{bei}_n' x).$

[Eq. 191 and 193, p. 170, Ref. 49.]

See also similar equations in $\text{ker}_n x$ and $\text{kei}_n x$, eq. 236 and 238, p. 172, Ref. 49.

840. $P_0(\mu) = 1$.

$P_1(\mu) = \mu$.

$P_2(\mu) = \dfrac{1}{2}(3\mu^2 - 1)$.

$P_3(\mu) = \dfrac{1}{2}(5\mu^3 - 3\mu)$.

$P_4(\mu) = \dfrac{1}{2\cdot 4}(5\cdot 7\mu^4 - 2\cdot 3\cdot 5\mu^2 + 1\cdot 3)$.

$P_5(\mu) = \dfrac{1}{2\cdot 4}(7\cdot 9\mu^5 - 2\cdot 5\cdot 7\mu^3 + 3\cdot 5\mu)$.

$P_6(\mu) = \dfrac{1}{2\cdot 4\cdot 6}(7\cdot 9\cdot 11\mu^6 - 3\cdot 5\cdot 7\cdot 9\mu^4$
$$+ 3\cdot 3\cdot 5\cdot 7\mu^2 - 1\cdot 3\cdot 5).$$

$P_7(\mu) = \dfrac{1}{2\cdot 4\cdot 6}(9\cdot 11\cdot 13\mu^7 - 3\cdot 7\cdot 9\cdot 11\mu^5$
$$+ 3\cdot 5\cdot 7\cdot 9\mu^3 - 3\cdot 5\cdot 7\mu).$$

. . . .

Note that the parentheses contain binomial coefficients as well as other factors. [Ref. 25, p. 956.]

841. $P_m(\mu) = \dfrac{(2m-1)(2m-3)\cdots 1}{m!}\left[\mu^m - \dfrac{m(m-1)}{2(2m-1)}\mu^{m-2}\right.$
$$\left. + \dfrac{m(m-1)(m-2)(m-3)}{2\cdot 4(2m-1)(2m-3)}\mu^{m-4} - \cdots\right].$$

The series terminates with the term involving μ if m is odd and with the term independent of μ if m is even. [Ref. 22, p. 145.]

842. $(m+1)P_{m+1}(\mu) - (2m+1)\mu P_m(\mu) - mP_{m-1}(\mu)$.

[Ref. 22, p. 151.]

843. $(\mu^2 - 1)P_m'(\mu) = m\mu P_m(\mu) - mP_{m-1}(\mu)$.

[Ref. 21, p. 137.]

844.   For large values of m,

$$P_m(\cos \theta) \approx \left(\frac{2}{m\pi \sin \theta} \right)^{1/2} \sin \left\{ \left(m + \frac{1}{2} \right) \theta + \frac{\pi}{4} \right\}.$$

[Ref. 21, p. 137.]

844.1. $P_m(x) = \dfrac{1}{2^m m!} \dfrac{d^m}{dx^m} (x^2 - 1)^m.$ [Ref. 22, p. 160, eq. 1.]

844.2. $P_m(1) = 1.$

844.3. $P_{2m}(-x) = P_{2m}(x).$

844.4. $P_{2m+1}(-x) = -P_{2m+1}(x).$

[Ref. 22, p. 150, eq. 5–7.]

845.   First Derivatives, $P_m{}'(\mu) = \dfrac{d}{d\mu} P_m(\mu).$

$P_0{}'(\mu) = 0.$

$P_1{}'(\mu) = 1.$

$P_2{}'(\mu) = 3\mu.$

$P_3{}'(\mu) = \dfrac{1}{2} (3 \cdot 5\mu^2 - 1 \cdot 3).$

$P_4{}'(\mu) = \dfrac{1}{2} (5 \cdot 7\mu^3 - 3 \cdot 5\mu).$

$P_5{}'(\mu) = \dfrac{1}{2 \cdot 4} (5 \cdot 7 \cdot 9\mu^4 - 2 \cdot 3 \cdot 5 \cdot 7\mu^2 + 1 \cdot 3 \cdot 5).$

$P_6{}'(\mu) = \dfrac{1}{2 \cdot 4} (7 \cdot 9 \cdot 11\mu^5 - 2 \cdot 5 \cdot 7 \cdot 9\mu^3 + 3 \cdot 5 \cdot 7\mu).$

$P_7{}'(\mu) = \dfrac{1}{2 \cdot 4 \cdot 6} (7 \cdot 9 \cdot 11 \cdot 13\mu^6 - 3 \cdot 5 \cdot 7 \cdot 9 \cdot 11\mu^4$
$$+ 3 \cdot 3 \cdot 5 \cdot 7 \cdot 9\mu^2 - 1 \cdot 3 \cdot 5 \cdot 7).$$

. . . .

Note that the parentheses contain binomial coefficients as well as other factors.   [Ref. 25, p. 957.]

For tables of numerical values see Ref. 22, pp. 278–281, Ref. 45, pp. 188–197, and Ref. 52, 53, and 54.

DEFINITE INTEGRALS

850.1. $\displaystyle\int_0^\infty x^{n-1}e^{-x}dx = \int_0^1 \left(\log\frac{1}{x}\right)^{n-1} dx = \Gamma(n).$

[See Table 1018.]

$\Gamma(n)$ is the **Gamma** function. The integral is finite when $n > 0$.

850.2. $\Gamma(n+1) = n\Gamma(n).$

850.3. $\Gamma(n)\Gamma(1-n) = \dfrac{\pi}{\sin n\pi}\cdot$

850.4. $\Gamma(n) = (n-1)!,$ when n is an integer > 0.

850.5. $\Gamma(1) = \Gamma(2) = 1.$ **850.6.** $\Gamma(\frac{1}{2}) = \sqrt{\pi}.$

850.7. $\Gamma(n+\frac{1}{2}) = 1\cdot3\cdot5\cdots(2n-3)(2n-1)\sqrt{\pi}/2^n,$

[n an integer > 0]. [Ref. 10, p. 301.]

851.1. $\log\Gamma(1+x) = -Cx + \dfrac{S_2x^2}{2} - \dfrac{S_3x^3}{3} + \dfrac{S_4x^4}{4} - \cdots,$

$$[x^2 < 1],$$

where C is Euler's constant,

$$C = \lim_{p\to\infty}\left[-\log p + 1 + \frac{1}{2} + \frac{1}{3} + \cdots + \frac{1}{p}\right] = 0.577\ 2157$$

and

$$S_p = 1 + \frac{1}{2^p} + \frac{1}{3^p} + \cdots.$$

851.2. $\log\Gamma(1+x) = \dfrac{1}{2}\log\dfrac{x\pi}{\sin x\pi} - Cx - \dfrac{S_3x^3}{3} - \dfrac{S_5x^5}{5} - \cdots.$

851.3. $\log\Gamma(1+x) = \dfrac{1}{2}\log\dfrac{x\pi}{\sin x\pi} - \dfrac{1}{2}\log\dfrac{1+x}{1-x}$

$$+ (1-C)x - (S_3-1)\frac{x^3}{3} - (S_5-1)\frac{x^5}{5} - \cdots.$$

Use **850.2** and **850.3** with these series for values of x greater than $\frac{1}{2}$.
[Ref. 7, par. 269–270 and Ref. 10, p. 303.]

851.4. $\Gamma(x+1) \approx x^x e^{-x} \sqrt{(2\pi x)} \left[1 + \dfrac{1}{12x} + \dfrac{1}{288x^2} \right.$

$$\left. - \frac{139}{51,840x^3} - \frac{571}{2,488,320x^4} + \cdots \right],$$

where $\approx$ denotes approximate equality. This gives an asymptotic expression for $x!$ when x is a large integer.

[Ref. 44, v. 1, p. 180.]   [See **11.**]

851.5. $\log \Gamma(x+1) \approx \tfrac{1}{2} \log (2\pi) - x + (x + \tfrac{1}{2}) \log x$

$$+ \frac{B_1}{1 \cdot 2x} - \frac{B_2}{3 \cdot 4x^3} + \frac{B_3}{5 \cdot 6x^5} - \cdots.$$

[See **45** and **47.1.**]

This is an asymptotic series. The absolute value of the error is less than the absolute value of the first term neglected.

[Ref. 42, pp. 153–154.]

Note that $B_1 = 1/6$, $B_2 = 1/30$, $B_3 = 1/42$, etc., as in **45**.

852.1. $\displaystyle\int_0^\infty e^{-x} \log x \, dx = -C,$

where $C = 0.577\ 2157$, as in **851.1.**

852.2. $\displaystyle\int_0^1 \log (\log x) dx = -C.$

852.3. $\displaystyle\int_0^1 \left(\frac{1}{\log x} + \frac{1}{1-x} \right) dx = C.$

852.4. $\displaystyle\int_0^\infty \frac{1}{x} \left(\frac{1}{1+x^2} - e^{-x} \right) dx = C.$

852.5. $\displaystyle\int_0^\infty \left(\frac{1}{e^x - 1} - \frac{1}{xe^x} \right) dx = C.$

853.1. $\Pi(n) = \Gamma(n + 1).$ [See **850** and Table 1018.]

$\Pi(n)$ is Gauss's Function.

853.2. If n is a positive integer, $\Pi(n) = n!$.

853.3. $\Pi(0) = 1.$

854.1. $\displaystyle\int_0^{\pi/2} \sin^m x \, dx = \int_0^{\pi/2} \cos^m x \, dx$

$$= \frac{2 \cdot 4 \cdot 6 \cdots (m-1)}{1 \cdot 3 \cdot 5 \cdots m},$$

[m an odd integer > 1],

$$= \frac{1 \cdot 3 \cdot 5 \cdots (m-1)}{2 \cdot 4 \cdot 6 \cdots m} \frac{\pi}{2},$$

[m an even integer],

$$= \frac{\sqrt{\pi}}{2} \frac{\Gamma\left(\dfrac{m+1}{2}\right)}{\Gamma\left(\dfrac{m}{2}+1\right)},$$

[m any value > -1].

855.1. $\displaystyle B(m, n) = \int_0^1 x^{m-1}(1-x)^{n-1} dx = \frac{\Gamma(m)\Gamma(n)}{\Gamma(m+n)}.$

$B(m, n)$ is called the Beta function.   [m and $n > 0.$]

855.2. $\displaystyle\int_0^\infty \frac{x^{m-1}}{(1+x)^{m+n}} \, dx = B(m, n).$

855.3. $\displaystyle\int_0^{\pi/2} \sin^m \theta \cos^n \theta \, d\theta = \frac{1}{2} B\left(\frac{m+1}{2}, \frac{n+1}{2}\right),$

[m and $n > -1$].   [Ref. 7, p. 259.]

855.4. $\displaystyle\int_0^1 x^m (1-x^2)^{(n-1)/2} dx = \frac{1}{2} B\left(\frac{m+1}{2}, \frac{n+1}{2}\right),$

[m and $n > -1$].   [Ref. 7, p. 259.]

855.5. $\displaystyle\int_0^1 \frac{x^{m-1} + x^{n-1}}{(1+x)^{m+n}} \, dx = B(m, n),$ [m and $n > 0$].

[Ref. 6, Art. 122.]

855.6. $\displaystyle\int_0^a x^{m-1}(a-x)^{n-1} dx = a^{m+n-1} B(m, n),$

[m and $n > 0$].   [Ref. 8, p. 133.]

855.7. $\displaystyle\int_0^\infty \frac{x^{m-1} dx}{(ax+b)^{m+n}} = \frac{B(m, n)}{a^m b^n},$ [m and $n > 0$].

[Ref. 6, Art. 122.]

856.1. $\displaystyle\int_1^\infty \frac{dx}{x^m} = \frac{1}{m-1},$ $\qquad\qquad\qquad\qquad [m > 1].$

[Ref. 20, p. 46, No. 107.]

856.2. $\displaystyle\int_0^\infty \frac{x^{p-1}dx}{1+x} = \frac{\pi}{\sin p\pi},$ $\qquad\qquad\qquad [0 < p < 1].$

[Ref. 7, p. 246.]

856.3. $\displaystyle\int_0^\infty \frac{dx}{(1+x)\sqrt{x}} = \pi.$

856.4. $\displaystyle\int_0^\infty \frac{dx}{(1+x)x^p} = \pi \csc p\pi,$ $\qquad\qquad\qquad [p < 1].$

[Ref. 16, p. 44.]

856.5. $\displaystyle\int_0^\infty \frac{dx}{(1-x)x^p} = -\pi \operatorname{ctn} p\pi,$ $\qquad\qquad [p < 1].$

[Ref. 16, p. 44.]

856.6. $\displaystyle\int_0^\infty \frac{x^{m-1}dx}{1+x^n} = \frac{\pi}{n \sin \dfrac{m\pi}{n}},$ $\qquad\qquad [0 < m < n].$

[Ref. 7, p. 246.]

856.7. $\displaystyle\int_0^\infty \frac{a\, dx}{a^2 + x^2} = \frac{\pi}{2},$ $\qquad\qquad\qquad\qquad [a > 0],$

$\qquad\qquad\qquad\quad = 0,$ $\qquad\qquad\qquad\qquad\quad [a = 0],$

$\qquad\qquad\qquad\quad = -\frac{\pi}{2},$ $\qquad\qquad\qquad\qquad [a < 0].$

[Ref. 5, No. 480.]

856.8. $\displaystyle\int_0^\infty \frac{dx}{(a^2 + x^2)(b^2 + x^2)} = \frac{\pi}{2ab(a+b)}.$

[Ref. 7, p. 73, No. 4.]

857.1. $\displaystyle\int_0^1 \frac{dx}{\sqrt{(1-x^{1/p})}} = \frac{p\sqrt{\pi}\,\Gamma(p)}{\Gamma(p+\frac{1}{2})}.$ [Ref. 6, p. 168, No. 5.]

857.2. $\displaystyle\int_0^1 \frac{dx}{1 + 2x\cos\varphi + x^2} = \frac{\varphi}{2\sin\varphi}.$

[Ref. 16, Table 6, No. 3.]

857.3. $\displaystyle\int_0^\infty \frac{dx}{1 + 2x\cos\varphi + x^2} = \frac{\varphi}{\sin\varphi}.$

[Ref. 40, p. 80, No. 10.]

858.1. $\displaystyle\int_0^\pi \sin mx \sin nx \, dx = 0,$ $[m \neq n;\ m,\ n = \text{integers}]$.

858.2. $\displaystyle\int_0^\pi \cos mx \cos nx \, dx = 0,$ $[m \neq n;\ m,\ n = \text{integers}]$.

                                               [Ref. 20, p. 46, No. 111.]

858.3. $\displaystyle\int_0^\pi \sin^2 x \, dx = \int_0^\pi \cos^2 x \, dx = \frac{\pi}{2}.$

858.4. $\displaystyle\int_0^\pi \sin^2 nx \, dx = \int_0^\pi \cos^2 nx \, dx = \frac{\pi}{2},$ $[n = \text{integer}]$.

                                               [Ref. 20, p. 46, No. 112.]

858.5. $\displaystyle\int_0^\infty \frac{\sin mx \, dx}{x} = \frac{\pi}{2},$ $[m > 0]$,

                       $= 0,$ $[m = 0]$,

                       $= -\frac{\pi}{2},$ $[m < 0]$.

                                           [Ref. 5, No. 484.]

858.51. $\displaystyle\int_0^\infty \frac{\cos ax - \cos bx}{x} \, dx = \log \frac{b}{a}.$

                                             [Ref. 7, p. 289, No. 8.]

858.6. $\displaystyle\int_0^\infty \frac{\sin x \, dx}{\sqrt{x}} = \int_0^\infty \frac{\cos x \, dx}{\sqrt{x}} = \sqrt{(\pi/2)}.$

                                             [Ref. 7, Art. 302.]

858.7. $\displaystyle\int_0^\infty \frac{\cos x \, dx}{x} = \infty.$

858.8. $\displaystyle\int_0^\infty \frac{\tan x}{x} \, dx = \frac{\pi}{2}.$

858.9. $\displaystyle\int_0^\infty \frac{\sin x \cos mx}{x} \, dx = 0,$ $[m^2 > 1]$,

                             $= \frac{\pi}{4},$ $[m = 1 \text{ or} -1]$,

                             $= \frac{\pi}{2},$ $[m^2 < 1]$.

859.1. $\displaystyle\int_0^{\pi/2} \frac{dx}{1 + a \cos x} = \frac{\cos^{-1} a}{\sqrt{(1 - a^2)}},$ $[a < 1]$.

                                             [Ref. 7, p. 22, No. 42.]

859.2. $\displaystyle\int_0^{2\pi} \frac{dx}{1 + a \cos x} = \frac{2\pi}{\sqrt{(1-a^2)}}$, $[a^2 < 1]$.

[Ref. 21, p. 111.]

859.21. $\displaystyle\int_0^{\pi} \frac{dx}{a - \cos x} = \frac{\pi}{\sqrt{(a^2-1)}}$, $[a > 1]$.

[Ref. 39, p. 191, No. 60.]

859.22. $\displaystyle\int_0^{\pi/2} \frac{dx}{(\sin x + \cos x)^2} = 1.$ [Ref. 40, Art. 88.]

859.3. $\displaystyle\int_0^{\infty} \frac{\cos mx}{1 + x^2}\, dx = \frac{\pi}{2} e^{-m}$, $[m > 0]$,

$\qquad\qquad\qquad\qquad\quad = \dfrac{\pi}{2} e^{m}$, $[m < 0]$.

[Ref. 7, par. 290.]

859.4. $\displaystyle\int_0^{\infty} \frac{\sin^2 x}{x^2}\, dx = \frac{\pi}{2}.$

859.5. $\displaystyle\int_0^{\infty} \sin(x^2)dx = \int_0^{\infty} \cos(x^2)dx = \tfrac{1}{2}\sqrt{(\pi/2)}.$

[Ref. 11, p. 156, Ex. 6.]

859.61. $\displaystyle\int_0^{\pi/2} \frac{\sin x\, dx}{\sqrt{(1 - k^2 \sin^2 x)}} = \frac{1}{2k} \log \frac{1+k}{1-k}$, $[k^2 < 1]$.

859.62. $\displaystyle\int_0^{\pi/2} \frac{\cos x\, dx}{\sqrt{(1 - k^2 \sin^2 x)}} = \frac{1}{k} \sin^{-1} k$, $[k^2 < 1]$.

[Ref. 16, Table 57, Nos. 2 and 3.]

859.63. $\displaystyle\int_0^{\pi/2} \frac{\sin^2 x\, dx}{\sqrt{(1 - k^2 \sin^2 x)}} = \frac{1}{k^2} (K - E)$,

[modulus k; $k^2 < 1$].

859.64. $\displaystyle\int_0^{\pi/2} \frac{\cos^2 x\, dx}{\sqrt{(1 - k^2 \sin^2 x)}} = \frac{1}{k^2} \{E - (1 - k^2)K\}$,

[modulus k; $k^2 < 1$].

[Ref. 16, Table 57, Nos. 5 and 7.]

860.1. $\displaystyle\int_0^{\pi} \frac{\cos mx\, dx}{1 - 2a \cos x + a^2} = \frac{\pi a^m}{1 - a^2}$,

[m an integer $\geqq 0$; $a^2 < 1$].

860.2. $\displaystyle\int_0^\pi \frac{(a - c\cos x)dx}{a^2 - 2ac\cos x + c^2} = \frac{\pi}{a},$ \hfill $[a > c],$

$$= 0, \qquad\qquad [a < c].$$

[Ref. 7, Art. 46.]

860.3. $\displaystyle\int_0^{\pi/2} \frac{\sin^2 x\, dx}{1 - 2a\cos 2x + a^2} = \frac{\pi}{4(1 + a)}.$

[Ref. 16, Table 50, No. 1.]

860.4. $\displaystyle\int_0^{\pi/2} \frac{\cos^2 x\, dx}{1 - 2a\cos 2x + a^2} = \frac{\pi}{4(1 - a)},$ \hfill $[a^2 < 1],$

$$= \frac{\pi}{4(a - 1)}, \qquad [a^2 > 1].$$

[Ref. 16, Table 50, No. 2.]

860.5. $\displaystyle\int_0^\pi \frac{dx}{\sqrt{(1 \pm 2a\cos x + a^2)}}$

$$= 2\int_0^{\pi/2} \frac{d\varphi}{\sqrt{(1 - a^2\sin^2\varphi)}} = 2K. \quad [a^2 < 1.]$$

[See **773.1** and Table 1040.]
[Ref. 16, Table 67, No. 5.]

861.1. $\displaystyle\int_0^\infty e^{-ax}dx = \frac{1}{a}.$

861.11. $\displaystyle\int_0^\infty \frac{e^{-ax} - e^{-bx}}{x}\, dx = \log\frac{b}{a}.$ \hfill [Ref. 7, par. 288.]

861.2. $\displaystyle\int_0^\infty x^n e^{-ax}dx = \frac{\Gamma(n + 1)}{a^{n+1}},$ \hfill $[n > -1, a > 0],$

$$= \frac{n!}{a^{n+1}}, \qquad [n = \text{positive integer}, a > 0].$$

861.3. $\displaystyle\int_0^\infty e^{-a^2x^2}dx = \frac{\sqrt{\pi}}{2a}.$ \hfill $[a > 0.]$ \quad [Ref. 7, Art. 272.]

861.4. $\displaystyle\int_0^\infty xe^{-x^2}dx = \frac{1}{2}.$

861.5. $\displaystyle\int_0^\infty x^2 e^{-x^2}dx = \frac{\sqrt{\pi}}{4}.$ \qquad **861.6.** $\displaystyle\int_{-\infty}^\infty x^2 e^{-x^2}dx = \frac{\sqrt{\pi}}{2}.$

861.7. $\displaystyle\int_0^\infty x^{2a}e^{-px^2}dx = \frac{1\cdot3\cdot5\cdots(2a-1)}{2^{a+1}p^a}\sqrt{\frac{\pi}{p}}.$

861.8. $\displaystyle\int_0^\infty e^{-x^p}dx = \frac{1}{p}\,\Gamma\left(\frac{1}{p}\right).$ $\qquad\qquad\qquad[p>0].$

862.1. $\displaystyle\int_0^\infty \frac{dx}{1+e^{px}} = \frac{1}{p}\log_\epsilon 2.$ $\qquad$ [Ref. 16, Table 27, No. 1.]

862.2. $\displaystyle\int_0^\infty \frac{x\,dx}{e^x-1} = \frac{\pi^2}{6}.$ $\qquad$ **862.3.** $\displaystyle\int_0^\infty \frac{x\,dx}{e^x+1} = \frac{\pi^2}{12}.$

863.1. $\displaystyle\int_0^\infty e^{-ax}\sin mx\,dx = \frac{m}{a^2+m^2},$ $\qquad\qquad[a>0].$

863.2. $\displaystyle\int_0^\infty e^{-ax}\cos mx\,dx = \frac{a}{a^2+m^2},$ $\qquad\qquad[a>0].$

[Ref. 7, Art. 291.]

863.3. $\displaystyle\int_0^\infty e^{-a^2x^2}\cos 2px\,dx = \frac{\sqrt{\pi}}{2a}e^{-p^2/a^2},$ $\qquad[a>0].$

[Ref. 7, Art. 283 and Ref. 20, p. 47, No. 119.]

863.4. $\displaystyle\int_0^\infty \frac{e^{-ax}\sin x}{x}\,dx = \operatorname{ctn}^{-1}a = \tan^{-1}\frac{1}{a},$ $\qquad[a>0].$

[Ref. 11, p. 154, Ex. 3.]

864.1. $\displaystyle\int_0^1 \frac{\log x}{1-x}\,dx = -\frac{\pi^2}{6}.$

[See **48.2** and Ref. 7, Art. 299.]

864.2. $\displaystyle\int_0^1 \frac{\log x}{1+x}\,dx = -\frac{\pi^2}{12}.$

864.3. $\displaystyle\int_0^1 \frac{\log x}{1-x^2}\,dx = -\frac{\pi^2}{8}.$

865.1. $\displaystyle\int_0^1 \frac{\log(1+x)}{x}\,dx = \frac{\pi^2}{12}.$

865.2. $\displaystyle\int_0^1 \frac{\log(1-x)}{x}\,dx = -\frac{\pi^2}{6}.$

865.3. $\displaystyle\int_0^1 x^{2a}\log(1+x)dx = \frac{2\log 2}{2a+1} + \frac{1}{2a+1}\sum_{n=1}^{2a+1}\frac{(-1)^n}{n},$

[a = integer].

865.4. $\displaystyle\int_0^1 x^{2a-1} \log\,(1+x)dx = \frac{1}{2a} \sum_{n=1}^{2a} \frac{(-1)^{n-1}}{n},$

[a = integer].

865.5. $\displaystyle\int_0^1 x^{a-1} \log\,(1-x)dx = -\frac{1}{a} \sum_{n=1}^{a} \frac{1}{n},$ [a = integer].

[Ref. 16, p. 152.]

865.6. $\displaystyle\int_0^1 \frac{\log\,(1+x)}{1+x^2}\, dx = \frac{\pi}{8} \log_e 2.$ [Ref. 7, Art. 51.]

866.1. $\displaystyle\int_0^1 \left(\log \frac{1}{x}\right)^{1/2} dx = \frac{\sqrt{\pi}}{2}.$

866.2. $\displaystyle\int_0^1 \left(\log \frac{1}{x}\right)^{-1/2} dx = \sqrt{\pi}.$

866.3. $\displaystyle\int_0^1 \left(\log \frac{1}{x}\right)^{p} dx = \Gamma(p+1),$ [$-1 < p < \infty$].

[Ref. 16, Table 30, No. 2.]

866.31. $\displaystyle\int_0^1 x^m \left(\log \frac{1}{x}\right)^p dx = \frac{\Gamma(p+1)}{(m+1)^{p+1}},$

[$m+1 > 0,\ p+1 > 0$].   [Ref. 40, Art. 97.]

866.4. $\displaystyle\int_0^1 (\log x)^p dx = (-1)^p p!.$ [Ref. 20, p. 47, No. 121.]

867.1. $\displaystyle\int_0^1 \log x \log\,(1+x)dx = 2 - 2\log 2 - \frac{\pi^2}{12}.$

867.2. $\displaystyle\int_0^1 \log x \log\,(1-x)dx = 2 - \frac{\pi^2}{6}.$

867.3. $\displaystyle\int_0^1 x \log\,(1+x)dx = \frac{1}{4}.$

867.4. $\displaystyle\int_0^1 x \log\,(1-x)dx = -\frac{3}{4}.$

867.5. $\displaystyle\int_0^1 x \log x \log\,(1+x)dx = \frac{\pi^2}{24} - \frac{1}{2}.$

867.6. $\displaystyle\int_0^1 x \log x \log\,(1-x)dx = 1 - \frac{\pi^2}{12}.$

867.7. $\displaystyle\int_0^1 (1+x) \log x \log\,(1+x)dx = \frac{3}{2} - 2\log 2 - \frac{\pi^2}{24}.$

867.8. $\displaystyle\int_0^1 (1-x) \log x \log (1-x) dx = 1 - \frac{\pi^2}{12}.$

868.1. $\displaystyle\int_0^{\pi/2} \log \sin x \, dx = \int_0^{\pi/2} \log \cos x \, dx = -\frac{\pi}{2} \log 2.$

[Ref. 7, Art. 51.]

868.2. $\displaystyle\int_0^{\pi} x \log \sin x \, dx = -\frac{\pi^2 \log 2}{2}.$ [Ref. 5, No. 522.]

868.3. $\displaystyle\int_0^{\pi/2} \sin x \log \sin x \, dx = \log 2 - 1.$

[Ref. 7, p. 74, No. 13.]

868.4. $\displaystyle\int_0^{\pi} \log (a \pm b \cos x) dx = \pi \log \left[\frac{a + \sqrt{(a^2 - b^2)}}{2} \right],$

$[a \geqq b].$ [Ref. 5, No. 523.]

868.5. $\displaystyle\int_0^{\pi/2} \log \tan x \, dx = 0.$ [Ref. 7, p. 74, No. 12.]

868.6. $\displaystyle\int_0^{\pi/4} \log (1 + \tan x) dx = \frac{\pi}{8} \log 2.$ [Ref. 7, Art. 51.]

868.7. $\displaystyle\int_0^{\pi} \log (a^2 - 2ab \cos x + b^2) dx = 2\pi \log a, \; [a \geqq b > 0],$
$= 2\pi \log b, \; [b \geqq a > 0].$

[Ref. 7, par. 292.]

869.1. $\displaystyle\int_0^{\infty} \frac{dx}{\cosh ax} = \frac{\pi}{2a}.$ [Ref. 20, p. 47, No. 120.]

875.1. $\displaystyle\int_0^{\infty} e^{-ax} J_0(bx) dx = \frac{1}{\sqrt{(a^2 + b^2)}}.$

875.2. $\displaystyle\int_0^{\infty} e^{-ax} J_n(bx) dx = \frac{1}{\sqrt{(a^2 + b^2)}} \left\{ \frac{\sqrt{(a^2 + b^2)} - a}{b} \right\}^n.$

[Ref. 12, p. 64, eq. (1) and (2).]

876. $\displaystyle\int_0^{\pi} \cos (n\varphi - x \sin \varphi) d\varphi = \pi J_n(x).$

where n is zero or any positive integer.

[Bessel's Integral. Ref. 12, p. 32, eq. (9).]

For very complete tables of definite integrals see References 15 and 16.

880. Simpson's Rule. When there are a number of values of $\overline{y = f(x)}$ for values of x at equal intervals, h, apart, an approximate numerical integration is given by

$$\int_{x=a}^{b} f(x)\, dx \approx \frac{h}{3}\left[y_0 + 4y_1 + 2y_2 + 4y_3 + 2y_4 + \cdots \\ + 4y_{2n-1} + y_{2n}\right]$$

where $h = x_1 - x_0 =$ the constant interval of x, so that $2nh = b - a$. The coefficients are alternately 4 and 2 as indicated. The approximation is in general more accurate as n is larger. In this way, a numerical result can often be obtained when the algebraic expression cannot be integrated in suitable form. This computation can be performed as one continuous operation on a manual calculating machine, using a table of $f(x)$.

881. An estimate of the error in the above approximate formula is

$$\frac{nh^5 f^{iv}(x)}{90} = \frac{(b-a)\, h^4 f^{iv}(x)}{180}$$

where the largest entry found in the fourth column of differences in the table of $f(x)$, in the range between a and b, may be used for the numerical value of $h^4 f^{iv}(x)$. See also pages 184–5 of "Methods of Advanced Calculus," by Philip Franklin (Refer. 39).

882. The following alternative formula is more accurate, with many functions, than No. 880. It also can be computed in one continuous operation on a manual calculating machine.

$$\int_{x=a}^{b} f(x)\, dx \approx \frac{h}{4.5}\left[1.4\,y_0 + 6.4\,y_1 + 2.4\,y_2 + 6.4\,y_3 + 2.8\,y_4 \\ + 6.4\,y_5 + 2.4\,y_6 + 6.4\,y_7 + 2.8\,y_8 \\ \cdots\cdots\cdots\cdots\cdots\cdots \\ + 6.4\,y_{4n-3} + 2.4\,y_{4n-2} + 6.4\,y_{4n-1} + 1.4\,y_{4n}\right]$$

where $4nh = b - a$.

DIFFERENTIAL EQUATIONS

890.1. Separation of the variables. If the equation can be put in the form $f_1(x)dx = f_2(y)dy$, each term may be integrated.

890.2. Separation of the variables by a substitution—Homogeneous equations. If the equation is of the form

$$f_1(x, y)dx + f_2(x, y)dy = 0,$$

where the functions are homogeneous in x and y and are of the same degree, let $y = ux$. Then

$$\frac{dx}{x} = -\frac{f_2(1, u)du}{f_1(1, u) + uf_2(1, u)}.$$

If more convenient let $x = uy$.

890.3. Separation of the variables by a substitution, for equations of the form

$$f_1(xy)y \, dx + f_2(xy)x \, dy = 0,$$

where f_1 and f_2 are any functions. Let $y = u/x$. Then

$$\frac{dx}{x} = \frac{f_2(u)du}{u\{f_2(u) - f_1(u)\}}.$$

890.4. An equation of the form

$$(ax + by + c)dx + (fx + gy + h)dy = 0$$

can be made homogeneous by putting $x = x' + m$ and $y = y' + n$. The quantities m and n can be found by solving the two simultaneous equations in m and n required to make the original equation homogeneous. This method does not apply if

$$\frac{ax + by}{fx + gy} = \text{a constant},$$

but we can then solve by substituting $ax + by = u$ and eliminating y or x.

890.5. Exact differential equations. If $M \, dx + N \, dy = 0$ is an equation in which

$$\frac{\partial M}{\partial y} = \frac{\partial N}{\partial x},$$

it is an exact differential equation.

Integrate $\int M\,dx$, regarding y as a constant and adding an unknown function of y, say $f(y)$; differentiate the result with respect to y and equate the new result to N; from the resulting equation determine the unknown function of y. The solution is then

$$\int M\,dx + f(y) + c = 0.$$

If more convenient, interchange M and N and also x and y in the above rule.

[See Ref. 32, *A Course in Mathematics*, by F. S. Woods and F. H. Bailey, vol. 2, ed. of 1909, p. 270.]

891.1. Linear equations of the first order. A differential equation is linear when it has only the first power of the function and of its derivatives. The linear equation of the first order is of the form

$$\frac{dy}{dx} + Py = Q \quad \text{or} \quad dy + Py\,dx = Q\,dx,$$

where P and Q are independent of y but may involve x.

Insert $e^{\int P\,dx}$ as an integrating factor. The solution is

$$y = e^{-\int P\,dx}\left[\int e^{\int P\,dx}Q\,dx + c\right].$$

891.2. Bernoulli's equation. If the equation is of the form

$$\frac{dy}{dx} + Py = Qy^{n},$$

where P and Q do not involve y, it can be made linear by substituting $1/y^{n-1} = u$. Divide the equation by y^{n} before making the substitution.

892. Equations of the first order but not of the first degree. Let

$$\frac{dy}{dx} = p.$$

If possible, solve the resulting equation for p. The equations given by putting p equal to the values so found may often be integrated, thus furnishing solutions of the given equation.

893.1. Equations of the second order, not containing y directly. Let $dy/dx = p$. The equation will become one of the first order in p and x. It may be possible to solve this by one of the methods of the preceding paragraphs.

893.2. Equations of the second order, not containing x directly.

Let

$$\frac{dy}{dx} = p.$$

Then

$$\frac{d^2y}{dx^2} = \frac{dp}{dy}\frac{dy}{dx} = p\frac{dp}{dy}.$$

The resulting equation is of the first order in p and y and it may be possible to solve it by one of the methods of the preceding paragraphs.

894. To solve

$$\frac{d^2y}{dx^2} + A\frac{dy}{dx} + By = 0,$$

where A and B are constants, find the roots of the auxiliary equation $p^2 + Ap + B = 0$. If the roots are real and unequal quantities a and b, the solution is $y = he^{ax} + ke^{bx}$, where h and k are constants.

If the roots are complex quantities $m + in$ and $m - in$,

$$y = e^{mx}(h \cos nx + k \sin nx).$$

If the roots are equal and are a, a,

$$y = e^{ax}(hx + k).$$

895. Equations of the nth order of the form

$$\frac{d^ny}{dx^n} + A\frac{d^{n-1}y}{dx^{n-1}} + B\frac{d^{n-2}y}{dx^{n-2}} + \cdots + Ky = 0,$$

where A, B, $\cdots$, K are constants. This is a linear differential equation.

For each distinct real root a of the auxiliary equation

$$p^n + Ap^{n-1} + Bp^{n-2} + \cdots + K = 0,$$

there is a term he^{ax} in the solution. The terms of the solution are to be added together.

When a occurs twice among the n roots of the auxiliary equation, the corresponding term is $e^{ax}(hx + k)$.

When a occurs three times, the corresponding term is

$$e^{ax}(hx^2 + kx + l),$$

and so forth.

When there is a pair of imaginary roots $m + in$ and $m - in$, there is a term in the solution

$$e^{mx}(h \cos nx + k \sin nx).$$

When the same pair occurs twice, the corresponding term in the solution is

$$e^{mx}\{(hx + k) \cos nx + (sx + t) \sin nx\}$$

and so forth.

896. Linear differential equations with constant coefficients.

$$\frac{d^n y}{dx^n} + A \frac{d^{n-1}y}{dx^{n-1}} + B \frac{d^{n-2}y}{dx^{n-2}} + \cdots + Ky = X$$

where X may involve x.

First solve the equation obtained by putting $X = 0$, as in **894** or **895**. Add to this solution a **particular integral** which satisfies the original equation and which need not contain constants of integration since n such constants have already been put in the solution.

897. The " homogeneous linear equation " of the second order,

$$x^2 \frac{d^2 y}{dx^2} + Ax \frac{dy}{dx} + By = f(x)$$

becomes a linear equation with constant coefficients

$$\frac{d^2 y}{dv^2} + (A - 1) \frac{dy}{dv} + By = f(e^v)$$

by substituting $x = e^v$.

[See *Elements of the Infinitesimal Calculus*, by G. H. Chandler. Ref. 8, Chaps. 44–45, or other textbooks.]

898. Linear partial differential equation of the first order,

$$P \frac{\partial z}{\partial x} + Q \frac{\partial z}{\partial y} = R.$$

To solve this, first solve the equations

$$\frac{dx}{P} = \frac{dy}{Q} = \frac{dz}{R},$$

and place the solution in the form $u = c_1$, $v = c_2$. Then

$$\varphi(u, v) = 0,$$

where φ is an arbitrary function, is the solution required.

[Ref. 11, p. 292.]

APPENDIX

A. Tables of Numerical Values

B. References

TABLE 1000—$\sqrt{a^2 + b^2}/a$

b/a	$\frac{\sqrt{a^2+b^2}}{a}$	b/a	$\frac{\sqrt{a^2+b^2}}{a}$	b/a	$\frac{\sqrt{a^2+b^2}}{a}$	b/a	$\frac{\sqrt{a^2+b^2}}{a}$	b/a	$\frac{\sqrt{a^2+b^2}}{a}$	b/a	$\frac{\sqrt{a^2+b^2}}{a}$	b/a	$\frac{\sqrt{a^2+b^2}}{a}$	b/a	$\frac{\sqrt{a^2+b^2}}{a}$
0	1.000	.175	1.015	.300	1.044	.350	1.059	.400	1.077	.450	1.097	.500	1.118	.550	1.141
.010	1.000	.180	1.016	.302	1.045	.352	1.060	.402	1.078	.452	1.097	.502	1.119	.552	1.142
.020	1.000	.185	1.017	.304	1.045	.354	1.061	.404	1.079	.454	1.098	.504	1.120	.554	1.143
.030	1.000	.190	1.018	.306	1.046	.356	1.061	.406	1.079	.456	1.099	.506	1.121	.556	1.144
.040	1.001	.195	1.019	.308	1.046	.358	1.062	.408	1.080	.458	1.100	.508	1.122	.558	1.145
.050	1.001	.200	1.020	.310	1.047	.360	1.063	.410	1.081	.460	1.101	.510	1.123	.560	1.146
.060	1.002	.205	1.021	.312	1.048	.362	1.064	.412	1.082	.462	1.102	.512	1.123	.562	1.147
.070	1.002	.210	1.022	.314	1.048	.364	1.065	.414	1.082	.464	1.102	.514	1.124	.564	1.148
.080	1.003	.215	1.023	.316	1.049	.366	1.065	.416	1.083	.466	1.103	.516	1.125	.566	1.149
.090	1.004	.220	1.024	.318	1.049	.368	1.066	.418	1.084	.468	1.104	.518	1.126	.568	1.150
.100	1.005	.225	1.025	.320	1.050	.370	1.066	.420	1.085	.470	1.105	.520	1.127	.570	1.151
.105	1.005	.230	1.026	.322	1.051	.372	1.067	.422	1.085	.472	1.106	.522	1.128	.572	1.152
.110	1.006	.235	1.027	.324	1.051	.374	1.068	.424	1.086	.474	1.107	.524	1.129	.574	1.153
.115	1.007	.240	1.028	.326	1.052	.376	1.068	.426	1.087	.476	1.108	.526	1.130	.576	1.154
.120	1.007	.245	1.030	.328	1.052	.378	1.069	.428	1.088	.478	1.108	.528	1.131	.578	1.155
.125	1.008	.250	1.031	.330	1.053	.380	1.070	.430	1.089	.480	1.109	.530	1.132	.580	1.156
.130	1.008	.255	1.032	.332	1.054	.382	1.070	.432	1.089	.482	1.110	.532	1.133	.582	1.157
.135	1.009	.260	1.033	.334	1.054	.384	1.071	.434	1.090	.484	1.111	.534	1.134	.584	1.158
.140	1.010	.265	1.035	.336	1.055	.386	1.072	.436	1.091	.486	1.112	.536	1.135	.586	1.159
.145	1.010	.270	1.036	.338	1.056	.388	1.073	.438	1.092	.488	1.113	.538	1.136	.588	1.160
.150	1.011	.275	1.037	.340	1.056	.390	1.073	.440	1.093	.490	1.114	.540	1.136	.590	1.161
.155	1.012	.280	1.038	.342	1.057	.392	1.074	.442	1.093	.492	1.114	.542	1.137	.592	1.162
.160	1.013	.285	1.040	.344	1.058	.394	1.075	.444	1.094	.494	1.115	.544	1.138	.594	1.163
.165	1.014	.290	1.041	.346	1.058	.396	1.076	.446	1.095	.496	1.116	.546	1.139	.596	1.164
.170	1.014	.295	1.043	.348	1.059	.398	1.076	.448	1.096	.498	1.117	.548	1.140	.598	1.165

TABLE 1000 (continued)—$\sqrt{(a^2 + b^2)}/a$

b/a	$\sqrt{(a^2+b^2)}$	b/a	$\sqrt{(a^2+b^2)}$	b/a	$\sqrt{(a^2+b^2)}$	b/a	$\sqrt{(a^2+b^2)}$	b/a	$\sqrt{(a^2+b^2)}$	b/a	$\sqrt{(a^2+b^2)}$	b/a	$\sqrt{(a^2+b^2)}$	b/a	$\sqrt{(a^2+b^2)}$
.600	1.166	.650	1.193	.700	1.221	.750	1.250	.800	1.281	.850	1.312	.900	1.345	.950	1.379
.602	1.167	.652	1.194	.702	1.222	.752	1.251	.802	1.282	.852	1.314	.902	1.347	.952	1.381
.604	1.168	.654	1.195	.704	1.223	.754	1.252	.804	1.283	.854	1.315	.904	1.348	.954	1.382
.606	1.169	.656	1.196	.706	1.224	.756	1.254	.806	1.284	.856	1.316	.906	1.349	.956	1.383
.608	1.170	.658	1.197	.708	1.225	.758	1.255	.808	1.286	.858	1.318	.908	1.351	.958	1.385
.610	1.171	.660	1.198	.710	1.226	.760	1.256	.810	1.287	.860	1.319	.910	1.352	.960	1.386
.612	1.172	.662	1.199	.712	1.228	.762	1.257	.812	1.288	.862	1.320	.912	1.353	.962	1.388
.614	1.173	.664	1.200	.714	1.229	.764	1.258	.814	1.289	.864	1.322	.914	1.355	.964	1.389
.616	1.175	.666	1.201	.716	1.230	.766	1.260	.816	1.291	.866	1.323	.916	1.356	.966	1.390
.618	1.176	.668	1.203	.718	1.231	.768	1.261	.818	1.292	.868	1.324	.918	1.357	.968	1.392
.620	1.177	.670	1.204	.720	1.232	.770	1.262	.820	1.293	.870	1.325	.920	1.359	.970	1.393
.622	1.178	.672	1.205	.722	1.233	.772	1.263	.822	1.294	.872	1.327	.922	1.360	.972	1.395
.624	1.179	.674	1.206	.724	1.235	.774	1.265	.824	1.296	.874	1.328	.924	1.362	.974	1.396
.626	1.180	.676	1.207	.726	1.236	.776	1.266	.826	1.297	.876	1.329	.926	1.363	.976	1.397
.628	1.181	.678	1.208	.728	1.237	.778	1.267	.828	1.298	.878	1.331	.928	1.364	.978	1.399
.630	1.182	.680	1.209	.730	1.238	.780	1.268	.830	1.300	.880	1.332	.930	1.366	.980	1.400
.632	1.183	.682	1.210	.732	1.239	.782	1.269	.832	1.301	.882	1.333	.932	1.367	.982	1.402
.634	1.184	.684	1.212	.734	1.240	.784	1.271	.834	1.302	.884	1.335	.934	1.368	.984	1.403
.636	1.185	.686	1.213	.736	1.242	.786	1.272	.836	1.303	.886	1.336	.936	1.370	.986	1.404
.638	1.186	.688	1.214	.738	1.243	.788	1.273	.838	1.305	.888	1.337	.938	1.371	.988	1.406
.640	1.187	.690	1.215	.740	1.244	.790	1.274	.840	1.306	.890	1.339	.940	1.372	.990	1.407
.642	1.188	.692	1.216	.742	1.245	.792	1.276	.842	1.307	.892	1.340	.942	1.374	.992	1.409
.644	1.189	.694	1.217	.744	1.246	.794	1.277	.844	1.309	.894	1.341	.944	1.375	.994	1.410
.646	1.191	.696	1.218	.746	1.248	.796	1.278	.846	1.310	.896	1.343	.946	1.377	.996	1.411
.648	1.192	.698	1.220	.748	1.249	.798	1.279	.848	1.311	.898	1.344	.948	1.378	.998	1.413
														1.000	1.414

$$\sqrt{(a^2 + b^2)} = a + \frac{b^2}{2a} - \frac{b^4}{8a^3} + \cdots \quad [b^2 < a^2].$$ The approximation $\sqrt{(a^2 + b^2)} = a + \dfrac{b^2}{2a}$ is correct within 1/1000 when $b/a < 0.3$.

TABLE 1010—TRIGONOMETRIC FUNCTIONS

[Characteristics of Logarithms omitted — determine by the usual rule from the value]

Radians	Degrees	Sine Value	Sine Log₁₀	Tangent Value	Tangent Log₁₀	Cotangent Value	Cotangent Log₁₀	Cosine Value	Cosine Log₁₀		
.0000	0° 00′	.0000	—	.0000	—			1.0000	.0000	90° 00′	1.5708
.0029	10	.0029	.4637	.0029	.4637	343.77	.5363	1.0000	.0000	50	1.5679
.0058	20	.0058	.7648	.0058	.7648	171.89	.2352	1.0000	.0000	40	1.5650
.0087	30	.0087	.9408	.0087	.9409	114.59	.0591	1.0000	.0000	30	1.5621
.0116	40	.0116	.0658	.0116	.0658	85.940	.9342	.9999	.0000	20	1.5592
.0145	50	.0145	.1627	.0145	.1627	68.750	.8373	.9999	.0000	10	1.5563
.0175	1° 00′	.0175	.2419	.0175	.2419	57.290	.7581	.9998	.9999	89° 00′	1.5533
.0204	10	.0204	.3088	.0204	.3089	49.104	.6911	.9998	.9999	50	1.5504
.0233	20	.0233	.3668	.0233	.3669	42.964	.6331	.9997	.9999	40	1.5475
.0262	30	.0262	.4179	.0262	.4181	38.188	.5819	.9997	.9999	30	1.5446
.0291	40	.0291	.4637	.0291	.4638	34.368	.5362	.9996	.9998	20	1.5417
.0320	50	.0320	.5050	.0320	.5053	31.242	.4947	.9995	.9998	10	1.5388
.0349	2° 00′	.0349	.5428	.0349	.5431	28.636	.4569	.9994	.9997	88° 00′	1.5359
.0378	10	.0378	.5776	.0378	.5779	26.432	.4221	.9993	.9997	50	1.5330
.0407	20	.0407	.6097	.0407	.6101	24.542	.3899	.9992	.9996	40	1.5301
.0436	30	.0436	.6397	.0437	.6401	22.904	.3599	.9990	.9996	30	1.5272
.0465	40	.0465	.6677	.0466	.6682	21.470	.3318	.9989	.9995	20	1.5243
.0495	50	.0494	.6940	.0495	.6945	20.206	.3055	.9988	.9995	10	1.5213
.0524	3° 00′	.0523	.7188	.0524	.7194	19.081	.2806	.9986	.9994	87° 00′	1.5184
.0553	10	.0552	.7423	.0553	.7429	18.075	.2571	.9985	.9993	50	1.5155
.0582	20	.0581	.7645	.0582	.7652	17.169	.2348	.9983	.9993	40	1.5126
.0611	30	.0610	.7857	.0612	.7865	16.350	.2135	.9981	.9992	30	1.5097
.0640	40	.0640	.8059	.0641	.8067	15.605	.1933	.9980	.9991	20	1.5068
.0669	50	.0669	.8251	.0670	.8261	14.924	.1739	.9978	.9990	10	1.5039
.0698	4° 00′	.0698	.8436	.0699	.8446	14.301	.1554	.9976	.9989	86° 00′	1.5010
.0727	10	.0727	.8613	.0729	.8624	13.727	.1376	.9974	.9989	50	1.4981
.0756	20	.0756	.8783	.0758	.8795	13.197	.1205	.9971	.9988	40	1.4952
.0785	30	.0785	.8946	.0787	.8960	12.706	.1040	.9969	.9987	30	1.4923
.0814	40	.0814	.9104	.0816	.9118	12.251	.0882	.9967	.9986	20	1.4893
.0844	50	.0843	.9256	.0846	.9272	11.826	.0728	.9964	.9985	10	1.4864
.0873	5° 00′	.0872	.9403	.0875	.9420	11.430	.0580	.9962	.9983	85° 00′	1.4835
.0902	10	.0901	.9545	.0904	.9563	11.059	.0437	.9959	.9982	50	1.4806
.0931	20	.0929	.9682	.0934	.9701	10.712	.0299	.9957	.9981	40	1.4777
.0960	30	.0958	.9816	.0963	.9836	10.385	.0164	.9954	.9980	30	1.4748
.0989	40	.0987	.9945	.0992	.9966	10.078	.0034	.9951	.9979	20	1.4719
.1018	50	.1016	.0070	.1022	.0093	9.7882	.9907	.9948	.9977	10	1.4690
.1047	6° 00′	.1045	.0192	.1051	.0216	9.5144	.9784	.9945	.9976	84° 00′	1.4661
.1076	10	.1074	.0311	.1080	.0336	9.2553	.9664	.9942	.9975	50	1.4632
.1105	20	.1103	.0426	.1110	.0453	9.0098	.9547	.9939	.9973	40	1.4603
.1134	30	.1132	.0539	.1139	.0567	8.7769	.9433	.9936	.9972	30	1.4573
.1164	40	.1161	.0648	.1169	.0678	8.5555	.9322	.9932	.9971	20	1.4544
.1193	50	.1190	.0755	.1198	.0786	8.3450	.9214	.9929	.9969	10	1.4515
.1222	7° 00′	.1219	.0859	.1228	.0891	8.1443	.9109	.9925	.9968	83° 00′	1.4486
.1251	10	.1248	.0961	.1257	.0995	7.9530	.9005	.9922	.9966	50	1.4457
.1280	20	.1276	.1060	.1287	.1096	7.7704	.8904	.9918	.9964	40	1.4428
.1309	30	.1305	.1157	.1317	.1194	7.5958	.8806	.9914	.9963	30	1.4399
.1338	40	.1334	.1252	.1346	.1291	7.4287	.8709	.9911	.9961	20	1.4370
.1367	50	.1363	.1345	.1376	.1385	7.2687	.8615	.9907	.9959	10	1.4341
.1396	8° 00′	.1392	.1436	.1405	.1478	7.1154	.8522	.9903	.9958	82° 00′	1.4312
.1425	10	.1421	.1525	.1435	.1569	6.9682	.8431	.9899	.9956	50	1.4283
.1454	20	.1449	.1612	.1465	.1658	6.8269	.8342	.9894	.9954	40	1.4254
.1484	30	.1478	.1697	.1495	.1745	6.6912	.8255	.9890	.9952	30	1.4224
.1513	40	.1507	.1781	.1524	.1831	6.5606	.8169	.9886	.9950	20	1.4195
.1542	50	.1536	.1863	.1554	.1915	6.4348	.8085	.9881	.9948	10	1.4166
.1571	9° 00′	.1564	.1943	.1584	.1997	6.3138	.8003	.9877	.9946	81° 00′	1.4137
		Value Log₁₀ COSINE		Value Log₁₀ COTANGENT		Value Log₁₀ TANGENT		Value Log₁₀ SINE		DEGREES	RADIANS

TABLE 1010 (*continued*)—TRIGONOMETRIC FUNCTIONS

[Characteristics of Logarithms omitted — determine by the usual rule from the value]

Radians	Degrees	Sine Value	Log₁₀	Tangent Value	Log₁₀	Cotangent Value	Log₁₀	Cosine Value	Log₁₀		
.1571	9° 00′	.1564	.1943	.1584	.1997	6.3138	.8003	.9877	.9946	81° 00′	1.4137
.1600	10	.1593	.2022	.1614	.2078	6.1970	.7922	.9872	.9944	50	1.4108
.1629	20	.1622	.2100	.1644	.2158	6.0844	.7842	.9868	.9942	40	1.4079
.1658	30	.1650	.2176	.1673	.2236	5.9758	.7764	.9863	.9940	30	1.4050
.1687	40	.1679	.2251	.1703	.2313	5.8708	.7687	.9858	.9938	20	1.4021
.1716	50	.1708	.2324	.1733	.2389	5.7694	.7611	.9853	.9936	10	1.3992
.1745	10° 00′	.1736	.2397	.1763	.2463	5.6713	.7537	.9848	.9934	80° 00′	1.3963
.1774	10	.1765	.2468	.1793	.2536	5.5764	.7464	.9843	.9931	50	1.3934
.1804	20	.1794	.2538	.1823	.2609	5.4845	.7391	.9838	.9929	40	1.3904
.1833	30	.1822	.2606	.1853	.2680	5.3955	.7320	.9833	.9927	30	1.3875
.1862	40	.1851	.2674	.1883	.2750	5.3093	.7250	.9827	.9924	20	1.3846
.1891	50	.1880	.2740	.1914	.2819	5.2257	.7181	.9822	.9922	10	1.3817
.1920	11° 00′	.1908	.2806	.1944	.2887	5.1446	.7113	.9816	.9919	79° 00′	1.3788
.1949	10	.1937	.2870	.1974	.2953	5.0658	.7047	.9811	.9917	50	1.3759
.1978	20	.1965	.2934	.2004	.3020	4.9894	.6980	.9805	.9914	40	1.3730
.2007	30	.1994	.2997	.2035	.3085	4.9152	.6915	.9799	.9912	30	1.3701
.2036	40	.2022	.3058	.2065	.3149	4.8430	.6851	.9793	.9909	20	1.3672
.2065	50	.2051	.3119	.2095	.3212	4.7729	.6788	.9787	.9907	10	1.3643
.2094	12° 00′	.2079	.3179	.2126	.3275	4.7046	.6725	.9781	.9904	78° 00′	1.3614
.2123	10	.2108	.3238	.2156	.3336	4.6382	.6664	.9775	.9901	50	1.3584
.2153	20	.2136	.3296	.2186	.3397	4.5736	.6603	.9769	.9899	40	1.3555
.2182	30	.2164	.3353	.2217	.3458	4.5107	.6542	.9763	.9896	30	1.3526
.2211	40	.2193	.3410	.2247	.3517	4.4494	.6483	.9757	.9893	20	1.3497
.2240	50	.2221	.3466	.2278	.3576	4.3897	.6424	.9750	.9890	10	1.3468
.2269	13° 00′	.2250	.3521	.2309	.3634	4.3315	.6366	.9744	.9887	77° 00′	1.3439
.2298	10	.2278	.3575	.2339	.3691	4.2747	.6309	.9737	.9884	50	1.3410
.2327	20	.2306	.3629	.2370	.3748	4.2193	.6252	.9730	.9881	40	1.3381
.2356	30	.2334	.3682	.2401	.3804	4.1653	.6196	.9724	.9878	30	1.3352
.2385	40	.2363	.3734	.2432	.3859	4.1126	.6141	.9717	.9875	20	1.3323
.2414	50	.2391	.3786	.2462	.3914	4.0611	.6086	.9710	.9872	10	1.3294
.2443	14° 00′	.2419	.3837	.2493	.3968	4.0108	.6032	.9703	.9869	76° 00′	1.3265
.2473	10	.2447	.3887	.2524	.4021	3.9617	.5979	.9696	.9866	50	1.3235
.2502	20	.2476	.3937	.2555	.4074	3.9136	.5926	.9689	.9863	40	1.3206
.2531	30	.2504	.3986	.2586	.4127	3.8667	.5873	.9681	.9859	30	1.3177
.2560	40	.2532	.4035	.2617	.4178	3.8208	.5822	.9674	.9856	20	1.3148
.2589	50	.2560	.4083	.2648	.4230	3.7760	.5770	.9667	.9853	10	1.3119
.2618	15° 00′	.2588	.4130	.2679	.4281	3.7321	.5719	.9659	.9849	75° 00′	1.3090
.2647	10	.2616	.4177	.2711	.4331	3.6891	.5669	.9652	.9846	50	1.3061
.2676	20	.2644	.4223	.2742	.4381	3.6470	.5619	.9644	.9843	40	1.3032
.2705	30	.2672	.4269	.2773	.4430	3.6059	.5570	.9636	.9839	30	1.3003
.2734	40	.2700	.4314	.2805	.4479	3.5656	.5521	.9628	.9836	20	1.2974
.2763	50	.2728	.4359	.2836	.4527	3.5261	.5473	.9621	.9832	10	1.2945
.2793	16° 00′	.2756	.4403	.2867	.4575	3.4874	.5425	.9613	.9828	74° 00′	1.2915
.2822	10	.2784	.4447	.2899	.4622	3.4495	.5378	.9605	.9825	50	1.2886
.2851	20	.2812	.4491	.2931	.4669	3.4124	.5331	.9596	.9821	40	1.2857
.2880	30	.2840	.4533	.2962	.4716	3.3759	.5284	.9588	.9817	30	1.2828
.2909	40	.2868	.4576	.2994	.4762	3.3402	.5238	.9580	.9814	20	1.2799
.2938	50	.2896	.4618	.3026	.4808	3.3052	.5192	.9572	.9810	10	1.2770
.2967	17° 00′	.2924	.4659	.3057	.4853	3.2709	.5147	.9563	.9806	73° 00′	1.2741
.2996	10	.2952	.4700	.3089	.4898	3.2371	.5102	.9555	.9802	50	1.2712
.3025	20	.2979	.4741	.3121	.4943	3.2041	.5057	.9546	.9798	40	1.2683
.3054	30	.3007	.4781	.3153	.4987	3.1716	.5013	.9537	.9794	30	1.2654
.3083	40	.3035	.4821	.3185	.5031	3.1397	.4969	.9528	.9790	20	1.2625
.3113	50	.3062	.4861	.3217	.5075	3.1084	.4925	.9520	.9786	10	1.2595
.3142	18° 00′	.3090	.4900	.3249	.5118	3.0777	.4882	.9511	.9782	72° 00′	1.2566
		Value Cosine	Log₁₀	Value Cotangent	Log₁₀	Value Tangent	Log₁₀	Value Sine	Log₁₀	Degrees	Radians

TABLE 1010 (continued)—TRIGONOMETRIC FUNCTIONS
[Characteristics of Logarithms omitted — determine by the usual rule from the value]

Radians	Degrees	Sine Value	Sine Log10	Tangent Value	Tangent Log10	Cotangent Value	Cotangent Log10	Cosine Value	Cosine Log10		
.3142	18° 00′	.3090	.4900	.3249	.5118	3.0777	.4882	.9511	.9782	72° 00′	1.2566
.3171	10	.3118	.4939	.3281	.5161	3.0475	.4839	.9502	.9778	50	1.2537
.3200	20	.3145	.4977	.3314	.5203	3.0178	.4797	.9492	.9774	40	1.2508
.3229	30	.3173	.5015	.3346	.5245	2.9887	.4755	.9483	.9770	30	1.2479
.3258	40	.3201	.5052	.3378	.5287	2.9600	.4713	.9474	.9765	20	1.2450
.3287	50	.3228	.5090	.3411	.5329	2.9319	.4671	.9465	.9761	10	1.2421
.3316	19° 00′	.3256	.5126	.3443	.5370	2.9042	.4630	.9455	.9757	71° 00′	1.2392
.3345	10	.3283	.5163	.3476	.5411	2.8770	.4589	.9446	.9752	50	1.2363
.3374	20	.3311	.5199	.3508	.5451	2.8502	.4549	.9436	.9748	40	1.2334
.3403	30	.3338	.5235	.3541	.5491	2.8239	.4509	.9426	.9743	30	1.2305
.3432	40	.3365	.5270	.3574	.5531	2.7980	.4469	.9417	.9739	20	1.2275
.3462	50	.3393	.5306	.3607	.5571	2.7725	.4429	.9407	.9734	10	1.2246
.3491	20° 00′	.3420	.5341	.3640	.5611	2.7475	.4389	.9397	.9730	70° 00′	1.2217
.3520	10	.3448	.5375	.3673	.5650	2.7228	.4350	.9387	.9725	50	1.2188
.3549	20	.3475	.5409	.3706	.5689	2.6985	.4311	.9377	.9721	40	1.2159
.3578	30	.3502	.5443	.3739	.5727	2.6746	.4273	.9367	.9716	30	1.2130
.3607	40	.3529	.5477	.3772	.5766	2.6511	.4234	.9356	.9711	20	1.2101
.3636	50	.3557	.5510	.3805	.5804	2.6279	.4196	.9346	.9706	10	1.2072
.3665	21° 00′	.3584	.5543	.3839	.5842	2.6051	.4158	.9336	.9702	69° 00′	1.2043
.3694	10	.3611	.5576	.3872	.5879	2.5826	.4121	.9325	.9697	50	1.2014
.3723	20	.3638	.5609	.3906	.5917	2.5605	.4083	.9315	.9692	40	1.1985
.3752	30	.3665	.5641	.3939	.5954	2.5386	.4046	.9304	.9687	30	1.1956
.3782	40	.3692	.5673	.3973	.5991	2.5172	.4009	.9293	.9682	20	1.1926
.3811	50	.3719	.5704	.4006	.6028	2.4960	.3972	.9283	.9677	10	1.1897
.3840	22° 00′	.3746	.5736	.4040	.6064	2.4751	.3936	.9272	.9672	68° 00′	1.1868
.3869	10	.3773	.5767	.4074	.6100	2.4545	.3900	.9261	.9667	50	1.1839
.3898	20	.3800	.5798	.4108	.6136	2.4342	.3864	.9250	.9661	40	1.1810
.3927	30	.3827	.5828	.4142	.6172	2.4142	.3828	.9239	.9656	30	1.1781
.3956	40	.3854	.5859	.4176	.6208	2.3945	.3792	.9228	.9651	20	1.1752
.3985	50	.3881	.5889	.4210	.6243	2.3750	.3757	.9216	.9646	10	1.1723
.4014	23° 00′	.3907	.5919	.4245	.6279	2.3559	.3721	.9205	.9640	67° 00′	1.1694
.4043	10	.3934	.5948	.4279	.6314	2.3369	.3686	.9194	.9635	50	1.1665
.4072	20	.3961	.5978	.4314	.6348	2.3183	.3652	.9182	.9629	40	1.1636
.4102	30	.3987	.6007	.4348	.6383	2.2998	.3617	.9171	.9624	30	1.1606
.4131	40	.4014	.6036	.4383	.6417	2.2817	.3583	.9159	.9618	20	1.1577
.4160	50	.4041	.6065	.4417	.6452	2.2637	.3548	.9147	.9613	10	1.1548
.4189	24° 00′	.4067	.6093	.4452	.6486	2.2460	.3514	.9135	.9607	66° 00′	1.1519
.4218	10	.4094	.6121	.4487	.6520	2.2286	.3480	.9124	.9602	50	1.1490
.4247	20	.4120	.6149	.4522	.6553	2.2113	.3447	.9112	.9596	40	1.1461
.4276	30	.4147	.6177	.4557	.6587	2.1943	.3413	.9100	.9590	30	1.1432
.4305	40	.4173	.6205	.4592	.6620	2.1775	.3380	.9088	.9584	20	1.1403
.4334	50	.4200	.6232	.4628	.6654	2.1609	.3346	.9075	.9579	10	1.1374
.4363	25° 00′	.4226	.6259	.4663	.6687	2.1445	.3313	.9063	.9573	65° 00′	1.1345
.4392	10	.4253	.6286	.4699	.6720	2.1283	.3280	.9051	.9567	50	1.1316
.4422	20	.4279	.6313	.4734	.6752	2.1123	.3248	.9038	.9561	40	1.1286
.4451	30	.4305	.6340	.4770	.6785	2.0965	.3215	.9026	.9555	30	1.1257
.4480	40	.4331	.6366	.4806	.6817	2.0809	.3183	.9013	.9549	20	1.1228
.4509	50	.4358	.6392	.4841	.6850	2.0655	.3150	.9001	.9543	10	1.1199
.4538	26° 00′	.4384	.6418	.4877	.6882	2.0503	.3118	.8988	.9537	64° 00′	1.1170
.4567	10	.4410	.6444	.4913	.6914	2.0353	.3086	.8975	.9530	50	1.1141
.4596	20	.4436	.6470	.4950	.6946	2.0204	.3054	.8962	.9524	40	1.1112
.4625	30	.4462	.6495	.4986	.6977	2.0057	.3023	.8949	.9518	30	1.1083
.4654	40	.4488	.6521	.5022	.7009	1.9912	.2991	.8936	.9512	20	1.1054
.4683	50	.4514	.6546	.5059	.7040	1.9768	.2960	.8923	.9505	10	1.1025
.4712	27° 00′	.4540	.6570	.5095	.7072	1.9626	.2928	.8910	.9499	63° 00′	1.0996
		Value Cosine	Log10	Value Cotangent	Log10	Value Tangent	Log10	Value Sine	Log10	Degrees	Radians

214

TABLE 1010 (*continued*)—TRIGONOMETRIC FUNCTIONS
[Characteristics of Logarithms omitted — determine by the usual rule from the value]

RADIANS	DEGREES	SINE Value	Log$_{10}$	TANGENT Value	Log$_{10}$	COTANGENT Value	Log$_{10}$	COSINE Value	Log$_{10}$		
.4712	27° 00′	.4540	.6570	.5095	.7072	1.9626	.2928	.8910	.9499	63° 00′	1.0996
.4741	10	.4566	.6595	.5132	.7103	1.9486	.2897	.8897	.9492	50	1.0966
.4771	20	.4592	.6620	.5169	.7134	1.9347	.2866	.8884	.9486	40	1.0937
.4800	30	.4617	.6644	.5206	.7165	1.9210	.2835	.8870	.9479	30	1.0908
.4829	40	.4643	.6668	.5243	.7196	1.9074	.2804	.8857	.9473	20	1.0879
.4858	50	.4669	.6692	.5280	.7226	1.8940	.2774	.8843	.9466	10	1.0850
.4887	28° 00′	.4695	.6716	.5317	.7257	1.8807	.2743	.8829	.9459	62° 00′	1.0821
.4916	10	.4720	.6740	.5354	.7287	1.8676	.2713	.8816	.9453	50	1.0792
.4945	20	.4746	.6763	.5392	.7317	1.8546	.2683	.8802	.9446	40	1.0763
.4974	30	.4772	.6787	.5430	.7348	1.8418	.2652	.8788	.9439	30	1.0734
.5003	40	.4797	.6810	.5467	.7378	1.8291	.2622	.8774	.9432	20	1.0705
.5032	50	.4823	.6833	.5505	.7408	1.8165	.2592	.8760	.9425	10	1.0676
.5061	29° 00′	.4848	.6856	.5543	.7438	1.8040	.2562	.8746	.9418	61° 00′	1.0647
.5091	10	.4874	.6878	.5581	.7467	1.7917	.2533	.8732	.9411	50	1.0617
.5120	20	.4899	.6901	.5619	.7497	1.7796	.2503	.8718	.9404	40	1.0588
.5149	30	.4924	.6923	.5658	.7526	1.7675	.2474	.8704	.9397	30	1.0559
.5178	40	.4950	.6946	.5696	.7556	1.7556	.2444	.8689	.9390	20	1.0530
.5207	50	.4975	.6968	.5735	.7585	1.7437	.2415	.8675	.9383	10	1.0501
.5236	30° 00′	.5000	.6990	.5774	.7614	1.7321	.2386	.8660	.9375	60° 00′	1.0472
.5265	10	.5025	.7012	.5812	.7644	1.7205	.2356	.8646	.9368	50	1.0443
.5294	20	.5050	.7033	.5851	.7673	1.7090	.2327	.8631	.9361	40	1.0414
.5323	30	.5075	.7055	.5890	.7701	1.6977	.2299	.8616	.9353	30	1.0385
.5352	40	.5100	.7076	.5930	.7730	1.6864	.2270	.8601	.9346	20	1.0356
.5381	50	.5125	.7097	.5969	.7759	1.6753	.2241	.8587	.9338	10	1.0327
.5411	31° 00′	.5150	.7118	.6009	.7788	1.6643	.2212	.8572	.9331	59° 00′	1.0297
.5440	10	.5175	.7139	.6048	.7816	1.6534	.2184	.8557	.9323	50	1.0268
.5469	20	.5200	.7160	.6088	.7845	1.6426	.2155	.8542	.9315	40	1.0239
.5498	30	.5225	.7181	.6128	.7873	1.6319	.2127	.8526	.9308	30	1.0210
.5527	40	.5250	.7201	.6168	.7902	1.6212	.2098	.8511	.9300	20	1.0181
.5556	50	.5275	.7222	.6208	.7930	1.6107	.2070	.8496	.9292	10	1.0152
.5585	32° 00′	.5299	.7242	.6249	.7958	1.6003	.2042	.8480	.9284	58° 00′	1.0123
.5614	10	.5324	.7262	.6289	.7986	1.5900	.2014	.8465	.9276	50	1.0094
.5643	20	.5348	.7282	.6330	.8014	1.5798	.1986	.8450	.9268	40	1.0065
.5672	30	.5373	.7302	.6371	.8042	1.5697	.1958	.8434	.9260	30	1.0036
.5701	40	.5398	.7322	.6412	.8070	1.5597	.1930	.8418	.9252	20	1.0007
.5730	50	.5422	.7342	.6453	.8097	1.5497	.1903	.8403	.9244	10	.9977
.5760	33° 00′	.5446	.7361	.6494	.8125	1.5399	.1875	.8387	.9236	57° 00′	.9948
.5789	10	.5471	.7380	.6536	.8153	1.5301	.1847	.8371	.9228	50	.9919
.5818	20	.5495	.7400	.6577	.8180	1.5204	.1820	.8355	.9219	40	.9890
.5847	30	.5519	.7419	.6619	.8208	1.5108	.1792	.8339	.9211	30	.9861
.5876	40	.5544	.7438	.6661	.8235	1.5013	.1765	.8323	.9203	20	.9832
.5905	50	.5568	.7457	.6703	.8263	1.4919	.1737	.8307	.9194	10	.9803
.5934	34° 00′	.5592	.7476	.6745	.8290	1.4826	.1710	.8290	.9186	56° 00′	.9774
.5963	10	.5616	.7494	.6787	.8317	1.4733	.1683	.8274	.9177	50	.9745
.5992	20	.5640	.7513	.6830	.8344	1.4641	.1656	.8258	.9169	40	.9716
.6021	30	.5664	.7531	.6873	.8371	1.4550	.1629	.8241	.9160	30	.9687
.6050	40	.5688	.7550	.6916	.8398	1.4460	.1602	.8225	.9151	20	.9657
.6080	50	.5712	.7568	.6959	.8425	1.4370	.1575	.8208	.9142	10	.9628
.6109	35° 00′	.5736	.7586	.7002	.8452	1.4281	.1548	.8192	.9134	55° 00′	.9599
.6138	10	.5760	.7604	.7046	.8479	1.4193	.1521	.8175	.9125	50	.9570
.6167	20	.5783	.7622	.7089	.8506	1.4106	.1494	.8158	.9116	40	.9541
.6196	30	.5807	.7640	.7133	.8533	1.4019	.1467	.8141	.9107	30	.9512
.6225	40	.5831	.7657	.7177	.8559	1.3934	.1441	.8124	.9098	20	.9483
.6254	50	.5854	.7675	.7221	.8586	1.3848	.1414	.8107	.9089	10	.9454
.6283	36° 00′	.5878	.7692	.7265	.8613	1.3764	.1387	.8090	.9080	54° 00′	.9425
		Value	Log$_{10}$	Value	Log$_{10}$	Value	Log$_{10}$	Value	Log$_{10}$	DEGREES	RADIANS
		COSINE		COTANGENT		TANGENT		SINE			

TABLE 1010 (continued)—TRIGONOMETRIC FUNCTIONS

[Characteristics of Logarithms omitted — determine by the usual rule from the value]

RADIANS	DEGREES	SINE Value	SINE Log$_{10}$	TANGENT Value	TANGENT Log$_{10}$	COTANGENT Value	COTANGENT Log$_{10}$	COSINE Value	COSINE Log$_{10}$		
.6283	36° 00′	.5878	.7692	.7265	.8613	1.3764	.1387	.8090	.9080	54° 00′	.9425
.6312	10	.5901	.7710	.7310	.8639	1.3680	.1361	.8073	.9070	50	.9396
.6341	20	.5925	.7727	.7355	.8666	1.3597	.1334	.8056	.9061	40	.9367
.6370	30	.5948	.7744	.7400	.8692	1.3514	.1308	.8039	.9052	30	.9338
.6400	40	.5972	.7761	.7445	.8718	1.3432	.1282	.8021	.9042	20	.9308
.6429	50	.5995	.7778	.7490	.8745	1.3351	.1255	.8004	.9033	10	.9279
.6458	37° 00′	.6018	.7795	.7536	.8771	1.3270	.1229	.7986	.9023	53° 00′	.9250
.6487	10	.6041	.7811	.7581	.8797	1.3190	.1203	.7969	.9014	50	.9221
.6516	20	.6065	.7828	.7627	.8824	1.3111	.1176	.7951	.9004	40	.9192
.6545	30	.6088	.7844	.7673	.8850	1.3032	.1150	.7934	.8995	30	.9163
.6574	40	.6111	.7861	.7720	.8876	1.2954	.1124	.7916	.8985	20	.9134
.6603	50	.6134	.7877	.7766	.8902	1.2876	.1098	.7898	.8975	10	.9105
.6632	38° 00′	.6157	.7893	.7813	.8928	1.2799	.1072	.7880	.8965	52° 00′	.9076
.6661	10	.6180	.7910	.7860	.8954	1.2723	.1046	.7862	.8955	50	.9047
.6690	20	.6202	.7926	.7907	.8980	1.2647	.1020	.7844	.8945	40	.9018
.6720	30	.6225	.7941	.7954	.9006	1.2572	.0994	.7826	.8935	30	.8988
.6749	40	.6248	.7957	.8002	.9032	1.2497	.0968	.7808	.8925	20	.8959
.6778	50	.6271	.7973	.8050	.9058	1.2423	.0942	.7790	.8915	10	.8930
.6807	39° 00′	.6293	.7989	.8098	.9084	1.2349	.0916	.7771	.8905	51° 00′	.8901
.6836	10	.6316	.8004	.8146	.9110	1.2276	.0890	.7753	.8895	50	.8872
.6865	20	.6338	.8020	.8195	.9135	1.2203	.0865	.7735	.8884	40	.8843
.6894	30	.6361	.8035	.8243	.9161	1.2131	.0839	.7716	.8874	30	.8814
.6923	40	.6383	.8050	.8292	.9187	1.2059	.0813	.7698	.8864	20	.8785
.6952	50	.6406	.8066	.8342	.9212	1.1988	.0788	.7679	.8853	10	.8756
.6981	40° 00′	.6428	.8081	.8391	.9238	1.1918	.0762	.7660	.8843	50° 00′	.8727
.7010	10	.6450	.8096	.8441	.9264	1.1847	.0736	.7642	.8832	50	.8698
.7039	20	.6472	.8111	.8491	.9289	1.1778	.0711	.7623	.8821	40	.8668
.7069	30	.6494	.8125	.8541	.9315	1.1708	.0685	.7604	.8810	30	.8639
.7098	40	.6517	.8140	.8591	.9341	1.1640	.0659	.7585	.8800	20	.8610
.7127	50	.6539	.8155	.8642	.9366	1.1571	.0634	.7566	.8789	10	.8581
.7156	41° 00′	.6561	.8169	.8693	.9392	1.1504	.0608	.7547	.8778	49° 00′	.8552
.7185	10	.6583	.8184	.8744	.9417	1.1436	.0583	.7528	.8767	50	.8523
.7214	20	.6604	.8198	.8796	.9443	1.1369	.0557	.7509	.8756	40	.8494
.7243	30	.6626	.8213	.8847	.9468	1.1303	.0532	.7490	.8745	30	.8465
.7272	40	.6648	.8227	.8899	.9494	1.1237	.0506	.7470	.8733	20	.8436
.7301	50	.6670	.8241	.8952	.9519	1.1171	.0481	.7451	.8722	10	.8407
.7330	42° 00′	.6691	.8255	.9004	.9544	1.1106	.0456	.7431	.8711	48° 00′	.8378
.7359	10	.6713	.8269	.9057	.9570	1.1041	.0430	.7412	.8699	50	.8348
.7389	20	.6734	.8283	.9110	.9595	1.0977	.0405	.7392	.8688	40	.8319
.7418	30	.6756	.8297	.9163	.9621	1.0913	.0379	.7373	.8676	30	.8290
.7447	40	.6777	.8311	.9217	.9646	1.0850	.0354	.7353	.8665	20	.8261
.7476	50	.6799	.8324	.9271	.9671	1.0786	.0329	.7333	.8653	10	.8232
.7505	43° 00′	.6820	.8338	.9325	.9697	1.0724	.0303	.7314	.8641	47° 00′	.8203
.7534	10	.6841	.8351	.9380	.9722	1.0661	.0278	.7294	.8629	50	.8174
.7563	20	.6862	.8365	.9435	.9747	1.0599	.0253	.7274	.8618	40	.8145
.7592	30	.6884	.8378	.9490	.9772	1.0538	.0228	.7254	.8606	30	.8116
.7621	40	.6905	.8391	.9545	.9798	1.0477	.0202	.7234	.8594	20	.8087
.7650	50	.6926	.8405	.9601	.9823	1.0416	.0177	.7214	.8582	10	.8058
.7679	44° 00′	.6947	.8418	.9657	.9848	1.0355	.0152	.7193	.8569	46° 00′	.8029
.7709	10	.6967	.8431	.9713	.9874	1.0295	.0126	.7173	.8557	50	.7999
.7738	20	.6988	.8444	.9770	.9899	1.0235	.0101	.7153	.8545	40	.7970
.7767	30	.7009	.8457	.9827	.9924	1.0176	.0076	.7133	.8532	30	.7941
.7796	40	.7030	.8469	.9884	.9949	1.0117	.0051	.7112	.8520	20	.7912
.7825	50	.7050	.8482	.9942	.9975	1.0058	.0025	.7092	.8507	10	.7883
.7854	45° 00′	.7071	.8495	1.0000	.0000	1.0000	.0000	.7071	.8495	45° 00′	.7854
		Value Log$_{10}$ COSINE		Value Log$_{10}$ COTANGENT		Value Log$_{10}$ TANGENT		Value Log$_{10}$ SINE		DEGREES	RADIANS

216

Table 1010, of Trigonometric Functions, is from *The Macmillan Mathematical Tables*, by E. R. Hedrick, Refer. 19, where there are also tables of 5-place values for every minute of angle, beginning on page 22.

Chart of the
Trigonometric Functions
and Their Inverses
Thus $y = \sin x$ gives $x = $ arc $\sin y$

TABLE 1015—RADIANS TO DEGREES, MINUTES, AND SECONDS

	RADIANS	TENTHS	HUNDREDTHS	THOUSANDTHS	TEN-THOUSANDTHS
1	57°17′44″.8	5°43′46″.5	0°34′22″.6	0° 3′26″.3	0° 0′20″.6
2	114°35′29″.6	11°27′53″.0	1° 8′45″.3	0° 6′52″.5	0° 0′41″.3
3	171°53′14″.4	17°11′19″.4	1°43′07″.9	0°10′18″.8	0° 1′01″.9
4	229°10′59″.2	22°55′05″.9	2°17′30″.6	0°13′45″.1	0° 1′22″.5
5	286°28′44″.0	28°38′52″.4	2°51′53″.2	0°17′11″.3	0° 1′43″.1
6	343°46′28″.8	34°22′38″.9	3°26′15″.9	0°20′37″.6	0° 2′03″.8
7	401° 4′13″.6	40° 6′25″.4	4° 0′38″.5	0°24′03″.9	0° 2′24″.4
8	458°21′58″.4	45°50′11″.8	4°35′01″.2	0°27′30″.1	0° 2′45″.0
9	515°39′43″.3	51°33′58″.3	5° 9′23″.8	0°30′56″.4	0° 3′05″.6

TABLE 1016—DEGREES, MINUTES, AND SECONDS TO RADIANS

Degrees						Minutes		Seconds	
0°	0.00000 00	60°	1.04719 76	120°	2.09439 51	0'	0.00000 00	0''	0.00000 00
1	0.01745 33	61	1.06465 08	121	2.11184 84	1	0.00029 09	1	0.00000 48
2	0.03490 66	62	1.08210 41	122	2.12930 17	2	0.00058 18	2	0.00000 97
3	0.05235 99	63	1.09955 74	123	2.14675 50	3	0.00087 27	3	0.00001 45
4	0.06981 32	64	1.11701 07	124	2.16420 83	4	0.00116 36	4	0.00001 94
5	0.08726 65	65	1.13446 40	125	2.18166 16	5	0.00145 44	5	0.00002 42
6	0.10471 98	66	1.15191 73	126	2.19911 49	6	0.00174 53	6	0.00002 91
7	0.12217 30	67	1.16937 06	127	2.21656 82	7	0.00203 62	7	0.00003 39
8	0.13962 63	68	1.18682 39	128	2.23402 14	8	0.00232 71	8	0.00003 88
9	0.15707 96	69	1.20427 72	129	2.25147 47	9	0.00261 80	9	0.00004 36
10	0.17453 29	70	1.22173 05	130	2.26892 80	10	0.00290 89	10	0.00004 85
11	0.19198 62	71	1.23918 38	131	2.28638 13	11	0.00319 98	11	0.00005 33
12	0.20943 95	72	1.25663 71	132	2.30383 46	12	0.00349 07	12	0.00005 82
13	0.22689 28	73	1.27409 04	133	2.32128 79	13	0.00378 15	13	0.00006 30
14	0.24434 61	74	1.29154 36	134	2.33874 12	14	0.00407 24	14	0.00006 79
15	0.26179 94	75	1.30899 69	135	2.35619 45	15	0.00436 33	15	0.00007 27
16	0.27925 27	76	1.32645 02	136	2.37364 78	16	0.00465 42	16	0.00007 76
17	0.29670 60	77	1.34390 35	137	2.39110 11	17	0.00494 51	17	0.00008 24
18	0.31415 93	78	1.36135 68	138	2.40855 44	18	0.00523 60	18	0.00008 73
19	0.33161 26	79	1.37881 01	139	2.42600 77	19	0.00552 69	19	0.00009 21
20	0.34906 59	80	1.39626 34	140	2.44346 10	20	0.00581 78	20	0.00009 70
21	0.36651 91	81	1.41371 67	141	2.46091 42	21	0.00610 87	21	0.00010 18
22	0.38397 24	82	1.43117 00	142	2.47836 75	22	0.00639 95	22	0.00010 67
23	0.40142 57	83	1.44862 33	143	2.49582 08	23	0.00669 04	23	0.00011 15
24	0.41887 90	84	1.46607 66	144	2.51327 41	24	0.00698 13	24	0.00011 64
25	0.43633 23	85	1.48352 99	145	2.53072 74	25	0.00727 22	25	0.00012 12
26	0.45378 56	86	1.50098 32	146	2.54818 07	26	0.00756 31	26	0.00012 61
27	0.47123 89	87	1.51843 64	147	2.56563 40	27	0.00785 40	27	0.00013 09
28	0.48869 22	88	1.53588 97	148	2.58308 73	28	0.00814 49	28	0.00013 57
29	0.50614 55	89	1.55334 30	149	2.60054 06	29	0.00843 58	29	0.00014 06
30	0.52359 88	90	1.57079 63	150	2.61799 39	30	0.00872 66	30	0.00014 54
31	0.54105 21	91	1.58824 96	151	2.63544 72	31	0.00901 75	31	0.00015 03
32	0.55850 54	92	1.60570 29	152	2.65290 05	32	0.00930 84	32	0.00015 51
33	0.57595 87	93	1.62315 62	153	2.67035 38	33	0.00959 93	33	0.00016 00
34	0.59341 19	94	1.64060 95	154	2.68780 70	34	0.00989 02	34	0.00016 48
35	0.61086 52	95	1.65806 28	155	2.70526 03	35	0.01018 11	35	0.00016 97
36	0.62831 85	96	1.67551 61	156	2.72271 36	36	0.01047 20	36	0.00017 45
37	0.64577 18	97	1.69296 94	157	2.74016 69	37	0.01076 29	37	0.00017 94
38	0.66322 51	98	1.71042 27	158	2.75762 02	38	0.01105 38	38	0.00018 42
39	0.68067 84	99	1.72787 60	159	2.77507 35	39	0.01134 46	39	0.00018 91
40	0.69813 17	100	1.74532 93	160	2.79252 68	40	0.01163 55	40	0.00019 39
41	0.71558 50	101	1.76278 25	161	2.80998 01	41	0.01192 64	41	0.00019 88
42	0.73303 83	102	1.78023 58	162	2.82743 34	42	0.01221 73	42	0.00020 36
43	0.75049 16	103	1.79768 91	163	2.84488 67	43	0.01250 82	43	0.00020 85
44	0.76794 49	104	1.81514 24	164	2.86234 00	44	0.01279 91	44	0.00021 33
45	0.78539 82	105	1.83259 57	165	2.87979 33	45	0.01309 00	45	0.00021 82
46	0.80285 15	106	1.85004 90	166	2.89724 66	46	0.01338 09	46	0.00022 30
47	0.82030 47	107	1.86750 23	167	2.91469 99	47	0.01367 17	47	0.00022 79
48	0.83775 80	108	1.88495 56	168	2.93215 31	48	0.01396 26	48	0.00023 27
49	0.85521 13	109	1.90240 89	169	2.94960 64	49	0.01425 35	49	0.00023 76
50	0.87266 46	110	1.91986 22	170	2.96705 97	50	0.01454 44	50	0.00024 24
51	0.89011 79	111	1.93731 55	171	2.98451 30	51	0.01483 53	51	0.00024 73
52	0.90757 12	112	1.95476 88	172	3.00196 63	52	0.01512 62	52	0.00025 21
53	0.92502 45	113	1.97222 21	173	3.01941 96	53	0.01541 71	53	0.00025 70
54	0.94247 78	114	1.98967 53	174	3.03687 29	54	0.01570 80	54	0.00026 18
55	0.95993 11	115	2.00712 86	175	3.05432 62	55	0.01599 89	55	0.00026 66
56	0.97738 44	116	2.02458 19	176	3.07177 95	56	0.01628 97	56	0.00027 15
57	0.99483 77	117	2.04203 52	177	3.08923 28	57	0.01658 06	57	0.00027 63
58	1.01229 10	118	2.05948 85	178	3.10668 61	58	0.01687 15	58	0.00028 12
59	1.02974 43	119	2.07694 18	179	3.12413 94	59	0.01716 24	59	0.00028 60
60	1.04719 76	120	2.09439 51	180	3.14159 27	60	0.01745 33	60	0.00029 09

TABLE 1018—GAMMA FUNCTION [See 850]

n	Γ(n)	Diff.	n	Γ(n)	Diff.	n	Γ(n)	Diff.	n	Γ(n)	Diff.
1.00	1.000 00	−567	1.25	.906 40	−200	1.50	.886 23	36	1.75	.919 06	231
1.01	.994 33	−549	1.26	.904 40	−190	1.51	.886 59	45	1.76	.921 37	239
1.02	.988 84	−529	1.27	.902 50	−178	1.52	.887 04	53	1.77	.923 76	247
1.03	.983 55	−511	1.28	.900 72	−168	1.53	.887 57	61	1.78	.926 23	254
1.04	.978 44	−494	1.29	.899 04	−157	1.54	.888 18	69	1.79	.928 77	261
1.05	.973 50	−476	1.30	.897 47	−147	1.55	.888 87	77	1.80	.931 38	270
1.06	.968 74	−459	1.31	.896 00	−136	1.56	.889 64	85	1.81	.934 08	277
1.07	.964 15	−442	1.32	.894 64	−126	1.57	.890 49	93	1.82	.936 85	284
1.08	.959 73	−427	1.33	.893 38	−116	1.58	.891 42	101	1.83	.939 69	292
1.09	.955 46	−411	1.34	.892 22	−107	1.59	.892 43	109	1.84	.942 61	300
1.10	.951 35	−395	1.35	.891 15	−97	1.60	.893 52	116	1.85	.945 61	308
1.11	.947 40	−381	1.36	.890 18	−87	1.61	.894 68	124	1.86	.948 69	315
1.12	.943 59	−366	1.37	.889 31	−77	1.62	.895 92	132	1.87	.951 84	323
1.13	.939 93	−351	1.38	.888 54	−69	1.63	.897 24	140	1.88	.955 07	331
1.14	.936 42	−338	1.39	.887 85	−59	1.64	.898 64	148	1.89	.958 38	339
1.15	.933 04	−324	1.40	.887 26	−50	1.65	.900 12	155	1.90	.961 77	346
1.16	.929 80	−310	1.41	.886 76	−40	1.66	.901 67	163	1.91	.965 23	354
1.17	.926 70	−297	1.42	.886 36	−32	1.67	.903 30	170	1.92	.968 77	363
1.18	.923 73	−284	1.43	.886 04	−23	1.68	.905 00	178	1.93	.972 40	370
1.19	.920 89	−272	1.44	.885 81	−15	1.69	.906 78	186	1.94	.976 10	378
1.20	.918 17	−259	1.45	.885 66	−6	1.70	.908 64	193	1.95	.979 88	386
1.21	.915 58	−247	1.46	.885 60	3	1.71	.910 57	201	1.96	.983 74	394
1.22	.913 11	−236	1.47	.885 63	12	1.72	.912 58	209	1.97	.987 68	403
1.23	.910 75	−223	1.48	.885 75	20	1.73	.914 67	216	1.98	.991 71	410
1.24	.908 52	−212	1.49	.885 95	28	1.74	.916 83	223	1.99	.995 81	419
									2.00	1.000 00	

For larger values of n, use this table and make successive applications of the following equation:

$$\Gamma(n + 1) = n\Gamma(n).$$

For more complete tables, see Ref. 6, p. 169, Ref. 44, v. 1, pp. 196–273, and Ref. 45, pp. 208–209.

TABLE 1020—LOGARITHMS TO BASE 10

N	0	1	2	3	4	5	6	7	8	9	1 2 3	4 5 6	7 8 9
10	0000	0043	0086	0128	0170	0212	0253	0294	0334	0374	4 8 12	17 21 25	29 33 37
11	0414	0453	0492	0531	0569	0607	0645	0682	0719	0755	4 8 11	15 19 23	26 30 34
12	0792	0828	0864	0899	0934	0969	1004	1038	1072	1106	3 7 10	14 17 21	24 28 31
13	1139	1173	1206	1239	1271	1303	1335	1367	1399	1430	3 6 10	13 16 19	23 26 29
14	1461	1492	1523	1553	1584	1614	1644	1673	1703	1732	3 6 9	12 15 18	21 24 27
15	1761	1790	1818	1847	1875	1903	1931	1959	1987	2014	3 6 8	11 14 17	20 22 25
16	2041	2068	2095	2122	2148	2175	2201	2227	2253	2279	3 5 8	11 13 16	18 21 24
17	2304	2330	2355	2380	2405	2430	2455	2480	2504	2529	2 5 7	10 12 15	17 20 22
18	2553	2577	2601	2625	2648	2672	2695	2718	2742	2765	2 5 7	9 12 14	16 19 21
19	2788	2810	2833	2856	2878	2900	2923	2945	2967	2989	2 4 7	9 11 13	16 18 20
20	3010	3032	3054	3075	3096	3118	3139	3160	3181	3201	2 4 6	8 11 13	15 17 19
21	3222	3243	3263	3284	3304	3324	3345	3365	3385	3404	2 4 6	8 10 12	14 16 18
22	3424	3444	3464	3483	3502	3522	3541	3560	3579	3598	2 4 6	8 10 12	14 16 17
23	3617	3636	3655	3674	3692	3711	3729	3747	3766	3784	2 4 6	7 9 11	13 15 17
24	3802	3820	3838	3856	3874	3892	3909	3927	3945	3962	2 4 5	7 9 11	12 14 16
25	3979	3997	4014	4031	4048	4065	4082	4099	4116	4133	2 4 5	7 9 10	12 14 16
26	4150	4166	4183	4200	4216	4232	4249	4265	4281	4298	2 3 5	7 8 10	11 13 15
27	4314	4330	4346	4362	4378	4393	4409	4425	4440	4456	2 3 5	6 8 9	11 12 14
28	4472	4487	4502	4518	4533	4548	4564	4579	4594	4609	2 3 5	6 8 9	11 12 14
29	4624	4639	4654	4669	4683	4698	4713	4728	4742	4757	1 3 4	6 7 9	10 12 13
30	4771	4786	4800	4814	4829	4843	4857	4871	4886	4900	1 3 4	6 7 9	10 11 13
31	4914	4928	4942	4955	4969	4983	4997	5011	5024	5038	1 3 4	5 7 8	10 11 12
32	5051	5065	5079	5092	5105	5119	5132	5145	5159	5172	1 3 4	5 7 8	9 11 12
33	5185	5198	5211	5224	5237	5250	5263	5276	5289	5302	1 3 4	5 7 8	9 11 12
34	5315	5328	5340	5353	5366	5378	5391	5403	5416	5428	1 2 4	5 6 8	9 10 11
35	5441	5453	5465	5478	5490	5502	5514	5527	5539	5551	1 2 4	5 6 7	9 10 11
36	5563	5575	5587	5599	5611	5623	5635	5647	5658	5670	1 2 4	5 6 7	8 10 11
37	5682	5694	5705	5717	5729	5740	5752	5763	5775	5786	1 2 4	5 6 7	8 9 11
38	5798	5809	5821	5832	5843	5855	5866	5877	5888	5899	1 2 3	5 6 7	8 9 10
39	5911	5922	5933	5944	5955	5966	5977	5988	5999	6010	1 2 3	4 5 7	8 9 10
40	6021	6031	6042	6053	6064	6075	6085	6096	6107	6117	1 2 3	4 5 6	8 9 10
41	6128	6138	6149	6160	6170	6180	6191	6201	6212	6222	1 2 3	4 5 6	7 8 9
42	6232	6243	6253	6263	6274	6284	6294	6304	6314	6325	1 2 3	4 5 6	7 8 9
43	6335	6345	6355	6365	6375	6385	6395	6405	6415	6425	1 2 3	4 5 6	7 8 9
44	6435	6444	6454	6464	6474	6484	6493	6503	6513	6522	1 2 3	4 5 6	7 8 9
45	6532	6542	6551	6561	6571	6580	6590	6599	6609	6618	1 2 3	4 5 6	7 8 9
46	6628	6637	6646	6656	6665	6675	6684	6693	6702	6712	1 2 3	4 5 6	7 7 8
47	6721	6730	6739	6749	6758	6767	6776	6785	6794	6803	1 2 3	4 5 6	7 7 8
48	6812	6821	6830	6839	6848	6857	6866	6875	6884	6893	1 2 3	4 5 6	7 7 8
49	6902	6911	6920	6928	6937	6946	6955	6964	6972	6981	1 2 3	4 4 5	6 7 8
50	6990	6998	7007	7016	7024	7033	7042	7050	7059	7067	1 2 3	3 4 5	6 7 8
51	7076	7084	7093	7101	7110	7118	7126	7135	7143	7152	1 2 3	3 4 5	6 7 8
52	7160	7168	7177	7185	7193	7202	7210	7218	7226	7235	1 2 3	3 4 5	6 7 7
53	7243	7251	7259	7267	7275	7284	7292	7300	7308	7316	1 2 2	3 4 5	6 6 7
54	7324	7332	7340	7348	7356	7364	7372	7380	7388	7396	1 2 2	3 4 5	6 6 7
N	0	1	2	3	4	5	6	7	8	9	1 2 2	4 5 6	7 8 9

The proportional parts are stated in full for every tenth at the right-hand side.
The logarithm of any number of four significant figures can be read directly by add-

220

TABLE 1020 (*continued*)—LOGARITHMS TO BASE 10

N	0	1	2	3	4	5	6	7	8	9	1 2 3	4 5 6	7 8 9
55	7404	7412	7419	7427	7435	7443	7451	7459	7466	7474	1 2 2	3 4 5	5 6 7
56	7482	7490	7497	7505	7513	7520	7528	7536	7543	7551	1 2 2	3 4 5	5 6 7
57	7559	7566	7574	7582	7589	7597	7604	7612	7619	7627	1 1 2	3 4 5	5 6 7
58	7634	7642	7649	7657	7664	7672	7679	7686	7694	7701	1 1 2	3 4 4	5 6 7
59	7709	7716	7723	7731	7738	7745	7752	7760	7767	7774	1 1 2	3 4 4	5 6 7
60	7782	7789	7796	7803	7810	7818	7825	7832	7839	7846	1 1 2	3 4 4	5 6 6
61	7853	7860	7868	7875	7882	7889	7896	7903	7910	7917	1 1 2	3 3 4	5 6 6
62	7924	7931	7938	7945	7952	7959	7966	7973	7980	7987	1 1 2	3 3 4	5 5 6
63	7993	8000	8007	8014	8021	8028	8035	8041	8048	8055	1 1 2	3 3 4	5 5 6
64	8062	8069	8075	8082	8089	8096	8102	8109	8116	8122	1 1 2	3 3 4	5 5 6
65	8129	8136	8142	8149	8156	8162	8169	8176	8182	8189	1 1 2	3 3 4	5 5 6
66	8195	8202	8209	8215	8222	8228	8235	8241	8248	8254	1 1 2	3 3 4	5 5 6
67	8261	8267	8274	8280	8287	8293	8299	8306	8312	8319	1 1 2	3 3 4	5 5 6
68	8325	8331	8338	8344	8351	8357	8363	8370	8376	8382	1 1 2	3 3 4	4 5 6
69	8388	8395	8401	8407	8414	8420	8426	8432	8439	8445	1 1 2	3 3 4	4 5 6
70	8451	8457	8463	8470	8476	8482	8488	8494	8500	8506	1 1 2	3 3 4	4 5 6
71	8513	8519	8525	8531	8537	8543	8549	8555	8561	8567	1 1 2	3 3 4	4 5 6
72	8573	8579	8585	8591	8597	8603	8609	8615	8621	8627	1 1 2	3 3 4	4 5 6
73	8633	8639	8645	8651	8657	8663	8669	8675	8681	8686	1 1 2	2 3 4	4 5 5
74	8692	8698	8704	8710	8716	8722	8727	8733	8739	8745	1 1 2	2 3 4	4 5 5
75	8751	8756	8762	8768	8774	8779	8785	8791	8797	8802	1 1 2	2 3 3	4 5 5
76	8808	8814	8820	8825	8831	8837	8842	8848	8854	8859	1 1 2	2 3 3	4 4 5
77	8865	8871	8876	8882	8887	8893	8899	8904	8910	8915	1 1 2	2 3 3	4 4 5
78	8921	8927	8932	8938	8943	8949	8954	8960	8965	8971	1 1 2	2 3 3	4 4 5
79	8976	8982	8987	8993	8998	9004	9009	9015	9020	9025	1 1 2	2 3 3	4 4 5
80	9031	9036	9042	9047	9053	9058	9063	9069	9074	9079	1 1 2	2 3 3	4 4 5
81	9085	9090	9096	9101	9106	9112	9117	9122	9128	9133	1 1 2	2 3 3	4 4 5
82	9138	9143	9149	9154	9159	9165	9170	9175	9180	9186	1 1 2	2 3 3	4 4 5
83	9191	9196	9201	9206	9212	9217	9222	9227	9232	9238	1 1 2	2 3 3	4 4 5
84	9243	9248	9253	9258	9263	9269	9274	9279	9284	9289	1 1 2	2 3 3	4 4 5
85	9294	9299	9304	9309	9315	9320	9325	9330	9335	9340	1 1 2	2 3 3	4 4 5
86	9345	9350	9355	9360	9365	9370	9375	9380	9385	9390	1 1 2	2 3 3	4 4 5
87	9395	9400	9405	9410	9415	9420	9425	9430	9435	9440	1 1 2	2 3 3	4 4 5
88	9445	9450	9455	9460	9465	9469	9474	9479	9484	9489	0 1 1	2 2 3	3 4 4
89	9494	9499	9504	9509	9513	9518	9523	9528	9533	9538	0 1 1	2 2 3	3 4 4
90	9542	9547	9552	9557	9562	9566	9571	9576	9581	9586	0 1 1	2 2 3	3 4 4
91	9590	9595	9600	9605	9609	9614	9619	9624	9628	9633	0 1 1	2 2 3	3 4 4
92	9638	9643	9647	9652	9657	9661	9666	9671	9675	9680	0 1 1	2 2 3	3 4 4
93	9685	9689	9694	9699	9703	9708	9713	9717	9722	9727	0 1 1	2 2 3	3 4 4
94	9731	9736	9741	9745	9750	9754	9759	9763	9768	9773	0 1 1	2 2 3	3 4 4
95	9777	9782	9786	9791	9795	9800	9805	9809	9814	9818	0 1 1	2 2 3	3 4 4
96	9823	9827	9832	9836	9841	9845	9850	9854	9859	9863	0 1 1	2 2 3	3 4 4
97	9868	9872	9877	9881	9886	9890	9894	9899	9903	9908	0 1 1	2 2 3	3 4 4
98	9912	9917	9921	9926	9930	9934	9939	9943	9948	9952	0 1 1	2 2 3	3 3 4
99	9956	9961	9965	9969	9974	9978	9983	9987	9991	9996	0 1 1	2 2 3	3 3 4
N	0	1	2	3	4	5	6	7	8	9	1 2 3	4 5 6	7 8 9

ing the proportional part corresponding to the fourth figure to the tabular number corresponding to the first three figures. There may be an error of 1 in the last place.

TABLE 1025—NATURAL LOGARITHMS

No.	0	1	2	3	4	5	6	7	8	9	Diff.
1.00	.0000	.0010	.0020	.0030	.0040	.0050	.0060	.0070	.0080	.0090	10
1.01	.0100	.0109	.0119	.0129	.0139	.0149	.0159	.0169	.0178	.0188	10–9
1.02	.0198	.0208	.0218	.0227	.0237	.0247	.0257	.0266	.0276	.0286	10–9
1.03	.0296	.0305	.0315	.0325	.0334	.0344	.0354	.0363	.0373	.0383	10–9
1.04	.0392	.0402	.0411	.0421	.0431	.0440	.0450	.0459	.0469	.0478	10–9
1.05	.0488	.0497	.0507	.0516	.0526	.0535	.0545	.0554	.0564	.0573	10–9
1.06	.0583	.0592	.0602	.0611	.0620	.0630	.0639	.0649	.0658	.0667	10–9
1.07	.0677	.0686	.0695	.0705	.0714	.0723	.0733	.0742	.0751	.0760	10–9
1.08	.0770	.0779	.0788	.0797	.0807	.0816	.0825	.0834	.0843	.0853	10–9
1.09	.0862	.0871	.0880	.0889	.0898	.0908	.0917	.0926	.0935	.0944	10–9
1.10	.0953	.0962	.0971	.0980	.0989	.0998	.1007	.1017	.1026	.1035	10–9
1.11	.1044	.1053	.1062	.1071	.1080	.1089	.1098	.1106	.1115	.1124	9–8
1.12	.1133	.1142	.1151	.1160	.1169	.1178	.1187	.1196	.1204	.1213	9–8
1.13	.1222	.1231	.1240	.1249	.1258	.1266	.1275	.1284	.1293	.1302	9–8
1.14	.1310	.1319	.1328	.1337	.1345	.1354	.1363	.1371	.1380	.1389	9–8
1.15	.1398	.1406	.1415	.1424	.1432	.1441	.1450	.1458	.1467	.1476	9–8
1.16	.1484	.1493	.1501	.1510	.1519	.1527	.1536	.1544	.1553	.1561	9–8
1.17	.1570	.1579	.1587	.1596	.1604	.1613	.1621	.1630	.1638	.1647	9–8
1.18	.1655	.1664	.1672	.1681	.1689	.1697	.1706	.1714	.1723	.1731	9–8
1.19	.1740	.1748	.1756	.1765	.1773	.1781	.1790	.1798	.1807	.1815	9–8
1.20	.1823	.1832	.1840	.1848	.1856	.1865	.1873	.1881	.1890	.1898	9–8
1.21	.1906	.1914	.1923	.1931	.1939	.1947	.1956	.1964	.1972	.1980	9–8
1.22	.1989	.1997	.2005	.2013	.2021	.2029	.2038	.2046	.2054	.2062	9–8
1.23	.2070	.2078	.2086	.2095	.2103	.2111	.2119	.2127	.2135	.2143	9–8
1.24	.2151	.2159	.2167	.2175	.2183	.2191	.2199	.2207	.2215	.2223	8
1.25	.2231	.2239	.2247	.2255	.2263	.2271	.2279	.2287	.2295	.2303	8
1.26	.2311	.2319	.2327	.2335	.2343	.2351	.2359	.2367	.2374	.2382	8–7
1.27	.2390	.2398	.2406	.2414	.2422	.2429	.2437	.2445	.2453	.2461	8–7
1.28	.2469	.2476	.2484	.2492	.2500	.2508	.2515	.2523	.2531	.2539	8–7
1.29	.2546	.2554	.2562	.2570	.2577	.2585	.2593	.2601	.2608	.2616	8–7
1.30	.2624	.2631	.2639	.2647	.2654	.2662	.2670	.2677	.2685	.2693	8–7
1.31	.2700	.2708	.2716	.2723	.2731	.2738	.2746	.2754	.2761	.2769	8–7
1.32	.2776	.2784	.2791	.2799	.2807	.2814	.2822	.2829	.2837	.2844	8–7
1.33	.2852	.2859	.2867	.2874	.2882	.2889	.2897	.2904	.2912	.2919	8–7
1.34	.2927	.2934	.2942	.2949	.2957	.2964	.2971	.2979	.2986	.2994	8–7
1.35	.3001	.3008	.3016	.3023	.3031	.3038	.3045	.3053	.3060	.3067	8–7
1.36	.3075	.3082	.3090	.3097	.3104	.3112	.3119	.3126	.3133	.3141	8–7
1.37	.3148	.3155	.3163	.3170	.3177	.3185	.3192	.3199	.3206	.3214	8–7
1.38	.3221	.3228	.3235	.3243	.3250	.3257	.3264	.3271	.3279	.3286	8–7
1.39	.3293	.3300	.3307	.3315	.3322	.3329	.3336	.3343	.3350	.3358	8–7
1.40	.3365	.3372	.3379	.3386	.3393	.3400	.3407	.3415	.3422	.3429	8–7
1.41	.3436	.3443	.3450	.3457	.3464	.3471	.3478	.3485	.3492	.3500	8–7
1.42	.3507	.3514	.3521	.3528	.3535	.3542	.3549	.3556	.3563	.3570	7
1.43	.3577	.3584	.3591	.3598	.3605	.3612	.3619	.3626	.3633	.3639	7–6
1.44	.3646	.3653	.3660	.3667	.3674	.3681	.3688	.3695	.3702	.3709	7
1.45	.3716	.3723	.3729	.3736	.3743	.3750	.3757	.3764	.3771	.3778	7–6
1.46	.3784	.3791	.3798	.3805	.3812	.3819	.3825	.3832	.3839	.3846	7–6
1.47	.3853	.3859	.3866	.3873	.3880	.3887	.3893	.3900	.3907	.3914	7–6
1.48	.3920	.3927	.3934	.3941	.3947	.3954	.3961	.3968	.3974	.3981	7–6
1.49	.3988	.3994	.4001	.4008	.4015	.4021	.4028	.4035	.4041	.4048	7–6

TABLE 1025 (*continued*)—NATURAL LOGARITHMS

No.	0	1	2	3	4	5	6	7	8	9	Diff.
1.50	.4055	.4061	.4068	.4075	.4081	.4088	.4095	.4101	.4108	.4114	7–6
1.51	.4121	.4128	.4134	.4141	.4148	.4154	.4161	.4167	.4174	.4181	7–6
1.52	.4187	.4194	.4200	.4207	.4213	.4220	.4226	.4233	.4240	.4246	7–6
1.53	.4253	.4259	.4266	.4272	.4279	.4285	.4292	.4298	.4305	.4311	7–6
1.54	.4318	.4324	.4331	.4337	.4344	.4350	.4357	.4363	.4370	.4376	7–6
1.55	.4383	.4389	.4395	.4402	.4408	.4415	.4421	.4428	.4434	.4440	7–6
1.56	.4447	.4453	.4460	.4466	.4472	.4479	.4485	.4492	.4498	.4504	7–6
1.57	.4511	.4517	.4523	.4530	.4536	.4543	.4549	.4555	.4562	.4568	7–6
1.58	.4574	.4581	.4587	.4593	.4600	.4606	.4612	.4618	.4625	.4631	7–6
1.59	.4637	.4644	.4650	.4656	.4662	.4669	.4675	.4681	.4688	.4694	7–6
1.60	.4700	.4706	.4713	.4719	.4725	.4731	.4737	.4744	.4750	.4756	7–6
1.61	.4762	.4769	.4775	.4781	.4787	.4793	.4800	.4806	.4812	.4818	7–6
1.62	.4824	.4830	.4837	.4843	.4849	.4855	.4861	.4867	.4874	.4880	7–6
1.63	.4886	.4892	.4898	.4904	.4910	.4916	.4923	.4929	.4935	.4941	7–6
1.64	.4947	.4953	.4959	.4965	.4971	.4977	.4983	.4990	.4996	.5002	7–6
1.65	.5008	.5014	.5020	.5026	.5032	.5038	.5044	.5050	.5056	.5062	6
1.66	.5068	.5074	.5080	.5086	.5092	.5098	.5104	.5110	.5116	.5122	6
1.67	.5128	.5134	.5140	.5146	.5152	.5158	.5164	.5170	.5176	.5182	6
1.68	.5188	.5194	.5200	.5206	.5212	.5218	.5224	.5230	.5235	.5241	6–5
1.69	.5247	.5253	.5259	.5265	.5271	.5277	.5283	.5289	.5295	.5300	6–5
1.70	.5306	.5312	.5318	.5324	.5330	.5336	.5342	.5347	.5353	.5359	6–5
1.71	.5365	.5371	.5377	.5382	.5388	.5394	.5400	.5406	.5412	.5417	6–5
1.72	.5423	.5429	.5435	.5441	.5446	.5452	.5458	.5464	.5470	.5475	6–5
1.73	.5481	.5487	.5493	.5499	.5504	.5510	.5516	.5522	.5527	.5533	6–5
1.74	.5539	.5545	.5550	.5556	.5562	.5568	.5573	.5579	.5585	.5590	6–5
1.75	.5596	.5602	.5608	.5613	.5619	.5625	.5630	.5636	.5642	.5647	6–5
1.76	.5653	.5659	.5664	.5670	.5676	.5682	.5687	.5693	.5698	.5704	6–5
1.77	.5710	.5715	.5721	.5727	.5732	.5738	.5744	.5749	.5755	.5761	6–5
1.78	.5766	.5772	.5777	.5783	.5789	.5794	.5800	.5805	.5811	.5817	6–5
1.79	.5822	.5828	.5833	.5839	.5844	.5850	.5856	.5861	.5867	.5872	6–5
1.80	.5878	.5883	.5889	.5895	.5900	.5906	.5911	.5917	.5922	.5928	6–5
1.81	.5933	.5939	.5944	.5950	.5955	.5961	.5966	.5972	.5977	.5983	6–5
1.82	.5988	.5994	.5999	.6005	.6010	.6016	.6021	.6027	.6032	.6038	6–5
1.83	.6043	.6049	.6054	.6060	.6065	.6070	.6076	.6081	.6087	.6092	6–5
1.84	.6098	.6103	.6109	.6114	.6119	.6125	.6130	.6136	.6141	.6146	6–5
1.85	.6152	.6157	.6163	.6168	.6173	.6179	.6184	.6190	.6195	.6200	6–5
1.86	.6206	.6211	.6217	.6222	.6227	.6233	.6238	.6243	.6249	.6254	6–5
1.87	.6259	.6265	.6270	.6275	.6281	.6286	.6291	.6297	.6302	.6307	6–5
1.88	.6313	.6318	.6323	.6329	.6334	.6339	.6345	.6350	.6355	.6360	6–5
1.89	.6366	.6371	.6376	.6382	.6387	.6392	.6397	.6403	.6408	.6413	6–5
1.90	.6419	.6424	.6429	.6434	.6440	.6445	.6450	.6455	.6461	.6466	6–5
1.91	.6471	.6476	.6481	.6487	.6492	.6497	.6502	.6508	.6513	.6518	6–5
1.92	.6523	.6528	.6534	.6539	.6544	.6549	.6554	.6560	.6565	.6570	6–5
1.93	.6575	.6580	.6586	.6591	.6596	.6601	.6606	.6611	.6617	.6622	6–5
1.94	.6627	.6632	.6637	.6642	.6647	.6653	.6658	.6663	.6668	.6673	6–5
1.95	.6678	.6683	.6689	.6694	.6699	.6704	.6709	.6714	.6719	.6724	6–5
1.96	.6729	.6735	.6740	.6745	.6750	.6755	.6760	.6765	.6770	.6775	6–5
1.97	.6780	.6785	.6790	.6796	.6801	.6806	.6811	.6816	.6821	.6826	6–5
1.98	.6831	.6836	.6841	.6846	.6851	.6856	.6861	.6866	.6871	.6876	5
1.99	.6881	.6886	.6891	.6896	.6901	.6906	.6911	.6916	.6921	.6926	5

TABLE 1025 (*continued*)—NATURAL LOGARITHMS

No.	0	1	2	3	4	5	6	7	8	9	Diff.
2.0	.6931	.6981	.7031	.7080	.7129	.7178	.7227	.7275	.7324	.7372	50–48
2.1	.7419	.7467	.7514	.7561	.7608	.7655	.7701	.7747	.7793	.7839	48–46
2.2	.7885	.7930	.7975	.8020	.8065	.8109	.8154	.8198	.8242	.8286	45–44
2.3	.8329	.8372	.8416	.8459	.8502	.8544	.8587	.8629	.8671	.8713	44–42
2.4	.8755	.8796	.8838	.8879	.8920	.8961	.9002	.9042	.9083	.9123	42–40
2.5	.9163	.9203	.9243	.9282	.9322	.9361	.9400	.9439	.9478	.9517	40–39
2.6	.9555	.9594	.9632	.9670	.9708	.9746	.9783	.9821	.9858	.9895	39–37
2.7	.9933	.9969	1.0006	1.0043	1.0080	1.0116	1.0152	1.0188	1.0225	1.0260	37–35
2.8	1.0296	1.0332	1.0367	1.0403	1.0438	1.0473	1.0508	1.0543	1.0578	1.0613	36–35
2.9	1.0647	1.0682	1.0716	1.0750	1.0784	1.0818	1.0852	1.0886	1.0919	1.0953	35–33
3.0	1.0986	1.1019	1.1053	1.1086	1.1119	1.1151	1.1184	1.1217	1.1249	1.1282	34–32
3.1	1.1314	1.1346	1.1378	1.1410	1.1442	1.1474	1.1506	1.1537	1.1569	1.1600	32–31
3.2	1.1632	1.1663	1.1694	1.1725	1.1756	1.1787	1.1817	1.1848	1.1878	1.1909	31–30
3.3	1.1939	1.1969	1.2000	1.2030	1.2060	1.2090	1.2119	1.2149	1.2179	1.2208	31–29
3.4	1.2238	1.2267	1.2296	1.2326	1.2355	1.2384	1.2413	1.2442	1.2470	1.2499	30–28
3.5	1.2528	1.2556	1.2585	1.2613	1.2641	1.2669	1.2698	1.2726	1.2754	1.2782	29–28
3.6	1.2809	1.2837	1.2865	1.2892	1.2920	1.2947	1.2975	1.3002	1.3029	1.3056	28–27
3.7	1.3083	1.3110	1.3137	1.3164	1.3191	1.3218	1.3244	1.3271	1.3297	1.3324	27–26
3.8	1.3350	1.3376	1.3403	1.3429	1.3455	1.3481	1.3507	1.3533	1.3558	1.3584	27–25
3.9	1.3610	1.3635	1.3661	1.3686	1.3712	1.3737	1.3762	1.3788	1.3813	1.3838	26–25
4.0	1.3863	1.3888	1.3913	1.3938	1.3962	1.3987	1.4012	1.4036	1.4061	1.4085	25–24
4.1	1.4110	1.4134	1.4159	1.4183	1.4207	1.4231	1.4255	1.4279	1.4303	1.4327	25–24
4.2	1.4351	1.4375	1.4398	1.4422	1.4446	1.4469	1.4493	1.4516	1.4540	1.4563	24–23
4.3	1.4586	1.4609	1.4633	1.4656	1.4679	1.4702	1.4725	1.4748	1.4770	1.4793	24–22
4.4	1.4816	1.4839	1.4861	1.4884	1.4907	1.4929	1.4951	1.4974	1.4996	1.5019	23–22
4.5	1.5041	1.5063	1.5085	1.5107	1.5129	1.5151	1.5173	1.5195	1.5217	1.5239	22
4.6	1.5261	1.5282	1.5304	1.5326	1.5347	1.5369	1.5390	1.5412	1.5433	1.5454	22–21
4.7	1.5476	1.5497	1.5518	1.5539	1.5560	1.5581	1.5602	1.5623	1.5644	1.5665	22–21
4.8	1.5686	1.5707	1.5728	1.5748	1.5769	1.5790	1.5810	1.5831	1.5851	1.5872	21–20
4.9	1.5892	1.5913	1.5933	1.5953	1.5974	1.5994	1.6014	1.6034	1.6054	1.6074	21–20
5.0	1.6094	1.6114	1.6134	1.6154	1.6174	1.6194	1.6214	1.6233	1.6253	1.6273	20–19
5.1	1.6292	1.6312	1.6332	1.6351	1.6371	1.6390	1.6409	1.6429	1.6448	1.6467	20–19
5.2	1.6487	1.6506	1.6525	1.6544	1.6563	1.6582	1.6601	1.6620	1.6639	1.6658	19
5.3	1.6677	1.6696	1.6715	1.6734	1.6752	1.6771	1.6790	1.6808	1.6827	1.6845	19–18
5.4	1.6864	1.6882	1.6901	1.6919	1.6938	1.6956	1.6974	1.6993	1.7011	1.7029	19–18
5.5	1.7047	1.7066	1.7084	1.7102	1.7120	1.7138	1.7156	1.7174	1.7192	1.7210	19–18
5.6	1.7228	1.7246	1.7263	1.7281	1.7299	1.7317	1.7334	1.7352	1.7370	1.7387	18–17
5.7	1.7405	1.7422	1.7440	1.7457	1.7475	1.7492	1.7509	1.7527	1.7544	1.7561	18–17
5.8	1.7579	1.7596	1.7613	1.7630	1.7647	1.7664	1.7681	1.7699	1.7716	1.7733	18–17
5.9	1.7750	1.7766	1.7783	1.7800	1.7817	1.7834	1.7851	1.7867	1.7884	1.7901	17–16
6.0	1.7918	1.7934	1.7951	1.7967	1.7984	1.8001	1.8017	1.8034	1.8050	1.8066	17–16
6.1	1.8083	1.8099	1.8116	1.8132	1.8148	1.8165	1.8181	1.8197	1.8213	1.8229	17–16
6.2	1.8245	1.8262	1.8278	1.8294	1.8310	1.8326	1.8342	1.8358	1.8374	1.8390	17–16
6.3	1.8405	1.8421	1.8437	1.8453	1.8469	1.8485	1.8500	1.8516	1.8532	1.8547	16–15
6.4	1.8563	1.8579	1.8594	1.8610	1.8625	1.8641	1.8656	1.8672	1.8687	1.8703	16–15
6.5	1.8718	1.8733	1.8749	1.8764	1.8779	1.8795	1.8810	1.8825	1.8840	1.8856	16–15
6.6	1.8871	1.8886	1.8901	1.8916	1.8931	1.8946	1.8961	1.8976	1.8991	1.9006	15
6.7	1.9021	1.9036	1.9051	1.9066	1.9081	1.9095	1.9110	1.9125	1.9140	1.9155	15–14
6.8	1.9169	1.9184	1.9199	1.9213	1.9228	1.9242	1.9257	1.9272	1.9286	1.9301	15–14
6.9	1.9315	1.9330	1.9344	1.9359	1.9373	1.9387	1.9402	1.9416	1.9430	1.9445	15–14

TABLE 1025 (*continued*)—NATURAL LOGARITHMS

No.	0	1	2	3	4	5	6	7	8	9	Diff.
7.0	1.9459	1.9473	1.9488	1.9502	1.9516	1.9530	1.9544	1.9559	1.9573	1.9587	15–14
7.1	1.9601	1.9615	1.9629	1.9643	1.9657	1.9671	1.9685	1.9699	1.9713	1.9727	14
7.2	1.9741	1.9755	1.9769	1.9782	1.9796	1.9810	1.9824	1.9838	1.9851	1.9865	14–13
7.3	1.9879	1.9892	1.9906	1.9920	1.9933	1.9947	1.9961	1.9974	1.9988	2.0001	14–13
7.4	2.0015	2.0028	2.0042	2.0055	2.0069	2.0082	2.0096	2.0109	2.0122	2.0136	14–13
7.5	2.0149	2.0162	2.0176	2.0189	2.0202	2.0215	2.0229	2.0242	2.0255	2.0268	14–13
7.6	2.0281	2.0295	2.0308	2.0321	2.0334	2.0347	2.0360	2.0373	2.0386	2.0399	14–13
7.7	2.0412	2.0425	2.0438	2.0451	2.0464	2.0477	2.0490	2.0503	2.0516	2.0528	13–12
7.8	2.0541	2.0554	2.0567	2.0580	2.0592	2.0605	2.0618	2.0631	2.0643	2.0656	13–12
7.9	2.0669	2.0681	2.0694	2.0707	2.0719	2.0732	2.0744	2.0757	2.0769	2.0782	13–12
8.0	2.0794	2.0807	2.0819	2.0832	2.0844	2.0857	2.0869	2.0882	2.0894	2.0906	13–12
8.1	2.0919	2.0931	2.0943	2.0956	2.0968	2.0980	2.0992	2.1005	2.1017	2.1029	13–12
8.2	2.1041	2.1054	2.1066	2.1078	2.1090	2.1102	2.1114	2.1126	2.1138	2.1150	13–12
8.3	2.1163	2.1175	2.1187	2.1199	2.1211	2.1223	2.1235	2.1247	2.1258	2.1270	12–11
8.4	2.1282	2.1294	2.1306	2.1318	2.1330	2.1342	2.1353	2.1365	2.1377	2.1389	12–11
8.5	2.1401	2.1412	2.1424	2.1436	2.1448	2.1459	2.1471	2.1483	2.1494	2.1506	12–11
8.6	2.1518	2.1529	2.1541	2.1552	2.1564	2.1576	2.1587	2.1599	2.1610	2.1622	12–11
8.7	2.1633	2.1645	2.1656	2.1668	2.1679	2.1691	2.1702	2.1713	2.1725	2.1736	12–11
8.8	2.1748	2.1759	2.1770	2.1782	2.1793	2.1804	2.1815	2.1827	2.1838	2.1849	12–11
8.9	2.1861	2.1872	2.1883	2.1894	2.1905	2.1917	2.1928	2.1939	2.1950	2.1961	12–11
9.0	2.1972	2.1983	2.1994	2.2006	2.2017	2.2028	2.2039	2.2050	2.2061	2.2072	12–11
9.1	2.2083	2.2094	2.2105	2.2116	2.2127	2.2138	2.2148	2.2159	2.2170	2.2181	11–10
9.2	2.2192	2.2203	2.2214	2.2225	2.2235	2.2246	2.2257	2.2268	2.2279	2.2289	11–10
9.3	2.2300	2.2311	2.2322	2.2332	2.2343	2.2354	2.2364	2.2375	2.2386	2.2396	11–10
9.4	2.2407	2.2418	2.2428	2.2439	2.2450	2.2460	2.2471	2.2481	2.2492	2.2502	11–10
9.5	2.2513	2.2523	2.2534	2.2544	2.2555	2.2565	2.2576	2.2586	2.2597	2.2607	11–10
9.6	2.2618	2.2628	2.2638	2.2649	2.2659	2.2670	2.2680	2.2690	2.2701	2.2711	11–10
9.7	2.2721	2.2732	2.2742	2.2752	2.2762	2.2773	2.2783	2.2793	2.2803	2.2814	11–10
9.8	2.2824	2.2834	2.2844	2.2854	2.2865	2.2875	2.2885	2.2895	2.2905	2.2915	11–10
9.9	2.2925	2.2935	2.2946	2.2956	2.2966	2.2976	2.2986	2.2996	2.3006	2.3016	11–10
10.0	2.3026										

x	$\mathrm{Log}_\epsilon\, x$	x	$\mathrm{Log}_\epsilon\, x$
10	2.3026	.1	$\overline{3}.6974$
100	4.6052	.01	$\overline{5}.3948$
1 000	6.9078	.001	$\overline{7}.0922$
10 000	9.2103	.000 1	$\overline{10}.7897$
100 000	11.5129	.000 01	$\overline{12}.4871$
1 000 000	13.8155	.000 001	$\overline{14}.1845$
· · · · · · · · · · ·		· · · · · · · · · · ·	

For a large table of natural logarithms, see Ref. 55d.

TABLE 1030—EXPONENTIAL AND HYPERBOLIC FUNCTIONS

x	e^x Value	e^x Log₁₀	e^{-x} Value	Sinh x Value	Sinh x Log₁₀	Cosh x Value	Cosh x Log₁₀	Tanh x Value
0.00	1.0000	.00000	1.0000	0.0000	− ∞	1.0000	.00000	.00000
0.01	1.0101	.00434	.99005	0.0100	.00001	1.0001	.00002	.01000
0.02	1.0202	.00869	.98020	0.0200	.30106	1.0002	.00009	.02000
0.03	1.0305	.01303	.97045	0.0300	.47719	1.0005	.00020	.02999
0.04	1.0408	.01737	.96079	0.0400	.60218	1.0008	.00035	.03998
0.05	1.0513	.02171	.95123	0.0500	.69915	1.0013	.00054	.04996
0.06	1.0618	.02606	.94176	0.0600	.77841	1.0018	.00078	.05993
0.07	1.0725	.03040	.93239	0.0701	.84545	1.0025	.00106	.06989
0.08	1.0833	.03474	.92312	0.0801	.90355	1.0032	.00139	.07983
0.09	1.0942	.03909	.91393	0.0901	.95483	1.0041	.00176	.08976
0.10	1.1052	.04343	.90484	0.1002	.00072	1.0050	.00217	.09967
0.11	1.1163	.04777	.89583	0.1102	.04227	1.0061	.00262	.10956
0.12	1.1275	.05212	.88692	0.1203	.08022	1.0072	.00312	.11943
0.13	1.1388	.05646	.87810	0.1304	.11517	1.0085	.00366	.12927
0.14	1.1503	.06080	.86936	0.1405	.14755	1.0098	.00424	.13909
9.15	1.1618	.06514	.86071	0.1506	.17772	1.0113	.00487	.14889
0.16	1.1735	.06949	.85214	0.1607	.20597	1.0128	.00554	.15865
0.17	1.1853	.07383	.84366	0.1708	.23254	1.0145	.00625	.16838
0.18	1.1972	.07817	.83527	0.1810	.25762	1.0162	.00700	.17808
0.19	1.2092	.08252	.82696	0.1911	.28136	1.0181	.00779	.18775
0.20	1.2214	.08686	.81873	0.2013	.30392	1.0201	.00863	.19738
0.21	1.2337	.09120	.81058	0.2115	.32541	1.0221	.00951	.20697
0.22	1.2461	.09554	.80252	0.2218	.34592	1.0243	.01043	.21652
0.23	1.2586	.09989	.79453	0.2320	.36555	1.0266	.01139	.22603
0.24	1.2712	.10423	.78663	0.2423	.38437	1.0289	.01239	.23550
0.25	1.2840	.10857	.77880	0.2526	.40245	1.0314	.01343	.24492
0.26	1.2969	.11292	.77105	0.2629	.41986	1.0340	.01452	.25430
0.27	1.3100	.11726	.76338	0.2733	.43663	1.0367	.01564	.26362
0.28	1.3231	.12160	.75578	0.2837	.45282	1.0395	.01681	.27291
0.29	1.3364	.12595	.74826	0.2941	.46847	1.0423	.01801	.28213
0.30	1.3499	.13029	.74082	0.3045	.48362	1.0453	.01926	.29131
0.31	1.3634	.13463	.73345	0.3150	.49830	1.0484	.02054	.30044
0.32	1.3771	.13897	.72615	0.3255	.51254	1.0516	.02187	.30951
0.33	1.3910	.14332	.71892	0.3360	.52637	1.0549	.02323	.31852
0.34	1.4049	.14766	.71177	0.3466	.53981	1.0584	.02463	.32748
0.35	1.4191	.15200	.70469	0.3572	.55290	1.0619	.02607	.33638
0.36	1.4333	.15635	.69768	0.3678	.56564	1.0655	.02755	.34521
0.37	1.4477	.16069	.69073	0.3785	.57807	1.0692	.02907	.35399
0.38	1.4623	.16503	.68386	0.3892	.59019	1.0731	.03063	.36271
0.39	1.4770	.16937	.67706	0.4000	.60202	1.0770	.03222	.37136
0.40	1.4918	.17372	.67032	0.4108	.61358	1.0811	.03385	.37995
0.41	1.5068	.17806	.66365	0.4216	.62488	1.0852	.03552	.38847
0.42	1.5220	.18240	.65705	0.4325	.63594	1.0895	.03723	.39693
0.43	1.5373	.18675	.65051	0.4434	.64677	1.0939	.03897	.40532
0.44	1.5527	.19109	.64404	0.4543	.65738	1.0984	.04075	.41364
0.45	1.5683	.19543	.63763	0.4653	.66777	1.1030	.04256	.42190
0.46	1.5841	.19978	.63128	0.4764	.67797	1.1077	.04441	.43008
0.47	1.6000	.20412	.62500	0.4875	.68797	1.1125	.04630	.43820
0.48	1.6161	.20846	.61878	0.4986	.69779	1.1174	.04822	.44624
0.49	1.6323	.21280	.61263	0.5098	.70744	1.1225	.05018	.45422
0.50	1.6487	.21715	.60653	0.5211	.71692	1.1276	.05217	.46212

TABLE 1030 (*continued*)—EXPONENTIAL AND HYPERBOLIC
FUNCTIONS

x	e^x Value	e^x Log$_{10}$	e^{-x} Value	Sinh x Value	Sinh x Log$_{10}$	Cosh x Value	Cosh x Log$_{10}$	Tanh x Value
0.50	1.6487	.21715	.60653	0.5211	.71692	1.1276	.05217	.46212
0.51	1.6653	.22149	.60050	0.5324	.72624	1.1329	.05419	.46995
0.52	1.6820	.22583	.59452	0.5438	.73540	1.1383	.05625	.47770
0.53	1.6989	.23018	.58860	0.5552	.74442	1.1438	.05834	.48538
0.54	1.7160	.23452	.58275	0.5666	.75330	1.1494	.06046	.49299
0.55	1.7333	.23886	.57695	0.5782	.76204	1.1551	.06262	.50052
0.56	1.7507	.24320	.57121	0.5897	.77065	1.1609	.06481	.50798
0.57	1.7683	.24755	.56553	0.6014	.77914	1.1669	.06703	.51536
0.58	1.7860	.25189	.55990	0.6131	.78751	1.1730	.06929	.52267
0.59	1.8040	.25623	.55433	0.6248	.79576	1.1792	.07157	.52990
0.60	1.8221	.26058	.54881	0.6367	.80390	1.1855	.07389	.53705
0.61	1.8404	.26492	.54335	0.6485	.81194	1.1919	.07624	.54413
0.62	1.8589	.26926	.53794	0.6605	.81987	1.1984	.07861	.55113
0.63	1.8776	.27361	.53259	0.6725	.82770	1.2051	.08102	.55805
0.64	1.8965	.27795	.52729	0.6846	.83543	1.2119	.08346	.56490
0.65	1.9155	.28229	.52205	0.6967	.84308	1.2188	.08593	.57167
0.66	1.9348	.28663	.51685	0.7090	.85063	1.2258	.08843	.57836
0.67	1.9542	.29098	.51171	0.7213	.85809	1.2330	.09095	.58498
0.68	1.9739	.29532	.50662	0.7336	.86548	1.2402	.09351	.59152
0.69	1.9937	.29966	.50158	0.7461	.87278	1.2476	.09609	.59798
0.70	2.0138	.30401	.49659	0.7586	.88000	1.2552	.09870	.60437
0.71	2.0340	.30835	.49164	0.7712	.88715	1.2628	.10134	.61068
0.72	2.0544	.31269	.48675	0.7838	.89423	1.2706	.10401	.61691
0.73	2.0751	.31703	.48191	0.7966	.90123	1.2785	.10670	.62307
0.74	2.0959	.32138	.47711	0.8094	.90817	1.2865	.10942	.62915
0.75	2.1170	.32572	.47237	0.8223	.91504	1.2947	.11216	.63515
0.76	2.1383	.33006	.46767	0.8353	.92185	1.3030	.11493	.64108
0.77	2.1598	.33441	.46301	0.8484	.92859	1.3114	.11773	.64693
0.78	2.1815	.33875	.45841	0.8615	.93527	1.3199	.12055	.65271
0.79	2.2034	.34309	.45384	0.8748	.94190	1.3286	.12340	.65841
0.80	2.2255	.34744	.44933	0.8881	.94846	1.3374	.12627	.66404
0.81	2.2479	.35178	.44486	0.9015	.95498	1.3464	.12917	.66959
0.82	2.2705	.35612	.44043	0.9150	.96144	1.3555	.13209	.67507
0.83	2.2933	.36046	.43605	0.9286	.96784	1.3647	.13503	.68048
0.84	2.3164	.36481	.43171	0.9423	.97420	1.3740	.13800	.68581
0.85	2.3396	.36915	.42741	0.9561	.98051	1.3835	.14099	.69107
0.86	2.3632	.37349	.42316	0.9700	.98677	1.3932	.14400	.69626
0.87	2.3869	.37784	.41895	0.9840	.99299	1.4029	.14704	.70137
0.88	2.4109	.38218	.41478	0.9981	.99916	1.4128	.15009	.70642
0.89	2.4351	.38652	.41066	1.0122	.00528	1.4229	.15317	.71139
0.90	2.4596	.39087	.40657	1.0265	.01137	1.4331	.15627	.71630
0.91	2.4843	.39521	.40252	1.0409	.01741	1.4434	.15939	.72113
0.92	2.5093	.39955	.39852	1.0554	.02341	1.4539	.16254	.72590
0.93	2.5345	.40389	.39455	1.0700	.02937	1.4645	.16570	.73059
0.94	2.5600	.40824	.39063	1.0847	.03530	1.4753	.16888	.73522
0.95	2.5857	.41258	.38674	1.0995	.04119	1.4862	.17208	.73978
0.96	2.6117	.41692	.38289	1.1144	.04704	1.4973	.17531	.74428
0.97	2.6379	.42127	.37908	1.1294	.05286	1.5085	.17855	.74870
0.98	2.6645	.42561	.37531	1.1446	.05864	1.5199	.18181	.75307
0.99	2.6912	.42995	.37158	1.1598	.06439	1.5314	.18509	.75736
1.00	2.7183	.43429	.36788	1.1752	.07011	1.5431	.18839	.76159

TABLE 1030 (continued)—EXPONENTIAL AND HYPERBOLIC
FUNCTIONS

x	e^x Value	e^x Log₁₀	e^{-x} Value	Sinh x Value	Sinh x Log₁₀	Cosh x Value	Cosh x Log₁₀	Tanh x Value
1.00	2.7183	.43429	.36788	1.1752	.07011	1.5431	.18839	.76159
1.01	2.7456	.43864	.36422	1.1907	.07580	1.5549	.19171	.76576
1.02	2.7732	.44298	.36060	1.2063	.08146	1.5669	.19504	.76987
1.03	2.8011	.44732	.35701	1.2220	.08708	1.5790	.19839	.77391
1.04	2.8292	.45167	.35345	1.2379	.09268	1.5913	.20176	.77789
1.05	2.8577	.45601	.34994	1.2539	.09825	1.6038	.20515	.78181
1.06	2.8864	.46035	.34646	1.2700	.10379	1.6164	.20855	.78566
1.07	2.9154	.46470	.34301	1.2862	.10930	1.6292	.21197	.78946
1.08	2.9447	.46904	.33960	1.3025	.11479	1.6421	.21541	.79320
1.09	2.9743	.47338	.33622	1.3190	.12025	1.6552	.21886	.79688
1.10	3.0042	.47772	.33287	1.3356	.12569	1.6685	.22233	.80050
1.11	3.0344	.48207	.32956	1.3524	.13111	1.6820	.22582	.80406
1.12	3.0649	.48641	.32628	1.3693	.13649	1.6956	.22931	.80757
1.13	3.0957	.49075	.32303	1.3863	.14186	1.7093	.23283	.81102
1.14	3.1268	.49510	.31982	1.4035	.14720	1.7233	.23636	.81441
1.15	3.1582	.49944	.31664	1.4208	.15253	1.7374	.23990	.81775
1.16	3.1899	.50378	.31349	1.4382	.15783	1.7517	.24346	.82104
1.17	3.2220	.50812	.31037	1.4558	.16311	1.7662	.24703	.82427
1.18	3.2544	.51247	.30728	1.4735	.16836	1.7808	.25062	.82745
1.19	3.2871	.51681	.30422	1.4914	.17360	1.7957	.25422	.83058
1.20	3.3201	.52115	.30119	1.5095	.17882	1.8107	.25784	.83365
1.21	3.3535	.52550	.29820	1.5276	.18402	1.8258	.26146	.83668
1.22	3.3872	.52984	.29523	1.5460	.18920	1.8412	.26510	.83965
1.23	3.4212	.53418	.29229	1.5645	.19437	1.8568	.26876	.84258
1.24	3.4556	.53853	.28938	1.5831	.19951	1.8725	.27242	.84546
1.25	3.4903	.54287	.28650	1.6019	.20464	1.8884	.27610	.84828
1.26	3.5254	.54721	.28365	1.6209	.20975	1.9045	.27979	.85106
1.27	3.5609	.55155	.28083	1.6400	.21485	1.9208	.28349	.85380
1.28	3.5966	.55590	.27804	1.6593	.21993	1.9373	.28721	.85648
1.29	3.6328	.56024	.27527	1.6788	.22499	1.9540	.29093	.85913
1.30	3.6693	.56458	.27253	1.6984	.23004	1.9709	.29467	.86172
1.31	3.7062	.56893	.26982	1.7182	.23507	1.9880	.29842	.86428
1.32	3.7434	.57327	.26714	1.7381	.24009	2.0053	.30217	.86678
1.33	3.7810	.57761	.26448	1.7583	.24509	2.0228	.30594	.86925
1.34	3.8190	.58195	.26185	1.7786	.25008	2.0404	.30972	.87167
1.35	3.8574	.58630	.25924	1.7991	.25505	2.0583	.31352	.87405
1.36	3.8962	.59064	.25666	1.8198	.26002	2.0764	.31732	.87639
1.37	3.9354	.59498	.25411	1.8406	.26496	2.0947	.32113	.87869
1.38	3.9749	.59933	.25158	1.8617	.26990	2.1132	.32495	.88095
1.39	4.0149	.60367	.24908	1.8829	.27482	2.1320	.32878	.88317
1.40	4.0552	.60801	.24660	1.9043	.27974	2.1509	.33262	.88535
1.41	4.0960	.61236	.24414	1.9259	.28464	2.1700	.33647	.88749
1.42	4.1371	.61670	.24171	1.9477	.28952	2.1894	.34033	.88960
1.43	4.1787	.62104	.23931	1.9697	.29440	2.2090	.34420	.89167
1.44	4.2207	.62538	.23693	1.9919	.29926	2.2288	.34807	.89370
1.45	4.2631	.62973	.23457	2.0143	.30412	2.2488	.35196	.89569
1.46	4.3060	.63407	.23224	2.0369	.30896	2.2691	.35585	.89765
1.47	4.3492	.63841	.22993	2.0597	.31379	2.2896	.35976	.89958
1.48	4.3929	.64276	.22764	2.0827	.31862	2.3103	.36367	.90147
1.49	4.4371	.64710	.22537	2.1059	.32343	2.3312	.36759	.90332
1.50	4.4817	.65144	.22313	2.1293	.32823	2.3524	.37151	.90515

TABLE 1030 (*continued*)—EXPONENTIAL AND HYPERBOLIC
FUNCTIONS

x	e^x Value	e^x Log₁₀	e^{-x} Value	Sinh x Value	Sinh x Log₁₀	Cosh x Value	Cosh x Log₁₀	Tanh x Value
1.50	4.4817	.65144	.22313	2.1293	.32823	2.3524	.37151	.90515
1.51	4.5267	.65578	.22091	2.1529	.33303	2.3738	.37545	.90694
1.52	4.5722	.66013	.21871	2.1768	.33781	2.3955	.37939	.90870
1.53	4.6182	.66447	.21654	2.2008	.34258	2.4174	.38334	.91042
1.54	4.6646	.66881	.21438	2.2251	.34735	2.4395	.38730	.91212
1.55	4.7115	.67316	.21225	2.2496	.35211	2.4619	.39126	.91379
1.56	4.7588	.67750	.21014	2.2743	.35686	2.4845	.39524	.91542
1.57	4.8066	.68184	.20805	2.2993	.36160	2.5073	.39921	.91703
1.58	4.8550	.68619	.20598	2.3245	.36633	2.5305	.40320	.91860
1.59	4.9037	.69053	.20393	2.3499	.37105	2.5538	.40719	.92015
1.60	4.9530	.69487	.20190	2.3756	.37577	2.5775	.41119	.92167
1.61	5.0028	.69921	.19989	2.4015	.38048	2.6013	.41520	.92316
1.62	5.0531	.70356	.19790	2.4276	.38518	2.6255	.41921	.92462
1.63	5.1039	.70790	.19593	2.4540	.38987	2.6499	.42323	.92606
1.64	5.1552	.71224	.19398	2.4806	.39456	2.6746	.42725	.92747
1.65	5.2070	.71659	.19205	2.5075	.39923	2.6995	.43129	.92886
1.66	5.2593	.72093	.19014	2.5346	.40391	2.7247	.43532	.93022
1.67	5.3122	.72527	.18825	2.5620	.40857	2.7502	.43937	.93155
1.68	5.3656	.72961	.18637	2.5896	.41323	2.7760	.44341	.93286
1.69	5.4195	.73396	.18452	2.6175	.41788	2.8020	.44747	.93415
1.70	5.4739	.73830	.18268	2.6456	.42253	2.8283	.45153	.93541
1.71	5.5290	.74264	.18087	2.6740	.42717	2.8549	.45559	.93665
1.72	5.5845	.74699	.17907	2.7027	.43180	2.8818	.45966	.93786
1.73	5.6407	.75133	.17728	2.7317	.43643	2.9090	.46374	.93906
1.74	5.6973	.75567	.17552	2.7609	.44105	2.9364	.46782	.94023
1.75	5.7546	.76002	.17377	2.7904	.44567	2.9642	.47191	.94138
1.76	5.8124	.76436	.17204	2.8202	.45028	2.9922	.47600	.94250
1.77	5.8709	.76870	.17033	2.8503	.45488	3.0206	.48009	.94361
1.78	5.9299	.77304	.16864	2.8806	.45948	3.0492	.48419	.94470
1.79	5.9895	.77739	.16696	2.9112	.46408	3.0782	.48830	.94576
1.80	6.0496	.78173	.16530	2.9422	.46867	3.1075	.49241	.94681
1.81	6.1104	.78607	.16365	2.9734	.47325	3.1371	.49652	.94783
1.82	6.1719	.79042	.16203	3.0049	.47783	3.1669	.50064	.94884
1.83	6.2339	.79476	.16041	3.0367	.48241	3.1972	.50476	.94983
1.84	6.2965	.79910	.15882	3.0689	.48698	3.2277	.50889	.95080
1.85	6.3598	.80344	.15724	3.1013	.49154	3.2585	.51302	.95175
1.86	6.4237	.80779	.15567	3.1340	.49610	3.2897	.51716	.95268
1.87	6.4883	.81213	.15412	3.1671	.50066	3.3212	.52130	.95359
1.88	6.5535	.81647	.15259	3.2005	.50521	3.3530	.52544	.95449
1.89	6.6194	.82082	.15107	3.2341	.50976	3.3852	.52959	.95537
1.90	6.6859	.82516	.14957	3.2682	.51430	3.4177	.53374	.95624
1.91	6.7531	.82950	.14808	3.3025	.51884	3.4506	.53789	.95709
1.92	6.8210	.83385	.14661	3.3372	.52338	3.4838	.54205	.95792
1.93	6.8895	.83819	.14515	3.3722	.52791	3.5173	.54621	.95873
1.94	6.9588	.84253	.14370	3.4075	.53244	3.5512	.55038	.95953
1.95	7.0287	.84687	.14227	3.4432	.53696	3.5855	.55455	.96032
1.96	7.0993	.85122	.14086	3.4792	.54148	3.6201	.55872	.96109
1.97	7.1707	.85556	.13946	3.5156	.54600	3.6551	.56290	.96185
1.98	7.2427	.85990	.13807	3.5523	.55051	3.6904	.56707	.96259
1.99	7.3155	.86425	.13670	3.5894	.55502	3.7261	.57126	.96331
2.00	7.3891	.86859	.13534	3.6269	.55953	3.7622	.57544	.96403

TABLE 1030 (*continued*)—EXPONENTIAL AND HYPERBOLIC
FUNCTIONS

x	e^x Value	e^x Log$_{10}$	e^{-x} Value	Sinh x Value	Sinh x Log$_{10}$	Cosh x Value	Cosh x Log$_{10}$	Tanh x Value
2.00	7.3891	.86859	.13534	3.6269	.55953	3.7622	.57544	.96403
2.01	7.4633	.87293	.13399	3.6647	.56403	3.7987	.57963	.96473
2.02	7.5383	.87727	.13266	3.7028	.56853	3.8355	.58382	.96541
2.03	7.6141	.88162	.13134	3.7414	.57303	3.8727	.58802	.96609
2.04	7.6906	.88596	.13003	3.7803	.57753	3.9103	.59221	.96675
2.05	7.7679	.89030	.12873	3.8196	.58202	3.9483	.59641	.96740
2.06	7.8460	.89465	.12745	3.8593	.58650	3.9867	.60061	.96803
2.07	7.9248	.89899	.12619	3.8993	.59099	4.0255	.60482	.96865
2.08	8 0045	.90333	.12493	3.9398	.59547	4.0647	.60903	.96926
2.09	8.0849	.90768	.12369	3.9806	.59995	4.1043	.61324	.96986
2.10	8.1662	.91202	.12246	4.0219	.60443	4.1443	.61745	.97045
2.11	8.2482	.91636	.12124	4.0635	.60890	4.1847	.62167	.97103
2.12	8.3311	.92070	.12003	4.1056	.61337	4.2256	.62589	.97159
2.13	8.4149	.92505	.11884	4.1480	.61784	4.2669	.63011	.97215
2.14	8.4994	.92939	.11765	4.1909	.62231	4.3085	.63433	.97269
2.15	8.5849	.93373	.11648	4.2342	.62677	4.3507	.63856	.97323
2.16	8.6711	.93808	.11533	4.2779	.63123	4.3932	.64278	.97375
2.17	8.7583	.94242	.11418	4.3221	.63569	4.4362	.64701	.97426
2.18	8.8463	.94676	.11304	4.3666	.64015	4.4797	.65125	.97477
2.19	8.9352	.95110	.11192	4.4116	.64460	4.5236	.65548	.97526
2.20	9.0250	.95545	.11080	4.4571	.64905	4.5679	.65972	.97574
2.21	9.1157	.95979	.10970	4.5030	.65350	4.6127	.66396	.97622
2.22	9.2073	.96413	.10861	4.5494	.65795	4.6580	.66820	.97668
2.23	9.2999	.96848	.10753	4.5962	.66240	4.7037	.67244	.97714
2.24	9.3933	.97282	.10646	4.6434	.66684	4.7499	.67668	.97759
2.25	9.4877	.97716	.10540	4.6912	.67128	4.7966	.68093	.97803
2.26	9.5831	.98151	.10435	4.7394	.67572	4.8437	.68518	.97846
2.27	9.6794	.98585	.10331	4.7880	.68016	4.8914	.68943	.97888
2.28	9.7767	.99019	.10228	4.8372	.68459	4.9395	.69368	.97929
2.29	9.8749	.99453	.10127	4.8868	.68903	4.9881	.69794	.97970
2.30	9.9742	.99888	.10026	4.9370	.69346	5.0372	.70219	.98010
2.31	10.074	.00322	.09926	4.9876	.69789	5.0868	.70645	.98049
2.32	10.176	.00756	.09827	5.0387	.70232	5.1370	.71071	.98087
2.33	10.278	.01191	.09730	5.0903	.70675	5.1876	.71497	.98124
2.34	10.381	.01625	.09633	5.1425	.71117	5.2388	.71923	.98161
2.35	10.486	.02059	.09537	5.1951	.71559	5.2905	.72349	.98197
2.36	10.591	.02493	.09442	5.2483	.72002	5.3427	.72776	.98233
2.37	10.697	.02928	.09348	5.3020	.72444	5.3954	.73203	.98267
2.38	10.805	.03362	.09255	5.3562	.72885	5.4487	.73630	.98301
2.39	10.913	.03796	.09163	5.4109	.73327	5.5026	.74056	.98335
2.40	11.023	04231	.09072	5.4662	.73769	5.5569	.74484	.98367
2.41	11.134	.04665	.08982	5.5221	.74210	5.6119	.74911	.98400
2.42	11.246	.05099	.08892	5.5785	.74652	5.6674	.75338	.98431
2.43	11.359	.05534	.08804	5.6354	.75093	5.7235	.75766	.98462
2.44	11.473	.05968	.08716	5 6929	.75534	5.7801	.76194	.98492
2.45	11.588	.06402	.08629	5.7510	.75975	5.8373	.76621	.98522
2.46	11.705	.06836	.08543	5.8097	.76415	5.8951	.77049	.98551
2.47	11.822	.07271	.08458	5.8689	.76856	5.9535	.77477	.98579
2.48	11.941	.07705	.08374	5.9288	.77296	6.0125	.77906	.98607
2.49	12.061	.08139	.08291	5.9892	.77737	6.0721	.78334	.98635
2.50	12.182	.08574	.08208	6.0502	.78177	6.1323	.78762	.98661

TABLE 1030 (*continued*)—EXPONENTIAL AND HYPERBOLIC
FUNCTIONS

x	e^x Value	e^x Log$_{10}$	e^{-x} Value	Sinh x Value	Sinh x Log$_{10}$	Cosh x Value	Cosh x Log$_{10}$	Tanh x Value
2.50	12.182	.08574	.08208	6.0502	.78177	6.1323	.78762	.98661
2.51	12.305	.09008	.08127	6.1118	.78617	6.1931	.79191	.98688
2.52	12.429	.09442	.08046	6.1741	.79057	6.2545	.79619	.98714
2.53	12.554	.09877	.07966	6.2369	.79497	6.3166	.80048	.98739
2.54	12.680	.10311	.07887	6.3004	.79937	6.3793	.80477	.98764
2.55	12.807	.10745	.07808	6.3645	.80377	6.4426	.80906	.98788
2.56	12.936	.11179	.07730	6.4293	.80816	6.5066	.81335	.98812
2.57	13.066	.11614	.07654	6.4946	.81256	6.5712	.81764	.98835
2.58	13.197	.12048	.07577	6.5607	.81695	6.6365	.82194	.98858
2.59	13.330	.12482	.07502	6.6274	.82134	6.7024	.82623	.98881
2.60	13.464	.12917	.07427	6.6947	.82573	6.7690	.83052	.98903
2.61	13.599	.13351	.07353	6.7628	.83012	6.8363	.83482	.98924
2.62	13.736	.13785	.07280	6.8315	.83451	6.9043	.83912	.98946
2.63	13.874	.14219	.07208	6.9008	.83890	6.9729	.84341	.98966
2.64	14.013	.14654	.07136	6.9709	.84329	7.0423	.84771	.98987
2.65	14.154	.15088	.07065	7.0417	.84768	7.1123	.85201	.99007
2.66	14.296	.15522	.06995	7.1132	.85206	7.1831	.85631	.99026
2.67	14.440	.15957	.06925	7.1854	.85645	7.2546	.86061	.99045
2.68	14.585	.16391	.06856	7.2583	.86083	7.3268	.86492	.99064
2.69	14.732	.16825	.06788	7.3319	.86522	7.3998	.86922	.99083
2.70	14.880	.17260	.06721	7.4063	.86960	7.4735	.87352	.99101
2.71	15.029	.17694	.06654	7.4814	.87398	7.5479	.87783	.99118
2.72	15.180	.18128	.06587	7.5572	.87836	7.6231	.88213	.99136
2.73	15.333	.18562	.06522	7.6338	.88274	7.6991	.88644	.99153
2.74	15.487	.18997	.06457	7.7112	.88712	7.7758	.89074	.99170
2.75	15.643	.19431	.06393	7.7894	.89150	7.8533	.89505	.99186
2.76	15.800	.19865	.06329	7.8683	.89588	7.9316	.89936	.99202
2.77	15.959	.20300	.06266	7.9480	.90026	8.0106	.90367	.99218
2.78	16.119	.20734	.06204	8.0285	.90463	8.0905	.90798	.99233
2.79	16.281	.21168	.06142	8.1098	.90901	8.1712	.91229	.99248
2.80	16.445	.21602	.06081	8.1919	.91339	8.2527	.91660	.99263
2.81	16.610	.22037	.06020	8.2749	.91776	8.3351	.92091	.99278
2.82	16.777	.22471	.05961	8.3586	.92213	8.4182	.92522	.99292
2.83	16.945	.22905	.05901	8.4432	.92651	8.5022	.92953	.99306
2.84	17.116	.23340	.05843	8.5287	.93088	8.5871	.93385	.99320
2.85	17.288	.23774	.05784	8.6150	.93525	8.6728	.93816	.99333
2.86	17.462	.24208	.05727	8.7021	.93963	8.7594	.94247	.99346
2.87	17.637	.24643	.05670	8.7902	.94400	8.8469	.94679	.99359
2.88	17.814	.25077	.05613	8.8791	.94837	8.9352	.95110	.99372
2.89	17.993	.25511	.05558	8.9689	.95274	9.0244	.95542	.99384
2.90	18.174	.25945	.05502	9.0596	.95711	9.1146	.95974	.99396
2.91	18.357	.26380	.05448	9.1512	.96148	9.2056	.96405	.99408
2.92	18.541	.26814	.05393	9.2437	.96584	9.2976	.96837	.99420
2.93	18.728	.27248	.05340	9.3371	.97021	9.3905	.97269	.99431
2.94	18.916	.27683	.05287	9.4315	.97458	9.4844	.97701	.99443
2.95	19.106	.28117	.05234	9.5268	.97895	9.5791	.98133	.99454
2.96	19.298	.28551	.05182	9.6231	.98331	9.6749	.98565	.99464
2.97	19.492	.28985	.05130	9.7203	.98768	9.7716	.98997	.99475
2.98	19.688	.29420	.05079	9.8185	.99205	9.8693	.99429	.99485
2.99	19.886	.29854	.05029	9.9177	.99641	9.9680	.99861	.99496
3.00	20.086	.30288	.04979	10.018	.00078	10.068	.00293	.99505

TABLE 1030 (*continued*)—EXPONENTIAL AND HYPERBOLIC FUNCTIONS

x	e^x Value	Log₁₀	e^{-x} Value	Sinh x Value	Log₁₀	Cosh x Value	Log₁₀	Tanh x Value
3.00	20.086	.30288	.04979	10.018	.00078	10.068	.00293	.99505
3.05	21.115	.32460	.04736	10.534	.02259	10.581	.02454	.99552
3.10	22.198	.34631	.04505	11.076	.04440	11.122	.04616	.99595
3.15	23.336	.36803	.04285	11.647	.06620	11.689	.06779	.99633
3.20	24.533	.38974	.04076	12.246	.08799	12.287	.08943	.99668
3.25	25.790	.41146	.03877	12.876	.10977	12.915	.11108	.99700
3.30	27.113	.43317	.03688	13.538	.13155	13.575	.13273	.99728
3.35	28.503	.45489	.03508	14.234	.15332	14.269	.15439	.99754
3.40	29.964	.47660	.03337	14.965	.17509	14.999	.17605	.99777
3.45	31.500	.49832	.03175	15.734	.19685	15.766	.19772	.99799
3.50	33.115	.52003	.03020	16.543	.21860	16.573	.21940	.99818
3.55	34.813	.54175	.02872	17.392	.24036	17.421	.24107	.99835
3.60	36.598	.56346	.02732	18.286	.26211	18.313	.26275	.99851
3.65	38.475	.58517	.02599	19.224	.28385	19.250	.28444	.99865
3.70	40.447	.60689	.02472	20.211	.30559	20.236	.30612	.99878
3.75	42.521	.62860	.02352	21.249	.32733	21.272	.32781	.99889
3.80	44.701	.65032	.02237	22.339	.34907	22.362	.34951	.99900
3.85	46.993	.67203	.02128	23.486	.37081	23.507	.37120	.99909
3.90	49.402	.69375	.02024	24.691	.39254	24.711	.39290	.99918
3.95	51.935	.71546	.01925	25.958	.41427	25.977	.41459	.99926
4.00	54.598	.73718	.01832	27.290	.43600	27.308	.43629	.99933
4.10	60.340	.78061	.01657	30.162	.47946	30.178	.47970	.99945
4.20	66.686	.82404	.01500	33.336	.52291	33.351	.52310	.99955
4.30	73.700	.86747	.01357	36.843	.56636	36.857	.56652	.99963
4.40	81.451	.91090	.01227	40.719	.60980	40.732	.60993	.99970
4.50	90.017	.95433	.01111	45.003	.65324	45.014	.65335	.99975
4.60	99.484	.99775	.01005	49.737	.69668	49.747	.69677	.99980
4.70	109.95	.04118	.00910	54.969	.74012	54.978	.74019	.99983
4.80	121.51	.08461	.00823	60.751	.78355	60.759	.78361	.99986
4.90	134.29	.12804	.00745	67.141	.82699	67.149	.82704	.99989
5.00	148.41	.17147	.00674	74.203	.87042	74.210	.87046	.99991
5.10	164.02	.21490	.00610	82.008	.91386	82.014	.91389	.99993
5.20	181.27	.25833	.00552	90.633	.95729	90.639	.95731	.99994
5.30	200.34	.30176	.00499	100.17	.00074	100.17	.00074	.99995
5.40	221.41	.34519	.00452	110.70	.04415	110.71	.04417	.99996
5.50	244.69	.38862	.00409	122.34	.08758	122.35	.08760	.99997
5.60	270.43	.43205	.00370	135.21	.13101	135.22	.13103	.99997
5.70	298.87	.47548	.00335	149.43	.17444	149.44	.17445	.99998
5.80	330.30	.51891	.00303	165.15	.21787	165.15	.21788	.99998
5.90	365.04	.56234	.00274	182.52	.26130	182.52	.26131	.99998
6.00	403.43	.60577	.00248	201.71	.30473	201.72	.30474	.99999
6.25	518.01	.71434	.00193	259.01	.41331	259.01	.41331	.99999
6.50	665.14	.82291	.00150	332.57	.52188	332.57	.52189	1.0000
6.75	854.06	.93149	.00117	427.03	.63046	427.03	.63046	1.0000
7.00	1096.6	.04006	.00091	548.32	.73903	548.32	.73903	1.0000
7.50	1808.0	.25721	.00055	904.02	.95618	904.02	.95618	1.0000
8.00	2981.0	.47436	.00034	1490.5	.17333	1490.5	.17333	1.0000
8.50	4914.8	.69150	.00020	2457.4	.39047	2457.4	.39047	1.0000
9.00	8103.1	.90865	.00012	4051.5	.60762	4051.5	.60762	1.0000
9.50	13360.	.12580	.00007	6679.9	.82477	6679.9	.82477	1.0000
10.00	22026.	.34294	.00005	11013.	.04191	11013.	.04191	1.0000

TABLE 1030 (*continued*)—EXPONENTIAL AND HYPERBOLIC
FUNCTIONS

For tables of exponential and hyperbolic functions, see
References 30, 45, 55b and 55c.

For large values of x use e^x = natural anti-logarithm of x,
which may be obtained from a table of natural logarithms.
When x is large, subtract multiples of 2.3026 from x. Note
also that

$$e^{-x} = 1/e^x$$

$$\sinh x = \tfrac{1}{2}(e^x - e^{-x})$$

$$\cosh x = \tfrac{1}{2}(e^x + e^{-x})$$

$$\tanh x = \frac{e^{2x} - 1}{e^{2x} + 1}$$

$$= 1 - \frac{2}{e^{2x}} + \frac{2}{e^{4x}} - \frac{2}{e^{6x}} + \cdots.$$

NOTE.—Tables 1010, 1015, 1016, 1020 and 1030 are from *The Macmillan
Mathematical Tables*.

TABLE 1040—COMPLETE ELLIPTIC INTEGRALS OF THE FIRST KIND

$$K = \int_0^{\pi/2} \frac{d\varphi}{\sqrt{(1 - \sin^2\theta \sin^2\varphi)}}$$ [See 773]

θ Deg.	K	Diff.	θ Deg.	K	Diff.	θ Deg.	K	Diff.	θ Deg.	K	Diff.	θ Deg.	K	Diff.	θ Deg.	K	Diff.
0	1.571	0	25	1.649	7	50	1.936	9	62.5	2.228	16	72.0	2.600	10	77.0	2.903	14
1	1.571	0	26	1.656	7	50.5	1.945	9	63	2.244	15	72.2	2.610	10	77.2	2.917	15
2	1.571	1	27	1.663	7	51	1.954	9	63.5	2.259	16	72.4	2.620	11	77.4	2.932	15
3	1.572	1	28	1.670	8	51.5	1.963	10	64	2.275	17	72.6	2.631	10	77.6	2.947	16
4	1.573	1	29	1.678	8	52	1.973	10	64.5	2.292	17	72.8	2.641	11	77.8	2.963	16
5	1.574	1	30	1.686	8	52.5	1.983	10	65	2.309	17	73.0	2.652	11	78.0	2.979	16
6	1.575	2	31	1.694	9	53	1.993	10	65.5	2.326	18	73.2	2.663	11	78.2	2.995	16
7	1.577	1	32	1.703	9	53.5	2.003	10	66	2.344	18	73.4	2.674	11	78.4	3.011	17
8	1.578	3	33	1.712	9	54	2.013	11	66.5	2.362	19	73.6	2.685	12	78.6	3.028	16
9	1.581	2	34	1.721	10	54.5	2.024	11	67	2.381	19	73.8	2.697	11	78.8	3.044	18
10	1.583	2	35	1.731	10	55	2.035	11	67.5	2.400	20	74.0	2.708	12	79.0	3.062	17
11	1.585	3	36	1.741	11	55.5	2.046	11	68	2.420	20	74.2	2.720	12	79.2	3.079	18
12	1.588	3	37	1.752	11	56	2.057	12	68.5	2.440	21	74.4	2.732	12	79.4	3.097	19
13	1.591	3	38	1.763	12	56.5	2.069	11	69	2.461	21	74.6	2.744	12	79.6	3.116	18
14	1.595	3	39	1.775	12	57	2.080	12	69.5	2.482	23	74.8	2.756	12	79.8	3.134	19
15	1.598	4	40	1.787	12	57.5	2.092	13	70.0	2.505	9	75.0	2.768	13	80.0	3.153	20
16	1.602	4	41	1.799	13	58	2.105	12	70.2	2.514	9	75.2	2.781	12	80.2	3.173	20
17	1.606	4	42	1.812	14	58.5	2.117	13	70.4	2.523	9	75.4	2.793	13	80.4	3.193	20
18	1.610	5	43	1.826	14	59	2.130	13	70.6	2.532	9	75.6	2.806	13	80.6	3.213	21
19	1.615	5	44	1.840	14	59.5	2.143	14	70.8	2.541	10	75.8	2.819	14	80.8	3.234	21
20	1.620	5	45	1.854	15	60	2.157	13	71.0	2.551	9	76.0	2.833	13	81.0	3.255	22
21	1.625	6	46	1.869	16	60.5	2.170	14	71.2	2.560	10	76.2	2.846	14	81.2	3.277	22
22	1.631	5	47	1.885	16	61	2.184	15	71.4	2.570	10	76.4	2.860	14	81.4	3.299	23
23	1.636	7	48	1.901	17	61.5	2.199	14	71.6	2.580	10	76.6	2.874	14	81.6	3.322	24
24	1.643	6	49	1.918	18	62	2.213	15	71.8	2.590	10	76.8	2.888	15	81.8	3.346	24

TABLE 1040 (continued)—COMPLETE ELLIPTIC INTEGRALS OF THE FIRST KIND

θ Degrees	K	Diff.	θ Degrees	K	Diff.
82.0	3.370	12	84.5	3.738	18
82.1	3.382	13	84.6	3.756	18
82.2	3.395	12	84.7	3.774	19
82.3	3.407	13	84.8	3.793	19
82.4	3.420	13	84.9	3.812	20
82.5	3.433	13	85.0	3.832	20
82.6	3.446	13	85.1	3.852	20
82.7	3.459	14	85.2	3.872	21
82.8	3.473	14	85.3	3.893	21
82.9	3.487	13	85.4	3.914	22
83.0	3.500	15	85.5	3.936	22
83.1	3.515	14	85.6	3.958	23
83.2	3.529	14	85.7	3.981	23
83.3	3.543	15	85.8	4.004	24
83.4	3.558	15	85.9	4.028	25
83.5	3.573	15	86.0	4.053	25
83.6	3.588	16	86.1	4.078	26
83.7	3.604	16	86.2	4.104	26
83.8	3.620	16	86.3	4.130	27
83.9	3.636	16	86.4	4.157	28
84.0	3.652	16	86.5	4.185	29
84.1	3.668	17	86.6	4.214	30
84.2	3.685	17	86.7	4.244	30
84.3	3.702	18	86.8	4.274	32
84.4	3.720	18	86.9	4.306	33

θ Degrees	Min.	K	Diff.
87.0		4.339	33
87.1		4.372	35
87.2		4.407	37
87.3		4.444	37
87.4		4.481	39
87.5		4.520	42
87.6		4.561	41
87.7		4.603	45
87.8		4.648	46
87.9		4.694	49
88.0		4.743	51
88.1		4.794	54
88.2		4.848	57
88.3		4.905	60
88.4		4.965	65
88.5		5.030	69
88.6		5.099	74
88.7		5.173	80
88.8		5.253	87
88.9		5.340	95
89	0	5.435	34
	2	5.469	35
	4	5.504	36
	6	5.540	38
	8	5.578	39

θ Degrees	Min.	K	Diff.
89	10	5.617	41
	12	5.658	42
	14	5.700	45
	16	5.745	46
	18	5.791	49
	20	5.840	51
	22	5.891	55
	24	5.946	57
	26	6.003	60
	28	6.063	65
	30	6.128	69
	32	6.197	74
	34	6.271	80
	36	6.351	87
	38	6.438	95
	40	6.533	51
	41	6.584	55
	42	6.639	57
	43	6.696	60
	44	6.756	65
	45	6.821	69
	46	6.890	74
	47	6.964	80
	48	7.044	87
	49	7.131	95

θ Degrees	Min.	K	Diff.
89	50	7.226	106
	51	7.332	117
	52	7.449	134
	53	7.583	154
	54	7.737	182
	55	7.919	224
	56	8.143	287
	57	8.430	406
	58	8.836	693
	59	9.529	
90	0	∞	

For values of θ greater than about 89° 50' it is often better to use series 773.3 than to interpolate from tables.

235

TABLE 1041—COMPLETE ELLIPTIC INTEGRALS OF THE SECOND KIND

$$E = \int_0^{\pi/2} \sqrt{(1 - \sin^2\theta \sin^2\varphi)}\,d\varphi \quad \text{[See 774]}$$

θ Degrees	E	Diff.	θ Degrees	E	Diff.	θ Degrees	E	Diff.
0	1.571	0	30	1.467	−6	60	1.211	−9
1	1.571	−1	31	1.461	−7	61	1.202	−10
2	1.570	0	32	1.454	*−7	62	1.192	−9
3	1.570	−1	33	1.447	−7	63	1.183	−10
4	1.569	−1	34	1.440	−8	64	1.173	−9
5	1.568	−2	35	1.432	−7	65	1.164	−9
6	1.566	−1	36	1.425	−8	66	1.155	−10
7	1.565	−2	37	1.417	−8	67	1.145	−9
8	1.563	−2	38	1.409	−8	68	1.136	−9
9	1.561	−2	39	1.401	−8	69	1.127	−9
10	1.559	−3	40	1.393	−8	70	1.118	−8
11	1.556	−2	41	1.385	−8	71	1.110	−9
12	1.554	−3	42	1.377	−9	72	1.101	−8
13	1.551	−3	43	1.368	−9	73	1.093	−9
14	1.548	−4	44	1.359	−8	74	1.084	−8
15	1.544	−3	45	1.351	−9	75	1.076	−7
16	1.541	−4	46	1.342	−9	76	1.069	−8
17	1.537	−4	47	1.333	−9	77	1.061	−7
18	1.533	−5	48	1.324	−9	78	1.054	−7
19	1.528	−4	49	1.315	−9	79	1.047	−7
20	1.524	−5	50	1.306	−10	80	1.040	−6
21	1.519	−5	51	1.296	−9	81	1.034	−6
22	1.514	−5	52	1.287	−9	82	1.028	−6
23	1.509	−5	53	1.278	−10	83	1.022	−5
24	1.504	−6	54	1.268	−9	84	1.017	−4
25	1.498	−6	55	1.259	−10	85	1.013	−4
26	1.492	−6	56	1.249	−9	86	1.009	−4
27	1.486	−6	57	1.240	−10	87	1.005	−2
28	1.480	−6	58	1.230	−9	88	1.003	−2
29	1.474	−7	59	1.221	−10	89	1.001	−1
						90	1.000	

For tables of elliptic integrals, see Ref. 47, 26, 31, 35, 36, 45, and 48.

TABLE 1045—NORMAL PROBABILITY INTEGRAL

$$\frac{1}{\sqrt{(2\pi)}} \int_{-x}^{x} e^{-t^2/2} dt \quad [\text{See 585}]$$

x	0	1	2	3	4	5	6	7	8	9	Diff.
.0	.0000	.0080	.0160	.0239	.0319	.0399	.0478	.0558	.0638	.0717	79–80
.1	.0797	.0876	.0955	.1034	.1113	.1192	.1271	.1350	.1428	.1507	78–79
.2	.1585	.1663	.1741	.1819	.1897	.1974	.2051	.2128	.2205	.2282	76–78
.3	.2358	.2434	.2510	.2586	.2661	.2737	.2812	.2886	.2961	.3035	73–76
.4	.3108	.3182	.3255	.3328	.3401	.3473	.3545	.3616	.3688	.3759	70–74
.5	.3829	.3899	.3969	.4039	.4108	.4177	.4245	.4313	.4381	.4448	67–70
.6	.4515	.4581	.4647	.4713	.4778	.4843	.4907	.4971	.5035	.5098	63–66
.7	.5161	.5223	.5285	.5346	.5407	.5467	.5527	.5587	.5646	.5705	58–62
.8	.5763	.5821	.5878	.5935	.5991	.6047	.6102	.6157	.6211	.6265	54–58
.9	.6319	.6372	.6424	.6476	.6528	.6579	.6629	.6680	.6729	.6778	49–53
1.0	.6827	.6875	.6923	.6970	.7017	.7063	.7109	.7154	.7199	.7243	44–48
1.1	.7287	.7330	.7373	.7415	.7457	.7499	.7540	.7580	.7620	.7660	39–43
1.2	.7699	.7737	.7775	.7813	.7850	.7887	.7923	.7959	.7995	.8029	34–38
1.3	.8064	.8098	.8132	.8165	.8198	.8230	.8262	.8293	.8324	.8355	30–34
1.4	.8385	.8415	.8444	.8473	.8501	.8529	.8557	.8584	.8611	.8638	26–30
1.5	.8664	.8690	.8715	.8740	.8764	.8789	.8812	.8836	.8859	.8882	22–26
1.6	.8904	.8926	.8948	.8969	.8990	.9011	.9031	.9051	.9070	.9090	19–22
1.7	.9109	.9127	.9146	.9164	.9181	.9199	.9216	.9233	.9249	.9265	16–19
1.8	.9281	.9297	.9312	.9328	.9342	.9357	.9371	.9385	.9399	.9412	13–16
1.9	.9426	.9439	.9451	.9464	.9476	.9488	.9500	.9512	.9523	.9534	11–13
2.0	.9545	.9556	.9566	.9576	.9586	.9596	.9606	.9615	.9625	.9634	9–11
2.1	.9643	.9651	.9660	.9668	.9676	.9684	.9692	.9700	.9707	.9715	7–9
2.2	.9722	.9729	.9736	.9743	.9749	.9756	.9762	.9768	.9774	.9780	6–7
2.3	.9786	.9791	.9797	.9802	.9807	.9812	.9817	.9822	.9827	.9832	4–6
2.4	.9836	.9840	.9845	.9849	.9853	.9857	.9861	.9865	.9869	.9872	4–5
2.5	.9876	.9879	.9883	.9886	.9889	.9892	.9895	.9898	.9901	.9904	3–4
2.6	.9907	.9909	.9912	.9915	.9917	.9920	.9922	.9924	.9926	.9929	2–3
2.7	.9931	.9933	.9935	.9937	.9939	.9940	.9942	.9944	.9946	.9947	1–2
2.8	.9949	.9950	.9952	.9953	.9955	.9956	.9958	.9959	.9960	.9961	1–2
2.9	.9963	.9964	.9965	.9966	.9967	.9968	.9969	.9970	.9971	.9972	1
3.0	.9973	.9974	.9975	.9976	.9976	.9977	.9978	.9979	.9979	.9980	0–1
3.1	.9981	.9981	.9982	.9983	.9983	.9984	.9984	.9985	.9985	.9986	0–1
3.2	.9986	.9987	.9987	.9988	.9988	.9988	.9989	.9989	.9990	.9990	0–1
3.3	.9990	.9991	.9991	.9991	.9992	.9992	.9992	.9992	.9993	.9993	0–1
3.4	.9993	.9994	.9994	.9994	.9994	.9994	.9995	.9995	.9995	.9995	0–1
3.5	.9995	.9996	.9996	.9996	.9996	.9997	.9996	.9996	.9997	.9997	0–1
3.6	.9997	.9997	.9997	.9997	.9997	.9997	.9997	.9998	.9998	.9998	0–1
3.7	.9998	.9998	.9998	.9998	.9998	.9998	.9998	.9998	.9998	.9998	0–1
3.8	.9999	.9999	.9999	.9999	.9999	.9999	.9999	.9999	.9999	.9999	0
3.9	.9999	.9999	.9999	.9999	.9999	.9999	.9999	.9999	.9999	.9999	0
4.0	.9999	.9999	.9999	.9999	.9999	.9999	1.0000	1.0000	1.0000	1.0000	0–1

For a large table of 15 decimal places, see Ref. 55e, "Tables of Probability Functions," Vol. II, A. N. Lowan, Technical Director, Work Projects Administration for the City of New York, 1942, sponsored by the National Bureau of Standards.

TABLE 1050—BESSEL FUNCTIONS

For tables of Bessel Functions of real arguments see References 12 and 50

$$\text{ber } x + i \text{ bei } x = J_0(xi\sqrt{i}) = I_0(x\sqrt{i})$$

x	ber x	bei x	ber$'$ x	bei$'$ x
0	$+1.0$	0	0	0
0.1	$+0.999\ 998\ 438$	$+0.002\ 500\ 000$	$-0.000\ 062\ 500$	$+0.049\ 999\ 974$
0.2	$+0.999\ 975\ 000$	$+0.009\ 999\ 972$	$-0.000\ 499\ 999$	$+0.099\ 999\ 167$
0.3	$+0.999\ 873\ 438$	$+0.022\ 499\ 684$	$-0.001\ 687\ 488$	$+0.149\ 993\ 672$
0.4	$+0.999\ 600\ 004$	$+0.039\ 998\ 222$	$-0.003\ 999\ 911$	$+0.199\ 973\ 334$
0.5	$+0.999\ 023\ 464$	$+0.062\ 493\ 218$	$-0.007\ 812\ 076$	$+0.249\ 918\ 621$
0.6	$+0.997\ 975\ 114$	$+0.089\ 979\ 750$	$-0.013\ 498\ 481$	$+0.299\ 797\ 507$
0.7	$+0.996\ 248\ 828$	$+0.122\ 448\ 939$	$-0.021\ 433\ 032$	$+0.349\ 562\ 345$
0.8	$+0.993\ 601\ 138$	$+0.159\ 886\ 230$	$-0.031\ 988\ 623$	$+0.399\ 146\ 758$
0.9	$+0.989\ 751\ 357$	$+0.202\ 269\ 363$	$-0.045\ 536\ 553$	$+0.448\ 462\ 528$
1.0	$+0.984\ 381\ 781$	$+0.249\ 566\ 040$	$-0.062\ 445\ 752$	$+0.497\ 396\ 511$
1.1	$+0.977\ 137\ 973$	$+0.301\ 731\ 269$	$-0.083\ 081\ 791$	$+0.545\ 807\ 563$
1.2	$+0.967\ 629\ 156$	$+0.358\ 704\ 420$	$-0.107\ 805\ 642$	$+0.593\ 523\ 499$
1.3	$+0.955\ 428\ 747$	$+0.420\ 405\ 966$	$-0.136\ 972\ 169$	$+0.640\ 338\ 102$
1.4	$+0.940\ 075\ 057$	$+0.486\ 733\ 934$	$-0.170\ 928\ 324$	$+0.686\ 008\ 176$
1.5	$+0.921\ 072\ 184$	$+0.557\ 560\ 062$	$-0.210\ 011\ 017$	$+0.730\ 250\ 674$
1.6	$+0.897\ 891\ 139$	$+0.632\ 725\ 677$	$-0.254\ 544\ 638$	$+0.772\ 739\ 922$
1.7	$+0.869\ 971\ 237$	$+0.712\ 037\ 292$	$-0.304\ 838\ 207$	$+0.813\ 104\ 947$
1.8	$+0.836\ 721\ 794$	$+0.795\ 261\ 955$	$-0.361\ 182\ 125$	$+0.850\ 926\ 951$
1.9	$+0.797\ 524\ 167$	$+0.882\ 122\ 341$	$-0.423\ 844\ 516$	$+0.885\ 736\ 950$
2.0	$+0.751\ 734\ 183$	$+0.972\ 291\ 627$	$-0.493\ 067\ 125$	$+0.917\ 013\ 613$
2.1	$+0.698\ 685\ 001$	$+1.065\ 388\ 161$	$-0.569\ 060\ 755$	$+0.944\ 181\ 339$
2.2	$+0.637\ 690\ 457$	$+1.160\ 969\ 944$	$-0.652\ 000\ 244$	$+0.966\ 608\ 614$
2.3	$+0.568\ 048\ 926$	$+1.258\ 528\ 975$	$-0.742\ 018\ 947$	$+0.983\ 606\ 691$
2.4	$+0.489\ 047\ 772$	$+1.357\ 485\ 476$	$-0.839\ 202\ 721$	$+0.994\ 428\ 643$
2.5	$+0.399\ 968\ 417$	$+1.457\ 182\ 044$	$-0.943\ 583\ 409$	$+0.998\ 268\ 847$
2.6	$+0.300\ 092\ 090$	$+1.556\ 877\ 774$	$-1.055\ 131\ 815$	$+0.994\ 262\ 944$
2.7	$+0.188\ 706\ 304$	$+1.655\ 742\ 407$	$-1.173\ 750\ 173$	$+0.981\ 488\ 365$
2.8	$+0.065\ 112\ 108$	$+1.752\ 850\ 564$	$-1.299\ 264\ 112$	$+0.958\ 965\ 456$
2.9	$-0.071\ 367\ 826$	$+1.847\ 176\ 116$	$-1.431\ 414\ 136$	$+0.925\ 659\ 305$
3.0	$-0.221\ 380\ 249$	$+1.937\ 586\ 785$	$-1.569\ 846\ 632$	$+0.880\ 482\ 324$
3.1	$-0.385\ 531\ 455$	$+2.022\ 839\ 042$	$-1.714\ 104\ 430$	$+0.822\ 297\ 688$
3.2	$-0.564\ 376\ 430$	$+2.101\ 573\ 388$	$-1.863\ 616\ 954$	$+0.749\ 923\ 691$
3.3	$-0.758\ 407\ 012$	$+2.172\ 310\ 131$	$-2.017\ 689\ 996$	$+0.662\ 139\ 131$
3.4	$-0.968\ 038\ 995$	$+2.233\ 445\ 750$	$-2.175\ 495\ 175$	$+0.557\ 689\ 801$
3.5	$-1.193\ 598\ 180$	$+2.283\ 249\ 967$	$-2.336\ 059\ 130$	$+0.435\ 296\ 178$
3.6	$-1.435\ 305\ 322$	$+2.319\ 863\ 655$	$-2.498\ 252\ 527$	$+0.293\ 662\ 421$
3.7	$-1.693\ 259\ 984$	$+2.341\ 297\ 714$	$-2.660\ 778\ 962$	$+0.131\ 486\ 760$
3.8	$-1.967\ 423\ 273$	$+2.345\ 433\ 061$	$-2.822\ 163\ 850$	$-0.052\ 526\ 621$
3.9	$-2.257\ 599\ 466$	$+2.330\ 021\ 882$	$-2.980\ 743\ 427$	$-0.259\ 654\ 097$
4.0	$-2.563\ 416\ 557$	$+2.292\ 690\ 323$	$-3.134\ 653\ 964$	$-0.491\ 137\ 441$
4.1	$-2.884\ 305\ 732$	$+2.230\ 942\ 780$	$-3.281\ 821\ 353$	$-0.748\ 166\ 860$
4.2	$-3.219\ 479\ 832$	$+2.142\ 167\ 987$	$-3.419\ 951\ 224$	$-1.031\ 862\ 169$
4.3	$-3.567\ 910\ 863$	$+2.023\ 647\ 069$	$-3.546\ 519\ 744$	$-1.343\ 251\ 997$
4.4	$-3.928\ 306\ 621$	$+1.872\ 563\ 796$	$-3.658\ 765\ 306$	$-1.683\ 250\ 947$
4.5	$-4.299\ 086\ 552$	$+1.686\ 017\ 204$	$-3.753\ 681\ 326$	$-2.052\ 634\ 662$
4.6	$-4.678\ 356\ 937$	$+1.461\ 036\ 836$	$-3.828\ 010\ 348$	$-2.452\ 013$
4.7	$-5.063\ 885\ 587$	$+1.194\ 600\ 797$	$-3.878\ 239\ 739$	$-2.881\ 799$
4.8	$-5.453\ 076\ 175$	$+0.883\ 656\ 854$	$-3.900\ 599\ 216$	$-3.342\ 181$
4.9	$-5.842\ 942\ 442$	$+0.525\ 146\ 811$	$-3.891\ 060\ 511$	$-3.833\ 085$
5.0	$-6.230\ 082\ 479$	$+0.116\ 034\ 382$	$-3.845\ 339\ 473$	$-4.354\ 141$

TABLE 1050 (*continued*)—BESSEL FUNCTIONS

x	ber x	bei x	ber$'$ x	bei$'$ x
5.1	−6.610 653 357	−0.346 663 218	−3.758 900 943	−4.904 641
5.2	−6.980 346 403	−0.865 839 727	−3.626 966 748	−5.483 505
5.3	−7.334 363 435	−1.444 260 151	−3.444 527 187	−6.089 232
5.4	−7.667 394 351	−2.084 516 693	−3.206 356 389	−6.719 859
5.5	−7.973 596 451	−2.788 980 155	−2.907 031 958	−7.372 913
5.6	−8.246 575 962	−3.559 746 593	−2.540 959 318	−8.015 365
5.7	−8.479 372 252	−4.398 579 111	−2.102 401 197	−8.733 576
5.8	−8.664 445 263	−5.306 844 640	−1.585 512 696	−9.433 252
5.9	−8.793 666 753	−6.285 445 623	−0.984 382 394	−10.139 389
6.0	−8.858 315 966	−7.334 746 541	−0.293 079 967	−10.846 224
6.1	−8.849 080 413	−8.454 495 269	+0.494 289 242	−11.547 179
6.2	−8.756 062 474	−9.643 739 286	+1.383 522 213	−12.234 815
6.3	−8.568 792 593	−10.900 736 825	+2.380 248 360	−12.900 779
6.4	−8.276 249 873	−12.222 863 128	+3.489 851 325	−13.535 755
6.5	−7.866 890 928	−13.606 512 001	+4.717 382 012	−14.129 423
6.6	−7.328 687 885	−15.046 992 991	+6.067 462 487	−14.670 413
6.7	−6.649 176 464	−16.538 424 538	+7.544 180 362	−15.146 266
6.8	−5.815 515 115	−18.073 623 609	+9.150 973 359	−15.543 406
6.9	−4.814 556 200	−19.643 992 365	+10.890 503 759	−15.847 109
7.0	−3.632 930 243	−21.239 402 580	+12.764 522 560	−16.041 489
7.1	−2.257 144 280	−22.848 078 597	+14.773 723 174	−16.109 484
7.2	−0.673 695 379	−24.456 479 797	+16.917 584 633	−16.032 856
7.3	+1.130 799 653	−26.049 183 639	+19.194 204 342	−15.792 207
7.4	+3.169 457 312	−27.608 770 523	+21.600 120 535	−15.367 001
7.5	+5.454 962 184	−29.115 711 867	+24.130 124 710	−14.735 602
7.6	+7.999 382 494	−30.548 262 965	+26.777 064 473	−13.875 334
7.7	+10.813 965 476	−31.882 362 359	+29.531 637 360	−12.762 551
7.8	+13.908 911 711	−33.091 539 670	+32.382 176 399	−11.372 739
7.9	+17.293 127 645	−34.146 833 988	+35.314 428 336	−9.680 623
8.0	+20.973 955 611	−35.016 725 165	+38.311 325 701	−7.660 318
8.1	+24.956 880 800	−35.667 080 514	+41.352 754 078	−5.285 490
8.2	+29.245 214 796	−36.061 119 681	+44.415 316 208	−2.529 555
8.3	+33.839 755 432	−36.159 400 616	+47.472 094 831	+0.634 098
8.4	+38.738 422 961	−35.919 829 830	+50.492 416 438	+4.231 841
8.5	+43.935 872 751	−35.297 700 300	+53.441 618 430	+8.289 519
8.6	+49.423 084 977	−34.245 760 640	+56.280 822 496	+12.832 116
8.7	+55.186 932 099	−32.714 319 308	+58.966 717 374	+17.883 387
8.8	+61.209 725 224	−30.651 387 879	+61.451 354 516	+23.465 444
8.9	+67.468 740 848	−28.002 867 538	+63.681 960 575	+29.598 302
9.0	+73.935 729 857	−24.712 783 168	+65.600 770 999	+36.299 384
9.1	+80.576 411 145	−20.723 569 533	+67.144 889 467	+43.582 976
9.2	+87.349 952 674	−15.976 414 197	+68.246 178 293	+51.459 634
9.3	+94.208 443 358	−10.411 661 917	+68.831 185 381	+59.935 547
9.4	+101.096 359 718	−3.969 285 324	+68.821 113 743	+69.011 850
9.5	+107.950 031 881	+3.410 573 282	+68.131 840 035	+78.683 888
9.6	+114.697 114 173	+11.786 984 189	+66.673 989 017	+88.940 434
9.7	+121.256 066 255	+21.217 531 810	+64.353 071 286	+99.762 855
9.8	+127.535 651 521	+31.757 530 896	+61.069 692 033	+111.124 240
9.9	+133.434 460 262	+43.459 152 933	+56.719 839 030	+122.988 479
10.0	+138.840 465 942	+56.370 458 554	+51.195 258 394	+135.309 302

For x up to 20, see Ref. 45 and 51.

TABLE 1050 (*continued*)—BESSEL FUNCTIONS
$$\ker x + i \kei x = K_0(x\sqrt{i})$$

x	ker x	kei x	ker$'$ x	kei$'$ x
0	$+ \infty$	$-0.785\ 398\ 2$	$- \infty$	0
0.1	$+2.420\ 474\ 0$	$-0.776\ 850\ 6$	$-9.960\ 959\ 3$	$+0.145\ 974\ 8$
0.2	$+1.733\ 142\ 7$	$-0.758\ 124\ 9$	$-4.922\ 948\ 5$	$+0.222\ 926\ 8$
0.3	$+1.337\ 218\ 6$	$-0.733\ 101\ 9$	$-3.219\ 865\ 2$	$+0.274\ 292\ 1$
0.4	$+1.062\ 623\ 9$	$-0.703\ 800\ 2$	$-2.352\ 069\ 9$	$+0.309\ 514\ 0$
0.5	$+0.855\ 905\ 9$	$-0.671\ 581\ 7$	$-1.819\ 799\ 8$	$+0.333\ 203\ 8$
0.6	$+0.693\ 120\ 7$	$-0.637\ 449\ 5$	$-1.456\ 538\ 6$	$+0.348\ 164\ 4$
0.7	$+0.561\ 378\ 3$	$-0.602\ 175\ 5$	$-1.190\ 943\ 3$	$+0.356\ 309\ 5$
0.8	$+0.452\ 882\ 1$	$-0.566\ 367\ 6$	$-0.987\ 335\ 1$	$+0.359\ 042\ 5$
0.9	$+0.362\ 514\ 8$	$-0.530\ 511\ 1$	$-0.825\ 868\ 7$	$+0.357\ 443\ 2$
1.0	$+0.286\ 706\ 2$	$-0.494\ 994\ 6$	$-0.694\ 603\ 9$	$+0.352\ 369\ 9$
1.1	$+0.222\ 844\ 5$	$-0.460\ 129\ 5$	$-0.585\ 905\ 3$	$+0.344\ 521\ 0$
1.2	$+0.168\ 945\ 6$	$-0.426\ 163\ 6$	$-0.494\ 643\ 2$	$+0.334\ 473\ 9$
1.3	$+0.123\ 455\ 4$	$-0.393\ 291\ 8$	$-0.417\ 227\ 4$	$+0.322\ 711\ 8$
1.4	$+0.085\ 126\ 0$	$-0.361\ 664\ 8$	$-0.351\ 055\ 1$	$+0.309\ 641\ 6$
1.5	$+0.052\ 934\ 9$	$-0.331\ 395\ 6$	$-0.294\ 181\ 6$	$+0.295\ 608\ 1$
1.6	$+0.026\ 029\ 9$	$-0.302\ 565\ 5$	$-0.245\ 114\ 7$	$+0.280\ 903\ 8$
1.7	$+0.003\ 691\ 1$	$-0.275\ 228\ 8$	$-0.202\ 681\ 8$	$+0.265\ 777\ 2$
1.8	$-0.014\ 696\ 1$	$-0.249\ 417\ 1$	$-0.165\ 942\ 4$	$+0.250\ 438\ 5$
1.9	$-0.029\ 661\ 4$	$-0.225\ 142\ 2$	$-0.134\ 128\ 2$	$+0.235\ 065\ 7$
2.0	$-0.041\ 664\ 5$	$-0.202\ 400\ 1$	$-0.106\ 601\ 0$	$+0.219\ 807\ 9$
2.1	$-0.051\ 106\ 5$	$-0.181\ 172\ 6$	$-0.082\ 823\ 4$	$+0.204\ 789\ 7$
2.2	$-0.058\ 338\ 8$	$-0.161\ 430\ 7$	$-0.062\ 337\ 3$	$+0.190\ 113\ 7$
2.3	$-0.063\ 670\ 5$	$-0.143\ 135\ 7$	$-0.044\ 747\ 9$	$+0.175\ 863\ 8$
2.4	$-0.067\ 373\ 5$	$-0.126\ 241\ 5$	$-0.029\ 712\ 3$	$+0.162\ 106\ 9$
2.5	$-0.069\ 688\ 0$	$-0.110\ 696\ 1$	$-0.016\ 929\ 8$	$+0.148\ 895\ 4$
2.6	$-0.070\ 825\ 7$	$-0.096\ 442\ 9$	$-0.006\ 135\ 8$	$+0.136\ 268\ 9$
2.7	$-0.070\ 973\ 6$	$-0.083\ 421\ 9$	$+0.002\ 904\ 3$	$+0.124\ 255\ 8$
2.8	$-0.070\ 296\ 3$	$-0.071\ 570\ 7$	$+0.010\ 399\ 0$	$+0.112\ 874\ 8$
2.9	$-0.068\ 939\ 0$	$-0.060\ 825\ 5$	$+0.016\ 534\ 2$	$+0.102\ 136\ 2$
3.0	$-0.067\ 029\ 2$	$-0.051\ 121\ 9$	$+0.021\ 476\ 2$	$+0.092\ 043\ 1$
3.1	$-0.064\ 678\ 6$	$-0.042\ 395\ 5$	$+0.025\ 373\ 8$	$+0.082\ 592\ 2$
3.2	$-0.061\ 984\ 8$	$-0.034\ 582\ 3$	$+0.028\ 360\ 3$	$+0.073\ 775\ 2$
3.3	$-0.059\ 032\ 9$	$-0.027\ 619\ 7$	$+0.030\ 555\ 4$	$+0.065\ 579\ 4$
3.4	$-0.055\ 896\ 6$	$-0.021\ 446\ 3$	$+0.032\ 066\ 2$	$+0.057\ 988\ 1$
3.5	$-0.052\ 639\ 3$	$-0.016\ 002\ 6$	$+0.032\ 988\ 6$	$+0.050\ 982\ 1$
3.6	$-0.049\ 315\ 6$	$-0.011\ 231\ 1$	$+0.033\ 408\ 7$	$+0.044\ 539\ 4$
3.7	$-0.045\ 971\ 7$	$-0.007\ 076\ 7$	$+0.033\ 403\ 0$	$+0.038\ 636\ 4$
3.8	$-0.042\ 646\ 9$	$-0.003\ 486\ 7$	$+0.033\ 040\ 0$	$+0.033\ 248\ 0$
3.9	$-0.039\ 373\ 61$	$-0.000\ 410\ 81$	$+0.032\ 380\ 46$	$+0.028\ 348\ 32$
4.0	$-0.036\ 178\ 85$	$+0.002\ 198\ 40$	$+0.031\ 478\ 49$	$+0.023\ 910\ 62$
4.1	$-0.033\ 084\ 40$	$+0.004\ 385\ 82$	$+0.030\ 381\ 79$	$+0.019\ 908\ 04$
4.2	$-0.030\ 107\ 58$	$+0.006\ 193\ 61$	$+0.029\ 132\ 42$	$+0.016\ 313\ 67$
4.3	$-0.027\ 261\ 77$	$+0.007\ 661\ 27$	$+0.027\ 767\ 30$	$+0.013\ 100\ 84$
4.4	$-0.024\ 556\ 89$	$+0.008\ 825\ 62$	$+0.026\ 318\ 68$	$+0.010\ 243\ 31$
4.5	$-0.021\ 999\ 88$	$+0.009\ 720\ 92$	$+0.024\ 814\ 54$	$+0.007\ 715\ 43$
4.6	$-0.019\ 595\ 03$	$+0.010\ 378\ 86$	$+0.023\ 279\ 08$	$+0.005\ 492\ 26$
4.7	$-0.017\ 344\ 41$	$+0.010\ 828\ 72$	$+0.021\ 733\ 00$	$+0.003\ 549\ 76$
4.8	$-0.015\ 248\ 19$	$+0.011\ 097\ 40$	$+0.020\ 193\ 91$	$+0.001\ 864\ 78$
4.9	$-0.013\ 304\ 90$	$+0.011\ 209\ 53$	$+0.018\ 676\ 61$	$+0.000\ 415\ 22$
5.0	$-0.011\ 511\ 73$	$+0.011\ 187\ 59$	$+0.017\ 193\ 40$	$-0.000\ 819\ 98$

TABLE 1050 (*continued*)—BESSEL FUNCTIONS

x	ker x	kei x	ker$'$ x	kei$'$ x
5.1	−0.009 864 74	+0.011 052 01	+0.015 754 36	−0.001 860 79
5.2	−0.008 359 11	+0.010 821 28	+0.014 367 57	−0.002 726 05
5.3	−0.006 989 28	+0.010 512 06	+0.013 039 35	−0.003 433 49
5.4	−0.005 749 13	+0.010 139 29	+0.011 774 46	−0.003 999 69
5.5	−0.004 632 16	+0.009 716 31	+0.010 576 33	−0.004 440 16
5.6	−0.003 631 56	+0.009 254 96	+0.009 447 17	−0.004 769 28
5.7	−0.002 740 38	+0.008 765 72	+0.008 388 18	−0.005 000 41
5.8	−0.001 951 58	+0.008 257 74	+0.007 399 67	−0.005 145 84
5.9	−0.001 258 12	+0.007 739 02	+0.006 481 21	−0.005 216 89
6.0	−0.000 653 04	+0.007 216 49	+0.005 631 71	−0.005 223 92
6.1	−0.000 129 53	+0.006 696 06	+0.004 849 57	−0.005 176 37
6.2	+0.000 319 05	+0.006 182 75	+0.004 132 75	−0.005 082 83
6.3	+0.000 699 12	+0.005 680 77	+0.003 478 86	−0.004 951 05
6.4	+0.001 016 83	+0.005 193 58	+0.002 885 23	−0.004 788 03
6.5	+0.001 278 080	+0.004 723 992	+0.002 348 995	−0.004 600 032
6.6	+0.001 488 446	+0.004 274 219	+0.001 867 130	−0.004 392 632
6.7	+0.001 653 215	+0.003 845 947	+0.001 436 521	−0.004 170 782
6.8	+0.001 777 354	+0.003 440 398	+0.001 053 999	−0.003 938 849
6.9	+0.001 865 512	+0.003 058 385	+0.000 716 382	−0.003 700 651
7.0	+0.001 922 022	+0.002 700 365	+0.000 420 510	−0.003 459 509
7.1	+0.001 950 901	+0.002 366 486	+0.000 163 267	−0.003 218 285
7.2	+0.001 955 861	+0.002 056 629	−0.000 058 386	−0.002 979 421
7.3	+0.001 940 312	+0.001 770 454	−0.000 247 403	−0.002 744 978
7.4	+0.001 907 373	+0.001 507 429	−0.000 406 628	−0.002 516 671
7.5	+0.001 859 888	+0.001 266 868	−0.000 538 787	−0.002 295 904
7.6	+0.001 800 431	+0.001 047 959	−0.000 646 478	−0.002 083 800
7.7	+0.001 731 326	+0.000 849 790	−0.000 732 165	−0.001 881 234
7.8	+0.001 654 654	+0.000 671 373	−0.000 798 170	−0.001 688 855
7.9	+0.001 572 275	+0.000 511 664	−0.000 846 677	−0.001 507 120
8.0	+0.001 485 834	+0.000 369 584	−0.000 879 724	−0.001 336 313
8.1	+0.001 396 782	+0.000 244 032	−0.000 899 210	−0.001 176 567
8.2	+0.001 306 386	+0.000 133 902	−0.000 906 891	−0.001 027 888
8.3	+0.001 215 743	+0.000 038 090	−0.000 904 388	−0.000 890 168
8.4	+0.001 125 797	−0.000 044 491	−0.000 893 190	−0.000 763 209
8.5	+0.001 037 349	−0.000 114 902	−0.000 874 656	−0.000 646 733
8.6	+0.000 951 070	−0.000 174 175	−0.000 850 022	−0.000 540 398
8.7	+0.000 867 511	−0.000 223 306	−0.000 820 407	−0.000 443 813
8.8	+0.000 787 120	−0.000 263 248	−0.000 786 819	−0.000 356 543
8.9	+0.000 710 249	−0.000 294 910	−0.000 750 159	−0.000 278 127
9.0	+0.000 637 164	−0.000 319 153	−0.000 711 231	−0.000 208 079
9.1	+0.000 568 055	−0.000 336 788	−0.000 670 745	−0.000 145 903
9.2	+0.000 503 046	−0.000 348 579	−0.000 629 326	−0.000 091 093
9.3	+0.000 442 203	−0.000 355 236	−0.000 587 517	−0.000 043 145
9.4	+0.000 385 540	−0.000 357 420	−0.000 545 789	−0.000 001 559
9.5	+0.000 333 029	−0.000 355 743	−0.000 504 544	+0.000 034 158
9.6	+0.000 284 604	−0.000 350 768	−0.000 464 122	+0.000 064 485
9.7	+0.000 240 168	−0.000 343 010	−0.000 424 806	+0.000 089 887
9.8	+0.000 199 598	−0.000 332 940	−0.000 386 830	+0.000 110 811
9.9	+0.000 162 751	−0.000 320 983	−0.000 350 379	+0.000 127 684
10.0	+0.000 129 466	−0.000 307 524	−0.000 315 597	+0.000 140 914

See Report of the British Assoc. for the Advancement of Science, 1912, p. 56; 1915, p. 36; and 1916, p. 122.

TABLE 1050 (*continued*)—BESSEL FUNCTIONS

$$\text{ber}_n x + i\,\text{bei}_n x = J_n(xi\sqrt{i}) = i^n I_n(x\sqrt{i})$$

$$\text{ber}_n' x = \frac{d}{dx}\,\text{ber}_n x$$

x	$\text{ber}_1 x$	$\text{bei}_1 x$	$\text{ber}_1' x$	$\text{bei}_1' x$
1	$-0.395\,868$	$+0.307\,557$	$-0.476\,664$	$+0.212\,036$
2	$-0.997\,078$	$+0.299\,775$	$-0.720\,532$	$-0.305\,845$
3	$-1.732\,64$	$-0.487\,45$	$-0.635\,99$	$-1.364\,13$
4	$-1.869\,25$	$-2.563\,82$	$+0.658\,74$	$-2.792\,83$
5	$+0.359\,78$	$-5.797\,91$	$+4.251\,33$	$-3.327\,80$
6	$+7.462\,20$	$-7.876\,68$	$+10.206\,52$	$+0.235\,45$
7	$+20.368\,9$	$-2.317\,2$	$+14.677\,5$	$+12.780\,7$
8	$+32.506\,9$	$+21.673\,5$	$+5.866\,4$	$+36.882\,2$
9	$+20.719\,2$	$+72.054\,3$	$-37.108\,0$	$+61.749\,0$
10	-59.478	$+131.879$	-132.087	$+45.127$

x	$\text{ber}_2 x$	$\text{bei}_2 x$	$\text{ber}_2' x$	$\text{bei}_2' x$
1	$+0.010\,411$	$-0.124\,675$	$+0.041\,623$	$-0.248\,047$
2	$+0.165\,279$	$-0.479\,225$	$+0.327\,788$	$-0.437\,789$
3	$+0.808\,37$	$-0.891\,02$	$+1.030\,93$	$-0.286\,47$
4	$+2.317\,85$	$-0.725\,36$	$+1.975\,73$	$+0.853\,82$
5	$+4.488\,43$	$+1.422\,10$	$+2.049\,97$	$+3.785\,30$
6	$+5.242\,91$	$+7.432\,44$	$-1.454\,56$	$+8.368\,74$
7	$-0.950\,4$	$+17.592\,4$	$-12.493\,0$	$+11.015\,1$
8	$-22.889\,0$	$+25.438\,9$	$-32.589\,1$	$+1.300\,6$
9	$-65.869\,2$	$+10.134\,8$	$-50.963\,2$	$-38.551\,6$
10	-111.779	-66.610	-28.840	-121.987

x	$\text{ber}_3 x$	$\text{bei}_3 x$	$\text{ber}_3' x$	$\text{bei}_3' x$
1	$+0.013\,788$	$+0.015\,629$	$+0.039\,433$	$+0.048\,634$
2	$+0.085\,612$	$+0.144\,210$	$+0.093\,575$	$+0.239\,418$
3	$+0.130\,44$	$+0.565\,38$	$+0.072\,00$	$+0.636\,27$
4	$-0.282\,63$	$+1.437\,76$	$-0.914\,09$	$+1.073\,55$
5	$-2.094\,35$	$+2.454\,41$	$-2.922\,76$	$+0.695\,57$
6	$-6.430\,04$	$+1.901\,46$	$-5.747\,81$	$-2.498\,96$
7	$-12.876\,5$	$-4.407\,2$	$-6.249\,2$	$-11.222\,9$
8	$-15.420\,4$	$-22.575\,0$	$+3.979\,6$	$-25.707\,4$
9	$+3.166\,6$	$-54.538\,7$	$+38.354\,6$	$-35.563\,4$
10	$+72.253$	-81.423	$+104.463$	-7.513

x	$\text{ber}_4 x$	$\text{bei}_4 x$	$\text{ber}_4' x$	$\text{bei}_4' x$
1	$-0.002\,60$	$-0.000\,13$	$-0.010\,40$	$-0.000\,78$
2	$-0.040\,97$	$-0.008\,30$	$-0.080\,56$	$-0.024\,83$
3	$-0.193\,27$	$-0.093\,02$	$-0.234\,32$	$-0.183\,52$
4	$-0.493\,10$	$-0.499\,85$	$-0.323\,71$	$-0.716\,65$
5	$-0.628\,67$	$-1.727\,62$	$+0.248\,34$	$-1.834\,36$
6	$+0.648\,3$	$-4.230\,2$	$+2.770\,0$	$-3.071\,1$
7	$+6.083\,5$	$-7.116\,9$	$+8.745\,2$	$-1.921\,9$
8	$+19.094\,7$	$-5.288\,8$	$+17.319\,5$	$+7.703\,5$
9	$+38.667$	$+14.082$	$+19.140$	$+34.545$
10	$+46.579$	$+70.500$	-12.148	$+80.465$

TABLE 1050 (*continued*)—BESSEL FUNCTIONS

x	$\mathrm{ber}_5 x$	$\mathrm{bei}_5 x$	$\mathrm{ber}_5' x$	$\mathrm{bei}_5' x$
1	$+0.000\ 19$	$-0.000\ 18$	$+0.000\ 97$	$-0.000\ 87$
2	$+0.006\ 80$	$-0.004\ 84$	$+0.017\ 84$	$-0.011\ 00$
3	$+0.058\ 59$	$-0.025\ 54$	$+0.104\ 78$	$-0.028\ 32$
4	$+0.273\ 08$	$-0.033\ 53$	$+0.360\ 76$	$+0.046\ 69$
5	$+0.851\ 04$	$+0.211\ 43$	$+0.815\ 11$	$+0.565\ 64$
6	$+1.830\ 5$	$+1.475\ 6$	$+1.007\ 4$	$+2.220\ 0$
7	$+2.209\ 0$	$+5.242\ 3$	$-0.847\ 2$	$+5.589\ 6$
8	$-1.821\ 3$	$+12.812\ 8$	$-8.623\ 9$	$+9.233\ 7$
9	-18.619	$+21.384$	-26.955	$+5.504$
10	-58.722	$+15.193$	-53.427	-24.511

TABLE 1050 (*continued*)—BESSEL FUNCTIONS

$$\mathrm{ker}_n x + i\,\mathrm{kei}_n x = i^{-n} K_n(x\sqrt{i})$$

x	$\mathrm{ker}_1 x$	$\mathrm{kei}_1 x$	$\mathrm{ker}_1' x$	$\mathrm{kei}_1' x$
1	$-0.740\ 322$	$-0.241\ 996$	$+0.887\ 604$	$+0.794\ 742$
2	$-0.230\ 806$	$+0.080\ 049$	$+0.287\ 983$	$+0.073\ 632$
3	$-0.049\ 898$	$+0.080\ 270$	$+0.100\ 178$	$-0.038\ 005$
4	$+0.005\ 351\ 3$	$+0.039\ 166\ 0$	$+0.022\ 690\ 0$	$-0.036\ 928\ 3$
5	$+0.012\ 737\ 4$	$+0.011\ 577\ 8$	$-0.002\ 318\ 3$	$-0.018\ 366\ 4$
6	$+0.007\ 676\ 09$	$+0.000\ 288\ 35$	$-0.005\ 920\ 41$	$-0.005\ 612\ 66$
7	$+0.002\ 743\ 59$	$-0.002\ 148\ 90$	$-0.003\ 660\ 46$	$+0.000\ 156\ 61$
8	$+0.000\ 322\ 857$	$-0.001\ 566\ 975$	$-0.001\ 352\ 336$	$+0.000\ 985\ 180$
9	$-0.000\ 355\ 78$	$-0.000\ 650\ 05$	$-0.000\ 185\ 34$	$+0.000\ 748\ 45$
10	$-0.000\ 322\ 80$	$-0.000\ 123\ 52$	$+0.000\ 158\ 19$	$+0.000\ 321\ 35$

	$\mathrm{ker}_2 x$	$\mathrm{kei}_2 x$	$\mathrm{ker}_2' x$	$\mathrm{kei}_2' x$
1	$+0.418\ 03$	$+1.884\ 20$	$-0.141\ 46$	$-4.120\ 77$
2	$+0.261\ 472$	$+0.309\ 001$	$-0.154\ 871$	$-0.528\ 809$
3	$+0.128\ 391$	$+0.036\ 804$	$-0.107\ 070$	$-0.116\ 579$
4	$+0.048\ 134\ 2$	$-0.017\ 937\ 6$	$-0.055\ 545\ 6$	$-0.014\ 941\ 8$
5	$+0.011\ 183\ 7$	$-0.018\ 064\ 9$	$-0.021\ 666\ 9$	$+0.008\ 046\ 0$
6	$-0.001\ 088\ 3$	$-0.009\ 093\ 7$	$-0.005\ 268\ 9$	$+0.008\ 255\ 2$
7	$-0.002\ 910\ 45$	$-0.002\ 820\ 51$	$+0.000\ 411\ 05$	$+0.004\ 265\ 37$
8	$-0.001\ 819\ 91$	$-0.000\ 149\ 65$	$+0.001\ 334\ 70$	$+0.001\ 373\ 73$
9	$-0.000\ 683\ 40$	$+0.000\ 477\ 20$	$+0.000\ 863\ 10$	$+0.000\ 102\ 03$
10	$-0.000\ 101\ 28$	$+0.000\ 370\ 64$	$+0.000\ 335\ 85$	$-0.000\ 215\ 04$

	$\mathrm{ker}_3 x$	$\mathrm{kei}_3 x$	$\mathrm{ker}_3' x$	$\mathrm{kei}_3' x$
1	$+4.887\ 27$	$-6.269\ 71$	$-16.289\ 7$	$+17.772\ 4$
2	$+0.298\ 022$	$-0.886\ 821$	$-0.850\ 418$	$+1.296\ 62$
3	$-0.036\ 451$	$-0.236\ 018$	$-0.080\ 360$	$+0.300\ 78$
4	$-0.052\ 071\ 1$	$-0.060\ 518\ 2$	$+0.017\ 701\ 2$	$+0.092\ 108\ 5$
5	$-0.029\ 282\ 9$	$-0.007\ 685\ 2$	$+0.022\ 435\ 5$	$+0.025\ 293\ 0$
6	$-0.011\ 449\ 9$	$+0.004\ 511\ 5$	$+0.012\ 924\ 7$	$+0.003\ 405\ 0$
7	$-0.002\ 707\ 2$	$+0.004\ 464\ 6$	$+0.005\ 212\ 6$	$-0.001\ 977\ 0$
8	$+0.000\ 267\ 67$	$+0.002\ 263\ 32$	$+0.001\ 292\ 32$	$-0.002\ 029\ 80$
9	$+0.000\ 720\ 5$	$+0.000\ 714\ 8$	$-0.000\ 094\ 4$	$-0.001\ 059\ 0$
10	$+0.000\ 456\ 3$	$+0.000\ 047\ 3$	$-0.000\ 327\ 3$	$-0.000\ 347\ 9$

TABLE 1050 (*continued*)—BESSEL FUNCTIONS

x	$\ker_4 x$	$\kei_4 x$	$\ker_4' x$	$\kei_4' x$
1	−47.753 1	+3.981 0	+191.990	−8.035
2	−2.774 90	+0.940 03	+5.966 15	−1.042 25
3	−0.410 62	+0.348 52	+0.740 16	−0.323 58
4	−0.057 09	+0.137 36	+0.136 71	−0.131 38
5	+0.007 143	+0.049 433	+0.020 426	−0.054 819
6	+0.012 375	+0.014 000	−0.003 344	−0.020 620
7	+0.007 257	+0.001 780	−0.005 361	−0.006 088
8	+0.002 878 3	−0.001 192 6	−0.003 228 8	−0.000 814 8
9	+0.000 680 7	−0.001 153 8	−0.001 317 5	+0.000 516 8
10	−0.000 072 2	−0.000 584 3	−0.000 327 2	+0.000 522 9
	$\ker_5 x$	$\kei_5 x$	$\ker_5' x$	$\kei_5' x$
1	+287.76	+253.88	−1407.9	−1306.0
2	+10.209 4	+6.076 6	−24.226 0	−17.818 4
3	+1.467 9	+0.353 1	−2.402 6	−1.125 3
4	+0.327 07	−0.052 99	−0.465 59	−0.071 26
5	+0.077 13	−0.056 32	−0.117 13	+0.026 42
6	+0.012 982	−0.029 378	−0.029 468	+0.023 332
7	−0.001 719	−0.011 767	−0.005 162	+0.011 279
8	−0.003 146 2	−0.003 455 3	+0.000 774 4	+0.005 038 1
9	−0.001 873 6	−0.000 417 5	+0.001 375 4	+0.001 529 2
10	−0.000 746 0	+0.000 324 1	+0.000 837 2	+0.000 200 1

[Ref. 14]

244

TABLE 1060—SOME NUMERICAL CONSTANTS

$$\sqrt{2} = 1.414\ 214$$
$$\sqrt{3} = 1.732\ 051$$
$$\sqrt{5} = 2.236\ 068$$
$$\sqrt{6} = 2.449\ 490$$
$$\sqrt{7} = 2.645\ 751$$
$$\sqrt{8} = 2.828\ 427$$
$$\sqrt{10} = 3.162\ 278$$
$$\pi = 3.141\ 592\ 654$$
$$\log_{10} \pi = 0.497\ 149\ 873$$
$$\pi^2 = 9.869\ 604\ 401$$
$$\frac{1}{\pi} = 0.318\ 309\ 886$$
$$\sqrt{\pi} = 1.772\ 453\ 851$$
$$\epsilon = 2.718\ 281\ 828$$
$$M = \log_{10} \epsilon = 0.434\ 294\ 482$$
$$1/M = \log_{\epsilon} 10 = 2.302\ 585\ 093$$
$$\log_{\epsilon} 2 = 0.693\ 147\ 181$$

TABLE 1070—GREEK ALPHABET

α	A	Alpha	ν	N	Nu
β	B	Beta	ξ	Ξ	Xi
γ	Γ	Gamma	o	O	Omicron
δ	Δ	Delta	π	Π	Pi
ϵ	E	Epsilon	ρ	P	Rho
ζ	Z	Zeta	$\sigma\ s$	Σ	Sigma
η	H	Eta	τ	T	Tau
$\theta\ \vartheta$	Θ	Theta	υ	Υ	Upsilon
ι	I	Iota	$\varphi\ \phi$	Φ	Phi
κ	K	Kappa	χ	X	Chi
λ	Λ	Lambda	ψ	Ψ	Psi
μ	M	Mu	ω	Ω	Omega

B. REFERENCES

1. *Integral Tables*, by Meyer Hirsch; Wm. Baynes & Son, London, 1823.
2. *Integraltafeln*, by F. Minding; C. Reimarus, Berlin, 1849.
3. *Synopsis of Elementary Results in Pure Mathematics*, by G. S. Carr; F. Hodgson, London, 1886.
4. *Sammlung von Formeln der Mathematik*, by W. Láska; F. Vieweg und Sohn, Braunschweig, 1894.
5. *A Short Table of Integrals*, by B. O. Peirce; Ginn & Co., Boston, 1929.
6. *Elementary Treatise on the Integral Calculus*, by Benj. Williamson; Longmans, Green & Co., London, 1896.
7. *A Treatise on the Integral Calculus*, by I. Todhunter; Macmillan & Co., London, 1921.
8. *Elements of the Infinitesimal Calculus*, by G. H. Chandler; J. Wiley & Sons, New York, 1907.
9. *Advanced Calculus*, by E. B. Wilson; Ginn & Co., Boston, 1912.
10. *Functions of a Complex Variable*, by J. Pierpont; Ginn & Co., Boston, 1914.
11. *Advanced Calculus*, by F. S. Woods; Ginn & Co., Boston, 1926.
12. *Treatise on Bessel Functions*, by Gray, Mathews and MacRobert; Macmillan & Co., London, 1931.
13. *Theory of Bessel Functions*, by G. N. Watson; Cambridge University Press, 1944.
14. *Bessel Functions for Alternating-Current Problems*, by H. B. Dwight; Transactions of American Institute of Electrical Engineers, July, 1929.
15. *Tables d'Intégrales Définies*, by B. de Haan; ed. of 1858–1864.
16. *Nouvelles Tables d'Intégrales Définies*, by B. de Haan; P. Engels, Leyden. 1867.
17. *Funktionentafeln mit Formeln und Kurven*, by E. Jahnke and F. Emde; 1933, 1938, and 1943.
18. *Elementary Integrals—A Short Table*, by T. J. I'a. Bromwich; Macmillan & Co., London, 1911.
19. *Logarithmic and Trigonometric Tables*, by E. R. Hedrick (*The Macmillan Mathematical Tables*); The Macmillan Co., New York, 1935.
20. *The Calculus*, by E. W. Davis and W. C. Brenke (*The Macmillan Mathematical Tables*); The Macmillan Co., New York, 1930.
21. *Synopsis of Applicable Mathematics*, by L. Silberstein; D. Van Nostrand Co., New York, 1923.
22. *Fourier's Series and Spherical, Cylindrical, and Ellipsoidal Harmonics*, by W. E. Byerly; Ginn & Co., Boston, 1893.
23. *Sag Calculations*, by J. S. Martin; ed. of 1931.
24. *Principles of Electric Power Transmission*, by L. F. Woodruff; J. Wiley & Sons, New York, 1938.
25. *The Magnetic Field of a Circular Cylindrical Coil*, by H. B. Dwight; Philosophical Magazine, Vol. XI, April, 1931.
26. *Scientific Paper 169 of the Bureau of Standards*, Washington, D. C., by E. B. Rosa and F. W. Grover; ed. of 1916, also published as Bulletin of the Bureau of Standards, Vol. 8, No. 1, 1912.
27. *Traité Élémentaire des Nombres de Bernoulli*, by N. Nielsen; Gauthier-Villars & Cie., Paris, 1923.

28. *American Standard Mathematical Symbols*, 1928, Report Z10 of American Engineering Standards Committee.

29. *Higher Trigonometry*, by J. B. Lock; Macmillan & Co., London, 1899.

30. *Smithsonian Mathematical Tables—Hyperbolic Functions*, by G. F. Becker and C. E. Van Orstrand; Smithsonian Institution, Washington, D. C., 1909.

31. *Smithsonian Mathematical Formulae and Tables of Elliptic Functions*, by E. P. Adams and R. L. Hippisley; Smithsonian Institution, Washington, D. C., 1922.

32. *A Course in Mathematics*, 2 vols., by F. S. Woods and F. H. Bailey; Ginn & Co., Boston, 1907–09.

33. *Elliptic Functions*, by A. Cayley; G. Bell & Sons, London, 1895.

34. *Introduction to the Theory of Infinite Series*, by T. J. I'a. Bromwich; Macmillan & Co., London, 1908.

35. *Theory of Elliptic Functions*, by H. Hancock; J. Wiley & Sons, New York, 1910.

36. *Elliptic Integrals*, by H. Hancock; J. Wiley & Sons, New York, 1917.

37. *Application of Hyperbolic Functions to Electrical Engineering Problems*, by A. E. Kennelly; University of London Press, 1912.

38. *Differential Equations for Electrical Engineers*, by P. Franklin; J. Wiley & Sons, New York, 1933.

39. *Methods of Advanced Calculus*, by P. Franklin; McGraw-Hill Book Co., New York, 1944.

40. *Integral Calculus*, by W. E. Byerly; Ginn & Co., Boston, 1898.

41. *College Algebra*, by H. B. Fine; Ginn & Co., Boston, 1905.

42. *Advanced Calculus*, by W. B. Fite; The Macmillan Co., New York, 1938.

43. *Application of the Method of Symmetrical Components*, by W. V. Lyon; McGraw-Hill Book Co., New York, 1937.

44. *Tables of the Higher Mathematical Functions*, by H. T. Davis; Principia Press, Bloomington, Indiana, Vol. 1, 1933, Vol. 2, 1935.

45. *Mathematical Tables of Elementary and Some Higher Mathematical Functions*, by H. B. Dwight; McGraw-Hill Book Co., New York, 1941.

46. *Electrical Coils and Conductors*, by H. B. Dwight; McGraw-Hill Book Co., New York, 1945.

47. *Traité des Fonctions Elliptiques*, Vol. 2, by A. M. Legendre; Huzard-Courcier, Paris, 1825.

48. *Ten-Figure Table of the Complete Elliptic Integrals*, by L. M. Milne-Thomson; Proc. Lond. Math. Soc., Ser. 2, Vol. 33, 1931.

49. *Bessel Functions for Engineers*, by N. W. McLachlan; Clarendon Press, Oxford, 1934.

50. *Bessel Functions*, Vol. VI, part I, Mathematical Tables of the British Association for the Advancement of Science; Cambridge University Press, 1937.

51. *Values of the Bessel Functions ber x and bei x and Their Derivatives*, by H. B. Dwight; Transactions of American Institute of Electrical Engineers, 1939, p. 787.

52. *Tables of the Spherical Function $P_n(x)$ and Its Derived Functions*, by H. Tallquist; Acta Soc. Sci. Fennicae, Finland, Vol. 32, 1906, p. 5, and Vol. 33, No. 9, 1908.

53. *Six-Place Tables of the 16 First Surface Zonal Harmonics $P_n(x)$*, by H. Tallquist; Acta Soc. Sci. Fennicae, Finland, 1937.

54. *Six-Place Tables of the 32 First Surface Zonal Harmonics $P_n(\cos\theta)$*, by H. Tallquist; Acta Soc. Sci. Fennicae, Finland, 1938.

55a. *Tables of Sines and Cosines for Radian Arguments*, 1940.
  b. *Tables of Circular and Hyperbolic Sines and Cosines for Radian Arguments*, 1939.
  c. *Tables of the Exponential Function*, 1939.
  d. *Table of Natural Logarithms*, Vols. I to IV, 1941.
  e. *Tables of Probability Functions*, Vols. I and II, 1941.
  f. *Tables of Sine, Cosine, and Exponential Integrals*, Vols. I and II, 1940. —A. N. Lowan, Technical Director; Work Projects Administration for the City of New York, sponsored by the National Bureau of Standards, Washington, D. C.
56. *Tafeln der Besselschen, Theta-, Kugel-, und anderer Funktionen*, by K. Hayashi; Julius Springer, Berlin, 1930.
57. *Tafeln für die Differenzenrechnung*, by K. Hayashi; Julius Springer, Berlin, 1933.
58. *Siebenstellige Werte der Trigonometrischen Funktionen von Tausendstel zu Tausendstel des Grades*, by J. Peters; C. P. Goerz, Berlin, 1918.
59. *Zehnstellige Logarithmen der Zahlen von 1 bis 100 000* (vol. 1), by J. Peters and J. Stein; Preussische Landesaufnahme, 1922.
60. *Zehnstellige Logarithmen der Trigonometrischen Funktionen von 0° bis 90° für jedes Tausendstel des Grades* (vol. 2), by J. Peters; Preussische Landesaufnahme, 1919.
61. *Transmission Circuits for Telephone Communication*, by K. S. Johnson; D. Van Nostrand Co., New York, 1939.
62. *Electromagnetic Theory*, by J. A. Stratton; McGraw-Hill Book Co., New York, 1941.
63. *Tables of the Bessel Functions of the First Kind*, by the Staff of the Harvard Computation Laboratory; Harvard University Press, 1947-48.
64. *An Index of Mathematical Tables*, by A. Fletcher, J. C. P. Miller, and L. Rosenhead; McGraw-Hill Book Company, New York, 1946.
65. *Barlow's Tables of Squares, Cubes, Square Roots, Cube Roots, and Reciprocals*, edited by L. J. Comrie. Chemical Pub. Co., New York, 1952.

INDEX